Nutrition and Growth

Edited by

Derrick B. Jelliffe

and

E. F. Patrice Jelliffe

University of California
Los Angeles, California

PLENUM PRESS · NEW YORK AND LONDON

Library of Congress Cataloging in Publication Data

Main entry under title:

Nutrition and growth.

  (Human nutrition; v. 2)
  Bibliography: p.
  Includes index.
  1. Nutrition. 2. Human growth. I. Jelliffe, Derrick Brian. II. Jelliffe, E. F. Patrice.
III. Series. [DNLM: 1. Growth. 2. Nutrition. 3. Monitoring, Physiologic. 4. Population
surveillance. QU145.3 H9183 v. 2]
QP141.H78 vol. 2   612'.3'08s [612'.3]   78-27076

ISBN-13: 978-1-4613-2918-3   e-ISBN-13: 978-1-4613-2916-9
DOI: 10.1007/978-1-4613-2916-9

©1979 Plenum Press, New York
Softcover reprint of the hardcover 1st edition 1979

A Division of Plenum Publishing Corporation
227 West 17th Street, New York, N.Y. 10011

*This volume is dedicated to all the
scientists whose work has advanced
our understanding of
how to apply knowledge
to benefit human nutrition.*

Contributors

Gordon B. Avery • Division of Neonatology, Children's Hospital, National Medical Center, Washington, D. C.

Stephen M. Bailey • Center for Human Growth and Development, University of Michigan, Ann Arbor, Michigan

Roy E. Brown • Department of Community Medicine, Mount Sinai School of Medicine, New York, New York

Paul C. Y. Chen • Department of Social and Preventive Medicine, University of Malaya, Kuala Lumpur, Malaysia

Hernan Delgado • Program of Population and Demography, Division of Human Development, Institute of Nutrition of Central America and Panama, Guatemala City, Guatemala

Stanley M. Garn • Center for Human Growth and Development, University of Michigan, Ann Arbor, Michigan

Mehari Gebre-Medhin • Institute of Nutrition, University of Uppsala, Uppsala, Sweden

C. Gopalan • Indian Council of Medical Research, New Delhi, India. Present affiliation: SEARO, WHO, New Delhi, India.

J. Michael Gurney • Caribbean Food and Nutrition Institute, Kingston, Jamaica, West Indies

Miguel A. Guzmán • Division of Statistics, Institute of Nutrition for Central America and Panama, Guatemala City, Guatemala

Jean-Pierre Habicht • Division of Nutritional Sciences, Cornell University, Ithaca, New York

Felix P. Heald • Division of Adolescent Medicine, University of Maryland Hospital, Baltimore, Maryland

Kamala S. Jaya Rao • National Institute of Nutrition, Hyderabad, India

Derrick B. Jelliffe • Division of Population, Family and International Health, School of Public Health, University of California, Los Angeles, California

E. F. Patrice Jelliffe • Division of Population, Family and International Health, School of Public Health, University of California, Los Angeles, California

Robert E. Klein • Division of Human Development, Institute of Nutrition of Central America and Panama, Guatemala City, Guatemala

J. Michael Lane • Nutrition Division, Center for Disease Control, Atlanta, Georgia

Aaron Lechtig • Program of Human Development, Division of Human Development, Institute of Nutrition of Central America and Panama, Guatemala City, Guatemala

June K. Lloyd • St. George's Hospital Medical School, University of London, London, England

Reynaldo Martorell • Food Research Institute, Stanford University, Stanford, California

David Morley • Institute of Child Health, University of London, London, England

Charlotte G. Neumann • Division of Population, Family and International Health, School of Public Health and Department of Pediatrics, University of California, Los Angeles, California

Milton Z. Nichaman • Nutrition Division, Center for Disease Control, Atlanta, Georgia

Meinhard Robinow • School of Medicine, University of Virginia, Charlottesville, Virginia

Adnan Shakir • Department of Pediatrics and Child Health, Baghdad University Medical College, Baghdad, Iraq

Bo Vahlquist • Deceased. Department of Pediatrics, University Hospital, Uppsala, Sweden

Napoleon Wolański • Department of Human Ecology, Polish Academy of Sciences, Warsaw, Poland

Charles Yarbrough • Computers for Marketing Corporation, Kenwood, California

Alfred J. Zerfas • Division of Population, Family and International Health, School of Public Health, University of California, Los Angeles, California

Foreword

The science of nutrition has advanced beyond expectation since Antoine Lavoisier as early as the 18th century showed that oxygen was necessary to change nutrients in foods to compounds which would become a part of the human body. He was also the first to measure metabolism and to show that oxidation within the body produces heat and energy. In the two hundred years that have elapsed, the essentiality of nitrogen-containing nutrients and of proteins for growth and maintenance of tissue has been established; the necessity for carbohydrates and certain types of fat for health has been documented; vitamins necessary to prevent deficiency diseases have been identified and isolated; and the requirement of many mineral elements for health has been demonstrated.

Further investigations have defined the role of these nutrients in metabolic processes and quantitated their requirements at various stages of development. Additional studies have involved their use in the possible prevention of, and therapy for, disease conditions.

This series of books was designed for the researcher or advanced student of nutritional science. The first volume is concerned with prenatal and postnatal nutrient requirements; the second volume with nutrient requirements for growth and development; the third with nutritional requirements of the adult; and the fourth with the role of nutrition in disease states. Our objectives were to review and evaluate that which is known and to point out those areas in which uncertainties and/or a lack of knowledge still exists with the hope of encouraging further research into the intricacies of human nutrition.

Roslyn B. Alfin-Slater
David Kritchevsky

Contents

Introduction: Perspectives and Needs 1

Part I • Influences on Growth

Chapter 1
Nutrient Needs
C. Gopalan and Kamala S. Jaya Rao

1. Energy 9
2. Protein 10
3. Fat 14
4. Vitamins 15
    4.1. Fat-Soluble Vitamins 16
    4.2. Water-Soluble Vitamins 19
5. Minerals 22
    5.1. Iron 22
    5.2. Calcium 24
6. Conclusions 25
7. References 26

Chapter 2
Genetic and Nutritional Interactions
Stanley M. Garn, Meinhard Robinow, and Stephen M. Bailey

1. Statistics of Growth Parameters 32
2. Population Differences and National Growth Norms 34
3. Nongenetic Determinants of the Size of the Newborn 36
4. The Multiple-Birth Model in the Study of Growth 38

5. The Genetics of Fatness and Obesity 39
6. Importance of Sample Size in Nutrition-Related Growth
   Research 40
7. New Strategies for Genetic Research 42
8. Separating Nutrition from Genetics and Vice Versa 44
9. References 45

Chapter 3
Nondietary Factors and Nutrition
Paul C. Y. Chen

1. The Influence of Infection on Growth 47
   1.1. Bacterial Infections 48
   1.2. Viral and Rickettsial Infections 49
   1.3. Protozoal Infections 50
   1.4. Helminthic Infections 50
   1.5. Diarrhea and Breast Feeding 50
   1.6. Weanling Diarrhea 51
2. The Influence of Food Practices on Growth 52
   2.1. The Meaning of Food 53
   2.2. Food Practices during the Period of Growth 55
3. The Influence of Socioeconomic and Other Social Factors on
   Growth 58
   3.1. Socioeconomic Influences 58
   3.2. Food Production 58
   3.3. Other Social Factors Influencing Growth 59
4. References 60

Chapter 4
Metabolic Anomalies, Nutrition, and Growth
Meinhard Robinow

1. Definitions 65
2. The Vulnerable Tissues 66
   2.1. Substrate Deficiency Disorders 66
   2.2. Disorders of Regulation 70
   2.3. End-Organ Failure 72
3. References 73

Part II • Ages of Man (Perspectives)

Chapter 5

Maternofetal Nutrition

Aaron Lechtig, Hernan Delgado, Reynaldo Martorell, Charles Yarbrough, and Robert E. Klein

1. Introduction 79
2. Influence of Maternal Nutrition on Fetal Growth 81
   2.1. Literature Review 81
   2.2. The INCAP Longitudinal Study 84
3. Influence of Nutrition on Other Maternal Characteristics 99
   3.1. Maternal Anthropometry during Pregnancy 99
   3.2. Birth Interval Components 103
   3.3. Blood Pressure, Edema, and Proteinuria during Pregnancy ... 113
4. Implications and Recommendations 118
5. Summary 120
6. References 122

Chapter 6

The Newborn

Gordon B. Avery

1. Nutrition of the Full-Term Newborn 129
   1.1. Breast Feeding 130
   1.2. Formula Feeding 131
   1.3. Vitamins 131
   1.4. Iron 132
   1.5. Fluoride 133
   1.6. Minerals 134
   1.7. Solid Foods 135
2. Nutrition of the Premature 138
   2.1. Special Problems of the Premature 138
   2.2. Strategies of Intake 140
   2.3. Parenteral Nutrition in the Newborn 142
   2.4. Nutritional Monitoring 146
   2.5. Necrotizing Enterocolitis 149
   2.6. Human Milk in the Premature 149
   2.7. The Small-for-Date Infant 151
   2.8. Summary 151
3. References 151

Chapter 7
The Young Child: Normal
Bo Vahlquist

1. Preschool Age—Definitions of Subgroups 153
2. Anthropometric Data (Swedish Section of the CIE
   Longitudinal Growth Study) 154
3. Effect of Low Birth Weight on Subsequent Growth
   Pattern 155
4. Body Composition 156
5. Psychomotor and Mental Development 158
6. Sexual Differences 158
7. Ethnic Differences 160
8. Forecasting of Growth and Development from
   Observations Made in Early Life 161
9. Physical Activity Pattern 162
10. "Normal" Disease Pattern 164
11. Home versus Institutional Environment 167
12. References 168

Chapter 8
The Young Child: Failure to Thrive
Roy E. Brown

1. Introduction 171
2. Causes of Growth Failure 172
   2.1. Congenital and Genetic Anomalies 173
   2.2. Enzymatic Defects 173
   2.3. Endocrine Deficiencies 173
   2.4. Chronic Diseases 173
   2.5. Miscellaneous Problems 173
3. Failure-to-Thrive Syndrome 173
   3.1. Nutritional Factors 174
   3.2. Infections 175
   3.3. Environmental Factors 175
4. How Does FTT Present? 176
5. The Investigation 177
6. Treatment and Management 179
7. Prevention 180
8. Summary 181
9. References 182

Chapter 9
The Young Child: Protein–Energy Malnutrition
J. Michael Gurney

1. Definitions 185
    1.1. Mild and Moderate PEM 186
    1.2. Severe Forms of PEM 188
    1.3. Interrelationships of PEM with Infections 190
    1.4. Classification of PEM 192
2. Epidemiology 195
    2.1. Nutritional Background 195
    2.2. Social and Economic Background of PEM 198
    2.3. Prevalence of PEM 202
3. Prevention and Treatment 205
    3.1. Primary Prevention 205
    3.2. Secondary Prevention 210
    3.3. Tertiary Prevention 211
4. References 212

Chapter 10
The Young Child: Obesity
June K. Lloyd

1. Prevalence 217
    1.1. Infancy 217
    1.2. Childhood and Adolescence 218
2. Natural History 219
    2.1. Infancy 219
    2.2. Childhood 221
3. Etiology 221
    3.1. Genetic Factors 222
    3.2. Energy Intake 223
    3.3. Energy Output 224
    3.4. Metabolic Factors 225
    3.5. Emotional Factors 226
4. Effects 226
    4.1. Growth and Puberty 226
    4.2. Metabolic Effects 227
    4.3. Cardiorespiratory Effects 228
    4.4. Emotional Effects 228
    4.5. Adipose Tissue Cellularity 229
5. Treatment 229
    5.1. Diet 230
    5.2. Physical Activity 232

     5.3. Psychiatric Treatment 232
  6. Prevention 232
     6.1. Prenatal Period 233
     6.2. Infancy 233
     6.3. Childhood 233
     6.4. Adolescence 234
  7. References 234

Chapter 11
The Adolescent
Felix P. Heald

  1. Introduction 239
  2. Secular Trends in Growth 240
  3. Menarche, Nutrition, and Growth 241
  4. Energy 241
  5. Protein 242
  6. Vitamins and Minerals 244
     6.1. Vitamin A 244
     6.2. Vitamin D 244
     6.3. Ascorbic Acid 244
     6.4. Folacin 245
     6.5. Vitamin B_{12} 245
     6.6. Niacin, Riboflavin, and Thiamine 245
     6.7. Vitamin B_6 245
     6.8. Minerals 245
  7. Anemia 246
     7.1. Iron Deficiency Anemia 247
     7.2. Megaloblastic Anemia 247
  8. Pregnancy 248
  9. Obesity 249
 10. References 251

Chapter 12
The Adult
Napoleon Wolański

  1. Adulthood—Stabilization or Initial Period of Regression 254
  2. Nutrition and Ecosensitivity 257
     2.1. Feeding, Ecosensitivity, and Nutritional Status 257
     2.2. Differences in Sex-Dependent Properties as a Criterion of
         Nutritional Status in Adults 260
  3. Rate of Regressive Changes, and Living Conditions (Including
     Nutrition) 261
  4. The Effects of Nutrition in Adults 263

5. Nutritional Requirements in Adulthood 265
6. Assessment of Nutritional Status 267
7. References 268

Part III • Growth Monitoring and Nutritional Assessment

Chapter 13
Optimal Nutritional Assessment
Stanley M. Garn

1. Introduction 273
2. Optimizing Length and Weight 274
3. Optimal Circumferences 277
4. The Optimal Use of Fatfolds 278
5. Optimal Use of Radiographic Information 280
6. Direct Photon Absorptiometry 284
7. Optimizing Food Intake Data 285
8. Optimizing Hematologic Determinations 287
9. Serum and Urinary Vitamins 289
10. Screening the At-Risk Group 291
11. Optimal, Necessary Follow-Up Studies 293
12. Optimal Steps in Data Recording, Reduction, and Analysis 294
13. Optimal Assessment 295
14. References 296

Chapter 14
Reference Data
Charlotte G. Neumann

1. Uses For Reference Data 299
2. Meaning of Reference Data 300
3. Criteria for Ideal Reference Data 301
4. International Reference Data 302
5. Locally Constructed Reference Standards 303
6. Genetic versus Environmental Influence on and the Use of
   Reference Data 304
7. Other Factors Affecting Reference Data 306
8. "Best" Available Height and Weight Reference Data for
   International Comparisons 306
9. Reference Data for Skinfold (Subcutaneous Fat) 307
10. Arm Circumference Reference Data 311
11. Reference Data for Head Circumference 312
12. Classification of PCM and Overnutrition 313

13. Commonly Used Classification Systems 314
    13.1. Weight for Age 314
    13.2. Combination of Weight for Age and Presence of Edema 315
    13.3. Weight for Height or Length 316
    13.4. Weight for Height Combined with Height for Age 318
14. Other Methods of Classification of PCM and Overnutrition 319
    14.1. Arm Circumference 319
    14.2. QUAC Stick: Arm Circumference for Height 320
    14.3. Skinfolds 320
    14.4. Sequential Nutritional Diagnosis 321
15. Conclusion 322
16. References 323

Chapter 15

Clinic Assessment

David Morley

 1. Introduction 329
 2. A Home-Based Record 330
 3. The Growth Chart in Undernutrition 331
 4. The Growth Chart in Recovery 333
 5. The Growth Chart, Breast Feeding, and Birth Interval 333
 6. Growth Charts and the At-Risk Child 334
 7. New Weighing Methods 334
 8. The Growth Chart in Surveillance 336
 9. Summary 336
10. References 337

Chapter 16

Anthropometric Field Methods: General

Alfred J. Zerfas

 1. Relevance of Anthropometry to Nutritional Status 339
 2. Surveys, Screening, and Surveillance 341
 3. Approach to Field Methodology 342
    3.1. Goals 343
    3.2. Objectives 343
    3.3. Scope of Work 343
    3.4. Constraints 344
 4. Application of Anthropometry with Other Measures 347
 5. Anthropometric Measures 348
    5.1. Weight 348
    5.2. Height and Length 351

    5.3. Mid-Upper-Arm Circumference 353
    5.4. Fatfold Thickness 353
  6. Methods of Measurement 356
    6.1. Weight 356
    6.2. Length 357
    6.3. Height 357
    6.4. Mid-Upper-Arm Circumference 358
    6.5. Triceps Fatfold 359
    6.6. Quality Control of Measurements 360
  7. References 362

Chapter 17
Anthropometric Field Methods: Criteria for Selection
Jean-Pierre Habicht, Charles Yarbrough, and Reynaldo Martorell

  1. Introduction 365
  2. Components of Sensitivity in Individuals 366
  3. Accuracy of Nutritional Status as a Description of Nutriture 368
  4. Dependability of an Indicator of Nutritional Status 370
  5. Precision of Measurement 372
  6. Interrelationships between Accuracy, Dependability, Precision, and
    Reliability 374
  7. Sensitivity and Sample Size 375
  8. Sensitivity in Populations 378
  9. Specificity of Response 380
 10. Feasibility 381
 11. Appendix 382
 12. References 386

Chapter 18
Anthropometric Field Methods: Simplified Methods
Adnan Shakir

  1. Discussion 389
  2. Conclusion 396
  3. References 397

Chapter 19
Presentation of Data
Miguel A. Guzmán, Charles Yarbrough, and Reynaldo Martorell

  1. Cross-Sectional Surveys 399
  2. Longitudinal Evaluations 401

3. Surveillance 402
4. General Comment 405
5. References 406

Chapter 20
Nutrition Surveillance in Developed Countries: The United States Experience
Milton Z. Nichaman and J. Michael Lane

1. Introduction 409
2. Development of Nutrition Surveillance Programs 411
   2.1. Identification of Population Nutrition Problems 411
   2.2. Identification of Surveillance Indices 412
   2.3. Identification of Data Sources 414
   2.4. Handling of Nutritional Surveillance Data 415
3. Examples from the Center for Disease Control Nutrition
   Surveillance System 416
4. Summary 429
5. References 429

Chapter 21
Nutrition Surveillance in Developing Countries, with Special Reference to Ethiopia
Mehari Gebre-Medhin

1. Principles of Nutritional Surveillance System for Developing
   Countries 432
2. Contents of the System 433
3. Operation of the System 433
   3.1. Reporting Stations 433
   3.2. Personnel 434
   3.3. Frequency of Surveillance 434
   3.4. Data Processing 435
   3.5. Administration 435
4. Preliminary Experience in Nutritional Surveillance in a Developing
   Country—The Recent Ethiopian Famine 435
5. Comments 438
6. References 440

Index 443

Introduction: Perspectives and Needs

Derrick B. Jelliffe and E. F. Patrice Jelliffe

As with all multiauthor publications, the present volume expresses many viewpoints, which is both inevitable and necessary in a subject so diffuse and with so many different perspectives as nutrition and growth. Each author has, of course, expressed his or her own views. In some cases, these views differ because some are concerned with individuals and others with communities. Often no absolute "answers" exist, but depend on the purpose of the measurements being made—for example, the levels (or standards) of reference used in assessing growth, or the dietary allowances considered as "normal."

Also, the word "growth" has various connotations and "nutrition" is an abstraction. As Tanner (1976) notes from the point of view of a research scientist "when a human auxologist talks of 'growth' without further qualification, he is thinking of growth in height. Weight he regards as a thoroughly unsatisfactory measurement, a hotch-potch of different tissues in varying proportions, in a sense the equivalent of the pulped brains of yesterday's developmental enzymologists." Conversely, serial changes in body weight are the main practical method for nutritional surveillance of young children in clinics in developing countries.

Much new knowledge has been accumulated in this field—for example, the concepts of "critical periods" as regards the lasting effects of abnormal nutrition (McCance, 1976) and the use of "growth velocity" to assess nutritional status (Tanner, 1976). Sometimes new information can add to the confusion of the applied field worker's interpretation of the situation or, still more difficult, with regard to the emphasis to be given in practical programs, including nutrition education. In turn, this can pose problems with the alteration of education of auxiliaries and of change of "messages" for mothers, especially in non-Western cultural contexts. For example, the alteration in nomenclature from "calorie" to "joule" and to the more mysterious SI units (WHO, 1977), and from "protein–calorie malnutrition" to "protein–energy malnutrition" (and other terms) may be laudable in terms of absolute logic, but has undoubtedly increased confusion in practice, at least temporarily (Jelliffe and Jelliffe, 1975).

Basic major issues still concern the interaction of nature and nurture in differing genetic groups of mankind, with possible adaptations by different communities to long-established dietary patterns and climates, and with widely varying forms, degrees, and timing of inadequate nutrition, usually intertwined with various conditioning infections and psychosocial stresses, with their hormonal consequences, such as the increased secretion of cortisol (Tanner, 1976). From these considerations stem, of course, such key issues as methods of disentangling heritability from environmental influences and, on a practical level (*Lancet,* 1978), the definition of optimal growth and hence body measurements, especially in children, which can be taken to indicate "eutrophy" for the particular genetic group, resulting from an ideal diet and protection from infections, parasitic burdens, and psychological stress. Which measurements are "ecosensitive" and which "heredosensitive" (Hiernaux, 1963)?

Should we strive for largeness as automatically best? Perhaps small is not only beautiful, but also healthier. It is still not decided whether the goals with nutrition in early childhood are concerned with producing "musclemen, geniuses, giants, dwarfs or Methuselahs" (Barness and György, 1962). The gross, whether under- or overnourished, is obvious and easy to detect. The Holy Grail of identifying the optimal still remains elusive, although the difficult task of measuring the ultimate functional effect of moderate undernutrition or overnutrition as suggested anthropometrically in early childhood has begun to be explored (Reddy *et al.*, 1976), with reference, for example, to long-term mortality, morbidity, ability to demonstrate "catch-up" or compensatory growth (McCance, 1976), and such special measurable functions as immunocompetence, endocrinological ability, intellectual development, degenerative diseases of adult life, and longevity.

The various complex and interacting community forces that affect the food availability and dietary pattern influence secular trends in growth, both upward and downward. The desiccated skeletons of pre-Conquistador Incans preserved in the Peruvian *altiplano* are notably taller than the present-day peasant population, perhaps in part due to the wider diet in the past, including the highly nutritious *quinoa (Chenopodium quinoa)*. At the present time, the upward movement of the secular trend in stature in the well-fed appears to have reached a genetic plateau, so that it seems unlikely that the hypernourished in the twenty-first century will be eight feet tall. However, with the current "urban avalanche" toward the cities in less developed Third World countries and, with suggestive evidence of apparent deterioration in young child nutrition in periurban slums, it may be that there is the danger of a stunted urban proletariat emerging, such as was found during military recruiting in Britain in World War I. Economic pressures and neocultural influences, including the cultural manipulation of advertising, are increasingly likely to have an impact on nutrition and hence growth.

In nutrition, as in any other aspects of life, fashions and trends appear, and then move aside. However, present-day focusing on the mother and fetus seems much more than this. Modern methods have enabled levels of reference for growth to be extended backwards into the fetal period, by measurements of well-dated abortions and by the use of ultrasound (McCance, 1976; Willcocks, 1977). In addition, the dyadic interaction and relationships between the pregnant and lactating mother and her fetus, young baby (exterogestate fetus), and weanling (transitional) seem increasingly to be recognized as a previously neglected continuum of profound significance to the nutrition and growth of the young human organism both *in utero*, at birth (Sterky and Mellander, 1978), and for at least the first few years after birth (Jelliffe and Jelliffe, 1978). Perhaps the useful, but biologically divisive, categorization of scientific medicine into pediatrics and obstetrics may be partly responsible for this relatively blind spot. The interrelationships between nutritional intake, body physique, fatness, maintenance of menstrual cycles, and fertility will be vital areas of concern, especially in developing countries (Frisch, 1977, 1978). Likewise, it seems clear that modern research into the nutritional needs of the fetus and the composition of human milk not only emphasizes the need to redefine recommended dietary allowances at this stage of life, but also indicates the difficulty of describing the giving of nourishment—a biological process—exclusively in chemical terms (Hall, 1978).

Both "nutrition" and the assessment of "growth" evoke many different images, from ultrasophisticated laboratory analyses to seemingly oversimplified applications in the field. In the present volume, it is hoped that these two wings have been represented and blended—ranging, for example, from the sophisticated statistical selection of measurements to the use of such valuable, but approximate, applied methods that can enable mothers themselves to weigh and chart the growth of their young children in Indonesian villages (Rohde *et al.*, 1975).

Finally, it is not the purpose of this book to concentrate overexclusively on the methods and problems of human growth that rightly preoccupy scientists in major centers in industrialized countries, as these have been covered admirably by others (Tanner *et al.*, 1976). Nor is it intended to present ready-made cookbook "how-to-do-it" solutions, methods, or dogma. The different authors are all widely experienced in their various fields, whether encompassing mainly basic science or field applications. It has been intended that they outline the present state of knowledge, with understandable reference to their own experience, and bring out the many areas of uncertainty in order to stimulate further thought and investigation. Above all, it is hoped that it has been possible to bridge between the sophisticated research approach and the village-level application. Both are valid facets of science. Both depend on the other for guidance or for validation. At the same time, application is the main world need. The measurement of the effects of good or inadequate nutrition on growth needs to be viewed principally in a practical context in relation to the development and evaluation of effective programs designed to improve the situation. If this is not the case, future observers may view our anthropometric nuances much as we do our own medieval ancestors' preoccupation with how many angels could sit on the head of a pin.

References

Barness, L. A., and György, P., 1962, Recent advances in infant nutrition, *World Rev. Nutr. Diet.* 3:1.

Frisch, R. E., 1977, Fatness and the onset and maintenance of menstrual cycles, *Res. Reproduct.* 9:1.

Frisch, R. E., 1978, Population, food intake and fertility, *Science* 179:22.

Hall, R. H., 1978, The fallacy in chemical thinking, *Entrophy Inst. Newslett.* 1:2(1).

Hiernaux, J., 1963, Heredity and environment, *Am. J. Phys. Anthropol.* 21:575.

Jelliffe, D. B., and Jelliffe, E. F. P., 1975, Kilo-what? *Br. Med. J.* 2:40.

Jelliffe, D. B., and Jelliffe, E. F. P., 1978, *Human Milk in the Modern World*, Oxford University Press, Oxford.

Jelliffe, D. B., and Jelliffe, E. F. P., 1979, *The Assessment of the Nutritional Status of the Community*, 2nd edition, Oxford University Press, Oxford.

Lancet, 1978, Nature and nurture in child growth, 1:322.

Reddy, V., Jagadeesan, V., Raghuramulu, N., Bhaskaran, C., and Srikantia, S. G., 1976, Functional significance of growth retardation in malnutrition, *Am. J. Clin. Nutr.* 29:1.

McCance, R. A., 1976, Critical periods of growth, *Proc. Nutr. Soc.* 35:309.

Rohde, J. E., Djauhar Ismail, and Rachmat Sutrisno, 1975, Mothers as weight watchers, *J. Trop. Pediatr. Environ. Child Health* 21:295.

Sterky, G., and Mellander, L. (eds.), 1978, Birthweight Distribution: An Indicator of Social Development, SAREC Report No. 8, Stockholm.

Tanner, J. M., 1976, Growth as a monitor of nutritional status, *Proc. Nutr. Soc.* 35:315.

Tanner, J. M., Whitehouse, R. H., Marshall, W. A., Healy, M. J. R., and Goldstein, H., 1976, *Assessment of Skeletal Maturity and Prediction of Adult Height*, Academic Press, New York.

World Health Organization, 1977, The SI for Health Professionals, WHO, Geneva.

Willcocks, J., 1977, The assessment of foetal growth, *Proc. Nutr. Soc.* 36:1.

I

Influences on Growth

Nutrient Needs

C. Gopalan and Kamala S. Jaya Rao

A child requires food not only for maintenance of body tissues but also for growth. The calorie and nutrient intakes vary widely from child to child and the daily fluctuations in the same child are considerable. Hence, it is easier to establish the quality of foods needed than to estimate the quantity needed to maintain health and growth. Besides, the quantity of nutrients needed to maintain good health varies from individual to individual; it is determined by various factors including the genetic make-up. Thus, the requirements for any nutrient vary over a range, for any particular physiological group. It is difficult to predict who among the group require high amounts and who will need lower amounts.

Advisable intakes of nutrients by a specific physiological group are set down in terms of recommended dietary allowances (RDA). RDA are the levels of intake considered to meet the requirements of practically all people in the specified group. RDA for infants and children are beset by the same limitations which apply to adult groups, and probably to a larger extent. Nevertheless, RDA do provide a sufficient margin of safety above minimal needs and provide important guidelines for understanding the nutrient needs of the group. RDA should not, however, be confused with actual requirements; they are always higher than the known needs of a majority of individuals.

In calculating RDA for infants and children, values considered to be consistent with satisfactory growth are taken into account. Doubts have been expressed as to whether growth alone is a sufficient criterion (Masek, 1976). It must be understood that the term satisfactory growth, by whatever definition, implies the concomitant presence of good health. This assumption, even if considered by some to be tenuous, has been fundamental to the setting down of RDA during infancy and childhood. In the absence of better criteria, growth

---

C. Gopalan • Indian Council of Medical Research, New Delhi, India. Present affiliation: SEARO, WHO, New Delhi, India. *Kamala S. Jaya Rao* • National Institute of Nutrition, Hyderabad, India.

will continue to be the yardstick for assessing nutrient needs in infants and children.

Fomon (1974) has listed five methods for estimating nutrient requirements in infants:

1. Analogy with the breast-fed infant.
2. Direct experimental evidence.
3. Extrapolation from evidence relating to human adults, infants, or growing animals.
4. Data on individuals developing deficiency signs.
5. Theoretically based calculations.

Data obtained from breast-fed infants and from those fed artificial formulas have been mostly employed for estimating nutrient requirements in infants. Here, the composition of mature human milk and the average daily feed volume have been taken into consideration. This method, however, does not give an indication of the minimum requirements but could, at best, hint at optimal requirements. It may be pertinent here to refer to a recent, interesting review (Alfin-Slater and Jelliffe, 1977) which, *inter alia*, highlights two seemingly simple and yet underappreciated pitfalls of this method.

The possible contribution of nutrient stores acquired during intrauterine life, to nutrient needs during the first 6 months of infancy, is generally not considered sufficiently. Thus, the nutrient needs of infants, computed on the basis of their nutrient intakes, may be underestimated, and conversely their actual dietary needs overestimated. The error, however, is mostly academic in nature. Normal infants are expected to be born with well-endowed nutrient stores and as long as their nutrient intake closely conforms to the composition of human milk, there can be no danger of any nutritional imbalance. On the other hand, this can have an important, practical bearing where infants born to undernourished mothers or premature infants are concerned. Such infants are known to be born with inadequate stores of several nutrients (Venkatachalam *et al.*, 1962; Iyengar and Apte, 1972) and a seemingly adequate output and composition of breast milk may still fail to satisfy the nutrient needs of these infants. This may be one of the reasons why no commercially available formula has been found to be suitable for premature infants (Fomon, 1974).

The assumption that present information on the composition of breast milk can help understand precise nutrient needs during infancy and childhood is questioned by Alfin-Slater and Jelliffe (1977) on one more ground. Most of the available information on breast milk composition has been cross-sectional in nature, for reasons that are not hard to find. At best, such studies provide information about interindividual differences but reveal nothing of intraindividual variations. Yet, the volume and composition of milk may not only show day-to-day variations but may be significantly influenced by circadian, hormonal, periprandial, and postprandial forces and by psychological factors. Thus, multiplying the value obtained at a single sampling by a fixed factor could grossly under- or overestimate the 24-hr intake of any nutrient by an infant.

From the foregoing discussion, it should be evident that our knowledge of nutrient needs during infancy is not only incomplete and inadequate, but probably to some extent misleading. The situation in regard to the nutrient requirements of older children is no better; in fact, it is even worse. There are virtually no studies in this direction, leaving a large lacuna in this important field. Most of the RDA for this group are therefore extrapolations from data on infants and adults. These limitations should be remembered as the requirements for different nutrients are discussed on the following pages.

1. Energy

Energy allowances are estimates of average needs of a group and are not recommended intakes for an individual. Energy needs of infants and children, expressed per unit body weight, are about two to three times those of adults. Energy is required by this group for basal metabolism, physical activity, and growth. Energy allowances for this group have remained fairly constant through the various revisions made by different national and international organizations.

Energy requirements for the first 6 months of infancy have been calculated from observations on infants with satisfactory growth. Gopalan and Belavady (1961) observed an average energy intake of 110 kcal/kg at birth. These observations were made on infants with low birth weights, born to poorly nourished mothers. It is doubtful whether this level of energy intake is sufficient to satisfy the energy requirements fully. Fomon and May (1958), studying infants bottle-fed pasteurized human milk, observed energy intake to fall from 140 kcal/kg body weight during the first 6 weeks of life to 94 kcal/kg during the 4- to 6-month period. An FAO Expert Committee (1957) had recommended 120 kcal/kg body weight during early infancy and about 110 kcal/kg during the latter part of infancy. The next committee, which met in 1973 (FAO/WHO, 1973), made no alterations in these figures.

It is estimated that about 50% of the total energy requirements are needed for basal activity and about 40% for physical activity and growth (Laupus, 1975). Rose and Mayer (1968) observed that physical activity is the most important variable in the energy balance of infants. They calculated that nearly 25% of the total energy is used up for physical activity. On the other hand, growth took up only about 7%: 5% for fat deposition and 1.8% for deposition of protein. Rose and Mayer, therefore, challenged the concept that the rate of growth influences calorie intake. Fomon *et al.* (1971) calculated that nearly one-third of the total energy consumed is utilized for growth, although according to the FAO/WHO Expert Committee (1973), this is true only in the first 3 months of life. Later, the proportion taken up for growth is less than 10% of the total energy uptake. However, such a division would appear to be more academic in nature. In reality it would not be wise to dissociate physical activity from growth, particularly in young children. These should be considered as interdependent variables.

Energy requirements for older children and adolescents are based on the data on consumption obtained from healthy children in the United States and the United Kingdom. Table I shows the energy requirements of children as computed by the FAO/WHO Expert Committee (1973). The requirements for children between 1 and 3 years of age are set down as 100 kcal/kg body weight. It may be noted that the figures for boys in the 16- to 19-year age group have been drastically reduced from earlier recommendations (FAO, 1957) and appear to be more realistic. It should also be noted that in communities in which a sizable proportion of the children are underweight, the requirements should be based on their age and not on their existing body weights. The higher energy intake in such cases is expected to facilitate catch-up growth.

2. Protein

Protein is needed in the diet to provide essential amino acids and nitrogen. Though it was believed that the requirement for protein could be equated with the requirement for amino acids, particularly essential amino acids, in actuality this is not so. In addition to supplying essential amino acids, protein is also needed for synthesis of other nitrogenous substances in the body.

Nitrogen is continually lost through urine, feces, perspiration, hair, skin, and other bodily secretions and excretions. Dietary protein is required to cover these obligatory losses. In addition, protein is also required for growth in children. Growth consists of deposition of new tissue and increased nitrogen

Table I. Energy Requirements during Infancy and Childhood[a]

Age (years)	Body weight	
	kcal/kg	kJ/kg
0–0.25	120	500
0.25–0.5	115	480
0.5–1	105–110	440–460
1–3	100	418
4–6	90	380
7–9	78	326
10–12		
Boys	71	297
Girls	62	259
13–15		
Boys	57	238
Girls	50	209

[a] From FAO/WHO (1973).

concentration in the body, the latter being referred to as chemical maturation. Though nitrogen retention occurs throughout the growth period, the rate is the greatest during infancy. In view of this, protein requirements are comparatively very high during infancy, particularly the first 6 months.

Essentially, two methods have been employed for deriving protein requirements of specific groups: nitrogen balance studies and the factorial approach. Of these, nitrogen balance studies are not suitable for infants, for two reasons. The studies have to be necessarily carried out for short periods, and since collection of the total losses is difficult, there is always a risk of nitrogen retention being overestimated. More importantly, the growing stage is one of nitrogen retention. With increasing protein intake, there is an increase in nitrogen retention in this group (Fomon, 1961) and data thus obtained give misleadingly high values.

The factorial approach depends on the computation of obligatory nitrogen losses observed on a minimum protein diet. This level of protein does not sustain growth and the FAO/WHO Expert Committee on Protein Requirements (1965) was of the opinion, justifiably, that an infant who is not growing but merely maintaining weight cannot be considered to be in a normal physiological state.

In view of these considerations, as with calories, protein intake compatible with a satisfactory rate of growth has been accepted as the basis for estimating protein requirements in infancy. It must therefore be emphasized again that values thus derived are not minimum requirements but are probably optimal levels. Gopalan (1956) observed a mean intake of 2 g protein/kg body weight during the first 4 weeks of life in breast-fed infants and about 1.0–1.2 g/kg body weight between 4 and 6 months of age. Fomon and May (1958) observed the intakes to be somewhat higher than this. Fomon (1960, 1961) also showed that though nitrogen retention increased with higher protein intakes, there was no additional increase in the rate of growth. Thus intakes above those observed in breast-fed infants appear to have no further known beneficial effect. Accordingly, the FAO/WHO Expert Group (1965) had recommended an intake of 2.3 g/kg body weight up to 3 months and 1.8 g/kg body weight from 3 to 6 months of life. The protein requirements during this period are expected to be met through human milk or through a suitable formula.

Beyond 6 months of age the allowances were reduced by 0.3 g/kg body weight for every 3-month period, the RDA being 1.5 g/kg for infants between 6 and 9 months of age and 1.2 g/kg for those between 9 and 12 months of age. These values were calculated from measurements of actual intakes found to promote satisfactory growth and good health. The safe level of protein intake during the last 6 months of infancy is now revised to a uniform 1.53 g/kg body weight (FAO/WHO, 1973).

Beyond 6 months of life, milk alone is not sufficient to satisfy growth needs and supplemental foods become necessary. Here, the quality of proteins fed also assumes equal significance. The FAO/WHO Expert Committee (1973) has also provided values for proteins having a lower score than egg or milk. The protein score is determined by the chemical score where the ratio of the concentration of the most limiting amino acid in the test protein to that in a reference protein is taken into account. When utilizing chemical score, the value for egg or human milk is set at 100. When the evaluation is in terms of net protein utilization (NPU), the ratio of the NPU of the test food to that of egg or human milk is to be employed. The Committee has provided intake values in terms of diets having protein qualities of 80%, 70%, and 60% relative to egg or milk. For example, for a protein with a score of 80, the recommended safe level of intake should be multiplied by a factor of 100/80.

In developing countries where supplemental foods are generally derived from plant sources, the quality and efficiency of utilization of the protein become more critical. In such situations it is recommended that the safe level of protein intake be set at 1.8 and 1.5 g/kg body weight, respectively, for the 6- to 9- and 9- to 12-month groups (Indian Council of Medical Research, 1968). These recommendations are made with the explicit understanding that only part of the total protein intake would be of vegetable origin, the rest coming through milk. The NPU of the vegetable protein is assumed to be not less than 65.

In children above 1 year of age the factorial approach is considered valid for computing protein requirements, since growth thereafter is not as rapid as during infancy. The obligatory loss of nitrogen in urine has been considered to be 1.4 mg/basal calorie and that in the feces, 0.4 mg/basal calorie (FAO/WHO, 1973). Making allowances for losses through other routes, the total obligatory nitrogen losses in children above 1 year of age is computed to be 2 mg/basal calorie each day. A provision of 7–16 mg nitrogen/kg body weight has been made for growth, the requirement for which decreases with increasing age. Comparison of nitrogen balance data with the factorial approach indicated that the values derived by the former method are consistently higher by about 30%. Though the exact reasons for this are not known, it is considered that this may be due to the fact that in children with optimal growth, the efficiency of nitrogen utilization is much lower than when the protein intake is low. Accordingly, the values obtained by the factorial method are multiplied by a factor of 1.3. The Committee was of the view that the additional 30% will also take care of interindividual and intraindividual variations in obligatory losses and in growth.

Essentially the same calculations as have been used for younger children have also been employed for computing protein requirements of older children. The requirement expressed on the basis of unit body weight decreases from 1.0 g at 4 years to 0.6 g during late adolescence. The safe levels of intake thus computed by the FAO/WHO Expert Committee (1973) are presented in Table II.

The quality of protein depends mainly on its amino acid composition, particularly the essential amino acids. For adults, eight essential amino acids have been identified: lysine, leucine, isoleucine, valine, methionine, phenylalanine, threonine, and tryptophan. The infant requires histidine also. For studying requirements of amino acids by the infant, data from Holt and Snyderman (1961) and from Fomon and Filer (1967) have been used; the lowest

Table II. Recommended Safe Levels of Protein Intake

Age (years)	g protein/kg body weight			
	Score 100[a]	Score 80	Score 70	Score 60
1	1.19	1.49	1.70	2.00
2	1.19	1.49	1.70	2.00
3	1.19	1.49	1.70	2.00
4–6	1.01	1.29	1.44	1.68
7–9	0.88	1.10	1.25	1.46
10–12				
Boys	0.81	1.00	1.16	1.35
Girls	0.72	0.90	1.03	1.20
13–15				
Boys	0.76	0.95	1.09	1.27
Girls	0.63	0.78	0.90	1.05

[a] For definition of score, see text.

levels of intake that maintained adequate growth were considered for study. From these data, amino acid requirements have been calculated for infants and expressed in terms of milligrams of protein. A similar method has been adopted for older children using the data from Nakagawa *et al.* (1964).

The intake recommended for any one nutrient presupposes that the requirements for all other nutrients and for energy are fully met. This is particularly so in the relationship between proteins and energy. Studies on experimental animals have clearly established the "protein-sparing" effects of energy in the diet (Munro, 1951). When energy is limiting, a part of the dietary protein will be diverted for energy purposes. It is increasingly recognized that the diets of preschool children or toddlers in the developing world, are grossly deficient in calories (Gopalan and Narasinga Rao, 1971; Payne, 1972; McLaren, 1974). Any attempt at meeting protein requirements in such a context will have little meaning and will be a very expensive way of supplying energy.

The NPU of diets based predominantly on cereals has been considered to be about 50. The RDA for protein for children subsisting on such diets are much higher than the recommendations made by the FAO/WHO Expert Group. The two recommendations are compared in Table III.

It has been argued on various occasions that provisions be made for extra needs during infections and minor stresses. Cuthbertson's classical studies have shown that a severe illness or injury can cause a serious loss of nitrogen (Cuthbertson, 1964). The RDA are set down for normal physiological states. The minor stresses enountered in such states and the variations they may cause in protein requirements contribute to the daily variations of protein requirements in individuals. Since provisions are already made for such daily variations, it is felt that extra allowances need not be made for minor stresses. The more seriously contended question is whether provisions should also be made for infections which are so common in childhood. The proponents of the

Table III. Recommended Safe Levels of Protein
Intake: Comparison of Protein with a Score of 100
Diets Having an NPU of 50

Age (years)	Protein score[a] 100	NPU of diet[b] 50
1	1.19	1.90
2	1.19	1.72
3	1.19	1.70
4–6	1.01	1.66
7–9	0.88	1.59
10–12		
Boys	0.81	1.48
Girls	0.72	1.44
13–15		
Boys	0.76	1.48
Girls	0.63	1.40

[a] FAO/WHO Expert Committee (1973).
[b] Predominantly cereal-based diets (Indian Council of Medical Research, 1968).

cause argue that extra protein should be provided as a buffer stock for impending infections. However, the general consensus is that there is no such entity as protein reserve (Holt and Snyderman, 1965; Chan, 1968; Waterlow, 1968). Even if the "labile protein pool"—the nitrogen lost when an individual is transferred from a high to a low protein diet—is considered to be a reserve, this hardly constitutes 1% of the total body protein (Waterlow, 1968; Young et al., 1968). Obviously there appears to be no need for an extra allowance for impending infections. This is not to deny the fact that protein requirements go up during serious illness and convalescence. However, it should be emphasized, as Holt and Snyderman (1965) maintained, that the requirements for a stress cannot be met in advance.

3. Fat

Fat is an important source of food energy. In addition, it serves two important functions in the body: it facilitates absorption of fat-soluble vitamins and provides essential fatty acids (EFA). Hansen and collaborators (1958) were the first to demonstrate the EFA need of the infant for growth and dermal integrity.

Spontaneous EFA deficiency has never been demonstrated in human infants. Only infants maintained on experimental diets providing extremely low amounts of linoleic acid have developed skin lesions (Hansen et al., 1958). On the other hand, phrynoderma is not infrequently encountered in older children. It is believed that EFA deficiency, in conjunction with deficiency of the B-complex group of vitamins, may be responsible for this (Gopalan, 1947; Srikantia and Pargaonkar, 1964).

Hansen's group (Hansen *et al.*, 1958) observed that the addition of linoleic acid to provide 1-2% of the total dietary calories could revert the skin to its normal pattern. Using the data of these workers, Holman *et al.* (1964) calculated the linoleic acid requirement of infants to be 1.4% of the total calories. The fat content and consequently the EFA concentration of human milk vary widely, the latter contributing 0.4-7% of the total dietary calories (Cuthbertson, 1976). Cuthbertson (1976) has calculated that 65 mg EFA/100 kcal can meet the daily needs of the infant with an ample margin of safety.

Studies on older children indicate that phrynoderma can develop at fat intakes which provide less than 3-4 g EFA (Srikantia and Belavady, 1961). The lesions could be cleared by administration of 7 g EFA. It is therefore suggested that diets of growing children should contain 5-10 g EFA.

A fear has been expressed that diets with low fat content will necessarily be high in carbohydrate and protein, the deleterious effects of which are not known (Fomon, 1974). The recommendations for EFA are minimal requirements needed to prevent skin lesions and to promote proper growth. It is well recognized that to provide the requisite calories and at the same time to maintain a low bulk, the diets of children and infants should contain a good amount of fat. Infant formulations may safely contain enough fat to provide 30-50% of the total calories. The safe upper limit for fat intake in older children is not known. High fat intakes are implicated in the etiology of atherosclerosis and the origins of the disease are believed to take root in early childhood. It has been suggested that the fat calories should not be more than 30% of total calories in adult dietaries (Indian Council of Medical Research, 1968; Food and Nutrition Board, 1974). Since children are more physically active, such restrictions on total fat intake are probably not necessary.

4. Vitamins

Vitamins are organic constituents of the diet, required in small amounts for the maintenance of the life and well-being of an organism. Vitamins are broadly classified into fat-soluble and water-soluble compounds. Initially, due to the absence of pure compounds, the dose of some vitamins was expressed in units defined in relation to the biological activity of the compound. This practice, which continues for some vitamins despite availability of the pure substances, should be discouraged.

Nutrition surveys and other relevant epidemiological data indicate a widespread deficiency of several vitamins, particularly in the developing countries and in underprivileged groups. Clinical signs attributable to hypovitaminosis A and to deficiencies of riboflavin and thiamine are well defined. These are encountered more frequently in children and pregnant and nursing mothers. There is also a body of opinion, though debatable, that vitamin deficiencies are fairly widespread even in industrialized, affluent communities.

Several methods can be employed to assess the minimum requirements of each vitamin.

1. *Methods based on dietary intakes.* These data, when obtained on healthy individuals, do not indicate minimum nor even optimum requirements. These may at best be considered as the upper limits consistent with good health. This is particularly so because of the widespread practice of fortification of a variety of foods with various vitamins and the equally widespread habit of consuming multivitamin preparations.

2. *Human depletion studies.* The minimum requirements of vitamins can be assessed by long-term depletion studies followed by determining the minimum intakes needed to reverse the signs or to prevent their appearance. Such studies yield information on the minimal requirements but do not give any idea of the optimum intakes which will provide for tissue saturation in the case of water-soluble vitamins or which may provide for reasonable storage in the case of fat-soluble vitamins.

3. *Studies on urinary excretion or tissue levels of vitamins.* Urinary excretion of certain vitamins decreases proportionately with intake up to a point below which changes in urinary excretion are minimal. Conversely, the excretion rises with increasing intakes. The return of the test dose or the dose retention studied 4 or 24 hr after the administration of a known dose of the vitamin is considered to indicate the degree of saturation of the tissues with the vitamin. On intakes which provide for full saturation of the tissues, return of a test dose would be very high.

As with proteins, the actual intakes of healthy, breast-fed infants with a normal pattern of growth have been used for setting down RDA of vitamins during infancy. Intake data on older children are woefully lacking for most of the vitamins. Extrapolations from intakes of infants and RDA for adults have been generally used for this group. Where a vitamin is known to be directly involved in energy metabolism, for example thiamine or riboflavin, recommendations have been made in relation to the recommended energy intake. In other instances, these have been set down in relation to body size.

4.1. Fat-Soluble Vitamins

4.1.1. Vitamin A

Vitamin A is present in the diet either as preformed vitamin A (retinol) or as its precursors, the most important of which nutritionally is β-carotene. The values of the vitamin are expressed in units of weight; the practice of expressing them in terms of international units is obsolete. The international unit is equivalent to 0.3 μg retinol (WHO, 1950).

From calculations of average milk intake per day and the average retinol content of human milk, it is estimated that breast-fed infants belonging to the well-nourished communities may receive 50–65 μg retinol/kg body weight daily. In undernourished communities the vitamin A content of human milk is low (Belavady and Gopalan, 1959) and it is computed that infants in these regions may receive at most 40 μg retinol/kg body weight (Gopalan and Belavady, 1961). Despite this, the growth of such infants is fairly satisfactory. However, it is not certain whether this low intake provides for adequate liver storage. Perhaps it does not, as many of these infants develop ocular signs of vitamin A deficiency soon after weaning. In fact hypovitaminosis A in children is one of the major public health problems in developing countries (McLaren, 1963; Oomen *et al.*, 1964; Srikantia, 1975). In view of this, it may be presumed that 40 μg retinol/kg may represent only the minimum requirement adequate to sustain growth. This assumption will hold good only provided the infants do not simultaneously draw upon their tissue reserves, a possibility that cannot entirely be ruled out. The FAO/WHO Expert Committee (1967) has recommended a total daily intake of 420 μg retinol during the first 6 months of infancy. This works out to about 65 μg/kg at the end of 3 months and 50 μg/kg at the end of 6 months of age.

Citing the review by Rodriguez and Irwin (1972), Anderson and Fomon (1974) have advocated an intake of 20 μg retinol/kg daily for infants. This was based on early studies in which weight gain and dark adaptation were taken as criteria for estimating requirements. Later studies (Lewis and Bodansky, 1943) indicated that these criteria were not as sensitive as plasma vitamin A levels, and based on these levels, an intake of 60 μg/kg had been advocated. The levels advised by Anderson and Fomon are much lower than those consumed by infants in developing countries, and the reasons for not considering the latter values as optimal have been discussed earlier. Hence the levels suggested by Anderson and Fomon appear to be unrealistically low.

Vitamin A requirements of children of other age groups have been computed from the RDA for infants and adults (FAO/WHO, 1967). The two points were joined by a curve based on body weight increments per kilogram of body weight at various ages. RDA have been computed for various age groups from this curve. These are presented in Table IV.

Barring the United States and other technologically advanced countries

Table IV. RDA for Retinol, Vitamin B₁₂, Folate, and Ascorbic Acid[a]

Age (years)	Retinol[b] (μg)	Vitamin B_{12} (μg)	Folic acid (μg)	Ascorbic acid[c] (μg)
0–0.5	420	0.3	40	20
0.5–1	300	0.3	60	20
1–3	250	0.9	100	20
4–6	300	1.5	100	20
7–9	400	1.5	100	20
10–12	575	2.0	100	20
13–15	725	2.0	200	30

[a] RDA set down by FAO/WHO (1967, 1970).
[b] Where a large part of dietary vitamin A is expected to be consumed as β-carotene, RDA must be multiplied by 4.
[c] Where consumption of foods in raw state (fruit, salads, etc.) is not common, RDA should be doubled.

where nearly half of the total dietary vitamin A is derived from animal foods as retinol, in the majority of the world dietary vitamin A is taken mainly in the form of β-carotene from tubers, vegetables, fruits, and so on (FAO/WHO, 1967). Though theoretically 1 μg β-carotene should equal 0.5 μg retinol, this may not be so because of the low availability of β-carotene. The FAO/WHO Expert Committee has assumed the absorption of β-carotene to be only 33%. Accepting a 50% conversion to retinol, the Committee suggested that 1 μg β-carotene be considered nutritionally equivalent to only 0.167 μg retinol. However, some recent studies (Lala and Reddy, 1970; Nageswara Rao and Narasinga Rao, 1970) show that β-carotene from fruits, greens, and cereal-based diets is absorbed to an extent of 70–90%. Accepting a minimum absorption of 50%, it is suggested that 1 μg β-carotene be considered equivalent to 0.25 μg retinol. In areas in which plant foods play a significant role in the diets of children, a significant part of dietary vitamin A will be available as β-carotene. Keeping in view the lower biological potency of β-carotene, the recommended intakes of vitamin A should be appropriately raised in such instances.

The fear that infants and children may not be receiving adequate vitamin A through their diets has led to the widespread practice of offering them supplements in the form of retinol esters and to fortification of infant foods with the vitamin. This carries an inherent danger of leading to hypervitaminosis A. It is therefore advocated that vitamin preparations containing high doses of retinol, say more than 1500 μg, should not be prescribed for routine use in growing children.

4.1.2. Vitamin D

Vitamin D is obtained through diet as ergocalciferol and cholecalciferol. It is also formed in the skin through the conversion of 7-dehydrocholesterol to cholecalciferol by the action of ultraviolet light.

It is difficult to assess the amount of cholecalciferol formed in the skin. It is also difficult to estimate dietary intake of the vitamin, due to lack of specific and sensitive methods of assay of the vitamin. Recent studies indicate that the active form of vitamin D in the body is not cholecalciferol but 1,25-dihydroxycholecalciferol. The transformation of cholecalciferol to its hydroxylated forms occurs in the liver and kidney (Omdahl and Deluca, 1973). It is now possible to measure blood levels of this active metabolite (Belsey *et al.*, 1974; Edelstein *et al.*, 1974) and this may aid in assessing more accurately the vitamin D status of a population group.

The Expert Committee of the FAO/WHO (1970) has reviewed data which showed that although 2.5 μg (100 IU) of vitamin D can prevent rickets in full-term infants, greater absorption of calcium and better growth can occur with intakes up to 10 μg. Since the availability of the vitamin through the action of sunlight is unknown and since in temperate and cold climates the exposure of infants to sunlight may be insufficient, the Committee has recommended an intake of 10 μg of vitamin D daily for infants.

In tropical countries where ample sunlight is available and where the exposure of infants to sunlight may also be adequate it is expected that endogenous synthesis of cholecalciferol may occur at a fairly good rate. However, as mentioned earlier, there is no way of assessing this. On the other hand, the fairly widespread prevalence of rickets in these regions suggests that endogenous synthesis may not meet the total requirements of infancy and early childhood. It is, however, felt that in such regions the RDA may be set at 5 μg and not at 10 μg (Indian Council of Medical Research, 1968). It is necessary to emphasize that this figure is purely arbitrary, as is the figure of 10 μg. Ten micrograms may be considered to be the upper limit and infants and children may not need more than this. Since rickets is most prevalent in the preschool age, the FAO/WHO Expert Committee has recommended the same intake of 10 μg/day for children up to 6 years of age. Thereafter the RDA is reduced to 2.5 μg.

4.2. Water-Soluble Vitamins

In view of the well-recognized role of thiamine, riboflavin, and niacin in energy metabolism, the requirements for these vitamins are considered in relation to energy intake. The biological variation among individuals is taken care of by the addition of 20% of the estimated minimum requirements. From birth to 6 months of age intakes through breast milk which are consistent with satisfactory growth have been considered. In the absence of adequate data, for children of all other age groups, the RDA for every 1000 kcal ingested have been accepted to be the same as those for adults. This underscores the sad lack of data for nutrient requirements during growth.

4.2.1. Thiamine

The FAO/WHO Expert Committee (1967) reviewed data providing information on levels of thiamine intake associated with clinical signs of thiamine deficiency and beriberi. From this, the Committee deduced that the minimum requirement of the vitamin would be 0.33 mg/1000 kcal. Allowing for a 20% individual variation, the RDA were set at 0.4 mg/1000 kcal.

Studies from India indicate a minimum requirement of 0.2 mg/1000 kcal for adults (Bamji, 1970). Thiamine losses during cooking are quite considerable. Milling of cereals, as is done with rice, is a major route of thiamine loss. In countries in which rice forms the staple food even for children, this important factor needs to be remembered. The practice of washing rice prior to cooking, as followed in some countries, also adds to the losses. In such situations RDA may actually have to be doubled for thiamine as obtained from raw foods. In view of this, it may be safer to accept a figure of 0.5 mg/1000 kcal (Indian Council of Medical Research, 1968). The Food and Nutrition Board of the United States (1974) has also recommended 0.5 mg thiamine/1000 kcal.

In infants above 6 months of age and in older children, the RDA for thiamine have been calculated in terms of recommended calorie intakes. Breast

milk supplies only about 0.20–0.25 g thiamine/1000 kcal (Belavady and Gopalan, 1959; Macy and Kelly, 1961; WHO, 1965). The FAO/WHO Expert
Committee (1967) therefore felt that infants under 6 months of age may not
have the requirement of 0.4 mg/1000 kcal, which is that of adults.

4.2.2. Riboflavin

The FAO/WHO Expert Committee (1967) has computed adult requirements for riboflavin from available data on diet surveys and controlled depletion–repletion studies. It was observed that clinical deficiency was associated
with intakes of 0.25 mg/1000 kcal (Hills et al., 1951; Pargaonkar and Srikantia,
1964). Urinary excretion was found to be low even at intakes of 0.3–0.4 mg/
1000 kcal (Williams et al., 1943; Horwitt et al., 1950; Pargaonkar and Srikantia,
1964). From these data, the Committee concluded that the minimum adult
requirement would be 0.44 mg/1000 kcal; allowing for individual variations, an
RDA of 0.55 mg/1000 kcal was arrived at.

Owing to the lack of data on children, the Committee considered that the
RDA for adults may be applied to all children above 6 months of age. The
FAO/WHO Expert Committee has made no specific recommendations for
infants below 6 months. From the data of Gopalan (1956) and of Belavady and
Gopalan (1959), the riboflavin intake of breast-fed infants appears to be above
0.25 mg/1000 kcal. Other data indicate a figure of about 0.5 mg/1000 kcal (Macy
and Kelly, 1961). It would therefore appear that the RDA of 0.55 mg/1000 kcal
recommended for adults may be accepted for infants also.

4.2.3. Niacin

Niacin, as its coenzymes the nicotinamide adenine dinucleotides, takes
part in glycolysis and in the electron transfer reactions of the respiratory chain;
hence, its requirement is related to energy metabolism. However, since tryptophan can be converted to niacin in the body, it also has a special relation to
protein metabolism. Therefore, in any consideration of niacin requirements,
dietary niacin and tryptophan should be taken together. It is generally agreed
that 60 mg tryptophan is equivalent to 1 mg niacin (Horwitt e al., 1956;
Goldsmith et al., 1961). Dietary allowances are made in terms of niacin equiv-
alents which are computed from the niacin and tryptophan contents of the
diets.

From data on depletion–repletion studies, the FAO/WHO Expert Committee (1967) arrived at an RDA of 6.6 niacin equiv/1000 kcal daily for adults,
as well as for children of all age groups.

Mature human milk contains on an average 150–180 µg niacin and about
15–18 mg tryptophan/100 ml (Deodhar and Ramakrishnan, 1959; Macy and
Kelly, 1961). This works out to about 6 niacin equiv/1000 kcal. Hence the
RDA for niacin for infants from birth to 6 months appears not to be different
from that of adults.

4.2.4. Vitamin B₁₂ and Folic Acid

One of the important biological processes in which both vitamin B_{12} and folic acid are involved is the synthesis of DNA. Deficiency of either one or both of these vitamins leads to defective DNA synthesis, which is strikingly evident in developing erythrocytes and leukocytes, manifesting as megaloblastic anemia.

Megaloblastic anemia of nutritional origin is not uncommon in infancy. Although it is generally believed that this is mostly due to folate deficiency, there are studies to show that insufficient vitamin B_{12} in breast milk can also be an important cause (Jadhav *et al.*, 1962; Srikantia and Reddy, 1967), particularly in developing countries where the intake of animal foods is inadequate or even negligible. In breast milk of mothers whose infants had megaloblastic anemia, the concentration of vitamin B_{12} was less than 10 ng/100 ml, and thus the infant received less than 100 ng vitamin B_{12} per day (Baker *et al.*, 1962; Srikantia and Reddy, 1967). Human milk normally supplies 300 ng of the vitamin per day (FAO/WHO, 1970). It is therefore felt that infants may require about 0.3 μg vitamin B_{12} daily. The RDA for other children have been arbitrarily fixed by the FAO/WHO Expert Committee (1970) based on their calorie requirements; these values are presented in Table IV.

Unlike vitamin B_{12}, which is present only in animal foods, dietary folates are present in both animal and plant foods, liver, meat, and leafy vegetables being good sources of the vitamin. Not many studies are available to determine folate requirements during growth. In one study it was observed that the infant's daily requirement may be between 20 and 50 μg (Sullivan *et al.*, 1966). The FAO/WHO Expert Committee (1970) has suggested an RDA of 40 μg for infants of 0–6 months and 60 μg for infants of 7–12 months. The requirements for children of other age groups have again been arbitrarily fixed, the values being 100 μg up to 13 years of age and 200 μg for adolescents (Table IV).

4.2.5. Vitamin B₆

Various clinical syndromes attributable to vitamin B_6 deficiency have been described in infants and children, well defined among which are a convulsive disorder and pyridoxine-responsive anemia (Coursin, 1964). However, knowledge regarding requirements of vitamin B_6 during growth is rather limited. Mature human milk supplies 10–20 μg pyridoxine/100 ml (Macy and Kelly, 1961). Thus infants under 6 months may be expected to receive up to 200 μg vitamin B_6 daily. Vitamin B_6 requirements are determined in relation to protein intake, in view of the vitamin's well-known role in transaminase reactions, tryptophan metabolism, and so on. The requirement is said to be around 10 μg/g protein and the RDA for infants under 6 months has been set down as 0.4 mg/day (Sauberlich, 1964; Anderson and Fomon, 1974). For older infants and children the RDA is 0.5–1.2 mg (Food and Nutrition Board, 1974). For adolescents the RDA is the same as for adults, 2 mg.

Besides the vitamins of the B group discussed here, several others, such as biotin, inositol, and pantothenic acid, are also required in human nutrition and growth. Specific deficiency symptoms directly attributable to the deficiencies of these vitamins have not been described in humans. It is possible that these vitamins are present in common foods in amounts sufficient to satisfy normal requirements. It is perhaps for this reason that specific recommendations for requirements are not made for these vitamins.

4.2.6. Ascorbic Acid

Ascorbic acid is known to take part in various metabolic processes but no definite coenzyme role has been ascribed to it. It is involved in connective tissue metabolism and wound healing, in the metabolism of certain amino acids such as tyrosine, in hemostasis, and so on. Its prophylactic role in warding off infections is severely contested. Ascorbic acid deficiency in infants manifests as scurvy, but scurvy is never seen in a breast-fed infant with satisfactory growth. It is seen in babies between 5 and 24 months of age with a peak incidence between 8 and 11 months (Woodruff, 1964; Anderson and Fomon, 1974).

The average daily intake of ascorbic acid of an infant through breast milk is about 30 mg daily (Woodruff, 1964). However, the FAO/WHO Expert Committee (1970) considered this to be only 20 mg and set the RDA at 20 mg for infants below 6 months. A daily supplement of 10 mg was found to be adequate to prevent scurvy (Woodruff, 1964), and allowing a sufficient safety margin, the RDA for the latter half of infancy can be considered to be 20 mg. In view of this, the RDA of 20 mg suggested for early infancy would also appear to be adequate. For older children the RDA have been arbitrarily fixed as 20 mg up to 12 years and as 30 mg for adolescents. Ascorbic acid losses during cooking are considerable and where raw foods are not generally included in the diet, it is suggested that the allowances be doubled (Indian Council of Medical Research, 1968).

5. Minerals

Of the various minerals required for human nutrition, only two major ones are being considered here, iron and calcium.

5.1. Iron

Iron plays an important role in respiratory processes. Almost all iron in the body is found bound to protein. Iron may be present complexed to porphyrin as heme compounds, the most important of which is hemoglobin. Others of importance are myoglobin, cytochromes, peroxidases, and catalases. Important among the nonheme, iron-containing proteins are ferritin, the storage form of iron; transferrin; the transport protein for iron; and lactoferrin in milk. Iron is also believed to play an important role in immune processes (Chandra, 1976).

Iron deficiency anemia is widely prevalent in children, particularly in developing countries (Gopalan and Srikantia, 1973). The prevalence rates reported from various regions are Africa 30–60%; Asia 50–90%; South America 15–50%; and the Middle East 25–70% (FAO/WHO, 1970). Children between 6 and 24 months of age in these regions have a very high incidence of iron deficiency anemia. The reasons for these are manifold. The infants are born with poor iron stores, since the mothers are generally themselves anemic. The intake of iron-rich foods and the bioavailability of iron from the diet are poor. Added to these are the high prevalence of infestation with intestinal parasites, malaria, and diarrheal diseases interfering with intestinal absorption.

Estimates of iron requirements of infants have been based on iron balance studies, calculations of increase in body iron content from birth to a specific period later on, and determining iron intakes necessary to maintain normal hematological values. Essentially similar methods have been used for estimating the requirements of preschool children and children between 6 and 12 years of age. These studies have been exhaustively reviewed (Bowering *et al.*, 1976).

At birth there is a surfeit of circulating hemoglobin. During the first 8 weeks of life this decreases and the iron released is mostly stored in the body. It is believed that mechanisms for absorption and utilization of exogenous iron are absent at birth and develop slowly (Hawkins, 1964). Thus the iron derived from maternal sources *in utero* and that present in breast milk are adequate to meet the requirements of the infant up to 6 months (Wadsworth, 1975). However, it should be noted that this pertains only to infants who have received adequate iron transplacentally and who are consequently born with adequate iron stores.

Iron requirements during growth are computed by taking into consideration the iron losses by using the factorial approach and the increase in body iron that occurs during growth. A normal full-term infant shows no increase in body iron before the end of the fourth month (FAO/WHO, 1970). Hawkins (1964) calculated the probable iron losses in adults and the probable iron content of the body at various ages. The FAO/WHO Expert Committee (1970) utilized these data to arrive at RDA for iron for older infants and children. The Committee suggested that the amount of absorbed iron should be 1.0 mg/day up to 12 years of age. The figure for boys of 13–16 years of age is 1.8 mg and for girls of the same age, 2.4 mg. The higher value for girls takes into account the additional losses due to menstrual blood loss.

While computing iron requirements two factors have to be taken into account: the biological availability of iron from foods and the presence of substances such as phytic acid which inhibit iron absorption. The absorption of iron from animal foods is not more than 20% (FAO/WHO, 1970). The availability of iron from plant foods *per se* is not less (Hawkins, 1964). However, cereal-based diets contain very high amounts of phytic acid which is a deterrent to iron absorption. Such diets are also poor in ascorbic acid, which facilitates iron absorption. The absorption of iron from diets predominantly based on cereals and other plant foods has been considered to be 10% (Indian Council of Medical Research, 1968; FAO/WHO, 1970) or even less (Narasinga Rao *et al.*, 1972). The RDA for iron suggested by the FAO/WHO Expert Committee (1970) to provide the requisite amount ultimately needed, depending on the type of diet consumed, are presented in Table V.

5.2. Calcium

During growth calcium is needed mainly to provide for bodily increments. It is estimated that from birth to adulthood about 1 kg or more of calcium is deposited in the body (Irwin and Kienholz, 1973), of which nearly 99% is in the skeleton. Disorders due to calcium deficiency have not been indisputably established and it is difficult to assess minimum requirements. People with wide ranges of intake are known to exist without any apparent disability. It has therefore been suggested that individuals or even population groups may adapt to low calcium intakes (Mitchell, 1944; Hegsted *et al.*, 1952).

Though calcium is needed for skeletal growth, rate of growth cannot be used for assessment of calcium requirements because growth is influenced by various factors. In fact it has been shown that in children with poor heights and weights due to undernutrition, calcium supplementation had no beneficial effect (Bansal *et al.*, 1964). This could be due to the overriding effect of calorie deficiency on growth retardation.

Calcium requirements during childhood have been estimated in most instances rather arbitrarily. Three types of data have been utilized—skeletal analysis and calculations of growth rate, balance studies, and dietary surveys. Most of the studies carried out to estimate calcium requirements of infants have been balance studies. These have been exhaustively reviewed (Irwin and Keinholz, 1973). Since calcium losses during infancy are believed to be neg-

Table V. Comparison of RDA for Iron Set Down by Various Groups (mg/day)

Age (years)	FAO/WHO (1970)[a]			FNB[b]	ICMR[c,d]
	10%	10–25%	25%		
0.5–1	10	7	5	15	1 (mg/kg)
1–3	10	7	5	15	15–20
4–6	10	7	5	10	15–20
7–9	10	7	5	10	15–20
10–12	10	7	5	18	15–20
13–15					
Boys	18	12	9	18	25
Girls	24	18	12	18	35

[a] Diets providing specified calories from animal foods.
[b] Food and Nutrition Board of the United States (1974).
[c] Indian Council of Medical Research (1968).
[d] Recommendations for cereal-based diets.

ligible compared to the requirements for growth (Hegsted *et al.*, 1952), it is felt that the increment in body content and the apparent retention of calcium will provide a reasonable estimate of the requirements. Fomon *et al.* (1963) observed that the retention of calcium through human milk during the first 4 months of infancy would be about 28 mg/kg daily or about 150 mg daily. Pratt (1957) considered 40–60 mg/kg daily as adequate. The availability of calcium through human milk is considered to be 50–70% and through cow's milk, about 35%. With an average calcium content of 340 mg/liter, human milk may be expected to supply 300 mg calcium daily, which may be considered to satisfy the needs of infants from birth to 6 months. For artificially fed infants and up to 12 months, the FAO/WHO Expert Committee (1962) has recommended 500–600 mg daily. It should be noted that in view of the paucity of data, the Committee felt that recommendation of "definite figures" is not justified and suggested only a range of "practical allowances" for all age groups.

For children of all other ages data on calcium accretion from birth to adulthood have been used. Though the rate of growth does not proceed at the same rate throughout childhood, enough data are not available to relate increases in body calcium to changes in weight and in height. In view of this, the FAO/WHO Expert Committee has suggested 400–500 mg daily for all children up to 9 years and 600–700 mg daily for all children from 10 to 15 years of age. It is necessary to note that these figures are suggested with the assumption that the needs for vitamin D are fully met. The values suggested by the Food and Nutrition Board of the United States (1974) are rather high and it would appear that calcium intakes of such magnitude are not necessary.

6. Conclusions

From the foregoing discussion it is evident that there is a considerable gap of knowledge regarding nutrient needs during human growth. An important reason for this is the difficulty in conducting studies in children to establish these needs, both qualitative and quantitative. In the field of trace elements particularly, one may say that there is virtually no data.

During the first half of infancy nutrient needs have mostly been computed by assessing the intakes, through breast milk, of infants with satisfactory growth rate. This latter term itself defies exact definition. Furthermore, this implies that the concentration of the nutrients in breast milk is not in excess of what the infant may need. This cannot be tested easily but it may assume significance when the composition of breast milk from different groups of women is being examined.

Much of the recommended allowances for all children beyond 6 months of age are extrapolations from data available on adults and infants. Balance studies are difficult to carry out in these age groups. On the other hand, data from dietary surveys will become less meaningful, due to the widespread practice of fortification of foods with various nutrients. Figures of actual intakes represent the range at which an individual shows no known signs of deficiency or of excess. Lately, however, there appears a tendency to consider the present intakes as "ideal requirements" (Cheraskin *et al.*, 1976). It is clearly time to examine the possible ill consequences, if any, of overnutrition not merely due to calorie excess or high fat intakes, but of high intakes of proteins, vitamins, and minerals as well. This bears scrutiny in the context of diseases such as atherosclerosis, diabetes, and cancer, whose etiopathogeneses are not well understood and whose primary manifestations are believed to take root possibly in childhood itself.

7. References

Alfin-Slater, R. B., and Jelliffe, D. B., 1977, Nutritional requirements with special reference to infancy, *Pediatr. Clin. North Am.* **24**:3.

Anderson, T. A., and Fomon, S. J., 1974, Vitamins, in: *Infant Nutrition* (S. J. Fomon, ed.), pp. 209–244, Saunders, Philadelphia.

Baker, S. J., Jacob, E., Rajan, K. T., and Swaminathan, S. P., 1962, Vitamin B_{12} deficiency in pregnancy and puerperium, *Br. Med. J.* **1**:1658.

Bamji, M. S., 1970, Transketolase activity and urinary excretion of thiamine in the assessment of thiamine nutrition status of Indians, *Am. J. Clin. Nutr.* **23**:52.

Bansal, P., Rau, P., Venkatachalam, P. S., and Gopalan, C., 1964, Effect of calcium supplementation on children in a rural community, *Indian J. Med. Res.* **52**:219.

Belavady, B., and Gopalan, C., 1959, Chemical composition of human milk in poor Indian women, *Indian J. Med. Res.* **47**:234.

Belsey, R. E., DeLuca, H. F., and Potts, J. T., Jr., 1974, A rapid assay for 25-OH-vitamin D_3 without preparative chromatography, *J. Clin. Endocrinol. Metab.* **38**:1046.

Bowering, J., Sanchez, M., and Irwin, M. I., 1976, A conspectus of research on iron requirements of man, *J. Nutr.* **106**:986.

Chan, H., 1968, Adaptation of urinary nitrogen excretion in infants to changes in protein intake, *Br. J. Nutr.* **22**:315.

Chandra, R. K., 1976, Iron and immunocompetence, *Nutr. Rev.* **34**:129.

Cheraskin, E., Ringsdorf, W. M., Jr., and Medford, F. H., 1976, The "ideal" daily niacin intake, *Int. J. Vitam. Nutr. Res.* **46**:58.

Coursin, D. B., 1964, Vitamin B_6 metabolism in infants and children, *Vitam. Horm.* **22**:756.

Cuthbertson, D. P., 1964, Physical injury and its effects on protein metabolism, in: *Mammalian Protein Metabolism* Vol. 2 (H. N. Munro and J. B. Allison, eds.), pp. 373–444, Academic Press, New York.

Cuthbertson, W. F. J., 1976, Essential fatty acid requirements in infancy, *Am J. Clin. Nutr.* **29**:559.

Deodhar, A. D., and Ramakrishnan, C. V., 1959, Studies on human lactation, *J. Trop. Pediatr.* **6**:44.

Edelstein, S., Charman, M., Lawson, D. E. M., and Kodicek, E., 1974, Competitive protein-binding assay for 25-hydroxycholecalciferol, *Clin. Sci. Mol. Med.* **46**:231.

FAO, 1957, Calorie Requirements, Report of the Second Committee on Calorie Requirements, Rome.

FAO/WHO, 1962, Calcium Requirements, Report of an FAO/WHO Expert Group, WHO Tech. Rep. Series No. 230, Geneva.

FAO/WHO, 1965, Protein Requirements, Report of a Joint FAO/WHO Expert Group, WHO Tech. Rep. Series No. 301, Geneva.

FAO/WHO, 1967, Requirements of Vitamin A, Thiamine, Riboflavine and Niacin, Report of a Joint FAO/WHO Expert Group, WHO Tech. Rep. Series No. 362, Geneva.

FAO/WHO, 1970, Requirements of Ascorbic Acid, Vitamin D, Vitamin B_{12}, Folate and Iron, Report of a Joint FAO/WHO Expert Group, WHO Tech. Rep. Series No. 452, Geneva.

FAO/WHO, 1973, Energy and Protein Requirements, Report of a Joint FAO/WHO Ad Hoc Expert Committee, WHO Tech. Rep. Series No. 522, Geneva.

Fomon, S. J., 1960, Comparative study of adequacy of protein from human milk and cow's milk in promoting nitrogen retention by normal full-term infants, *Pediatrics* 26:51.

Fomon, S. J., 1961, Nitrogen balance studies with normal full-term infants receiving high intakes of protein, *Pediatrics* 28:347.

Fomon, S. J., 1974, *Infant Nutrition,* pp. 109–117, Saunders, Philadelphia.

Fomon, S. J., and Filer, L. J., Jr., 1967, Amino acid requirements for normal growth, in: *Amino Acid Metabolism and Genetic Variation* (W. L. Nyhan, ed.), pp. 391–401, McGraw-Hill, New York.

Fomon, S. J., and May, C. D., 1958, Metabolic studies of normal full-term infants fed pasteurized human milk, *Pediatrics* 22:101.

Fomon, S. J., Owen, G. M., Jensen, R. L., and Thomas, L. N., 1963, Calcium and phosphorus balance studies with normal full-term infants fed pooled human milk or various formulas, *Am. J. Clin. Nutr.* 12:346.

Fomon, S. J., Thomas, L. N., Filer, L. J., Jr., Zeigler, E. E., and Leonard, M. T., 1971, Food consumption and growth of normal infants fed milk-based formulas, *Acta Pediatr. Scand.,* Suppl. No. 223.

Food and Nutrition Board, National Research Council, U.S., 1974, *Recommended Dietary Allowances,* 8th rev. ed., National Academy of Sciences, Washington, D.C.

Goldsmith, G. A., Miller, O. N., and Unglaub, W. G., 1961, Efficiency of tryptophan as a niacin precursor in man, *J. Nutr.* 73:172.

Gopalan, C., 1947, The aetiology of "phrynoderma," *Indian Med. Gaz.* 82:16.

Gopalan, C., 1956, Protein intake of breast-fed poor Indian infants, *J. Trop. Pediatr.* 2:89.

Gopalan, C., and Belavady, B., 1961, Nutrition and lactation, *Fed. Proc.* 20 (Suppl. 7):177.

Gopalan, C., and Narasinga Rao, B. S., 1971, Nutritional constraints on growth and development in current Indian dietaries, *Indian J. Med. Res.* 59(6):111.

Gopalan, C., and Srikantia, S. G., 1973, Nutrition and disease, *World Rev. Nutr. Diet.* 16:98.

Hansen, A. E., Haggard, M. E., Boelsche, A. N., Adam, D. J. D., and Wiese, H. F., 1958, Essential fatty acids in infant nutrition, *J. Nutr.* 66:565.

Hawkins, W. W., 1964, Iron, copper and cobalt, in: *Nutrition: A Comprehensive Treatise,* Vol. I (G. H. Beaton and E. W. McHenry, eds.), pp. 309–372. Academic Press, New York.

Hegsted, D. M., Moscoso, I., and Collazos Ch. C., 1952, A study of the minimum calcium requirements of adult men, *J. Nutr.* 46:181.

Hills, O. W., Liebert, E., Steinberg, D. L., and Horwitt, M. K., 1951, Clinical aspects of dietary depletion of riboflavin, *Arch. Intern. Med.* 87:682.

Holman, R. T., Caster, W. O., and Wiese, H. F., 1964, The essential fatty acid requirements of infants and the assessment of their dietary intake of linoleate by serum fatty acid analysis, *Am. J. Clin. Nutr.* 14:70.

Holt, L. E., Jr., and Snyderman, S. E., 1961, The amino acid requirements of infants, *J. Am. Med. Assoc.* 175:100.

Holt, L. E., Jr., and Snyderman, S. E., 1965, Protein and amino acid requirements of infants and children, *Nutr. Abstr. Rev.* 35:1.

Horwitt, M. K., Harvey, C. C., Hills, O. W., and Liebert, E., 1950, Correlation of urinary excretion of riboflavin with dietary intake and symptoms of ariboflavinosis, *J. Nutr.* 41:247.

Horwitt, M. K., Harvey, C. C., Rothwell, W. S., Cutler, J. L., and Haffron, D., 1956, Tryptophan–niacin relationships in man, *J. Nutr.* 60 (Suppl. 1):30.

Indian Council of Medical Research, 1968, *Dietary Allowances for Indians* (C. Gopalan and B. S. Narasinga Rao, eds.), National Institute of Nutrition, Hyderabad, India.

Irwin, M. I., and Kienholz, E. W., 1973, A conspectus of research on calcium requirements of man, *J. Nutr.* 103:1019.

Iyengar, L., and Apte, S. V., 1972, Nutrient stores in human foetal livers, *Br. J. Nutr.* 27:313.

Jadhav, M., Webb, J. K. G., Vaishnava, S., and Baker, S. J., 1962, Vitamin B_{12} deficiency in Indian infants, *Lancet* 2:903.

Lala, V. R., and Reddy, V., 1970, Absorption of β-carotene from green leafy vegetables in undernourished children, *Am. J. Clin. Nutr.* **23**:110.

Laupus, W. E., 1975, Nutritional requirements, in: *Nelson Textbook of Pediatrics* (V. C. Vaughan III, R. J. McKay, and W. E. Nelson, eds.), pp. 146–162, Saunders, Philadelphia.

Lewis, J. M., and Bodansky, O., 1943, Minimum vitamin A requirements in infants as described by vitamin A concentration in blood, *Proc. Soc. Exp. Biol. Med.* **52**:265.

Macy, I. G., and Kelly, H. J., 1961, Human milk and cow's milk in infant nutrition, in: *Milk: The Mammary Gland and Its Secretion,* Vol. II (S. K. Kon and A. T. Cowie, eds.), pp. 265–304, Academic Press, New York.

Masek, J., 1976, Recommended nutrient allowances, *World Rev. Nutr. Diet.* **25**:1.

McLaren, D. S., 1963, *Malnutrition and the Eye,* pp. 216–229, Academic Press, New York.

McLaren, D. S., 1974, The great protein fiasco, *Lancet* **2**:93.

Mitchell, H. H., 1944, Adaptation to undernutrition, *J. Am. Diet. Assoc.* **20**:511.

Munro, H. H., 1951, Carbohydrate and fat as factors in protein utilization and metabolism, *Physiol. Rev.* **31**:449.

Nageswara Rao, C., and Narasinga Rao, B. S., 1970, Absorption of dietary carotenes in human subjects, *Am. J. Clin. Nutr.* **23**:105.

Nakagawa, I., Takahashi, T., Suzuki, T., and Kobayashi, K., 1964, Amino acid requirements of children: Nitrogen balance at the minimal level of essential amino acids, *J. Nutr.* **83**:115.

Narasinga Rao, B. S., Prasad, S., and Apte, S. V., 1972, Iron absorption in Indians studied by whole body counting: A comparison of iron compounds used in salt fortification, *Br. J. Haematol.* **22**:281.

Omdahl, J. L., and DeLuca, H. F., 1973, Regulation of vitamin D metabolism and function, *Physiol. Rev.* **53**:327.

Oomen, H. A. P. C., McLaren, D. S., and Escapini, H., 1964, Epidemiology and public health aspects of hypovitaminosis A, *Trop. Geogr. Med.* **16**:271.

Pargaonkar, V. U., and Srikantia, S. G., 1964, Evaluation of riboflavin nutriture by urinary and red cell riboflavin, *Indian J. Med. Res.* **52**:1202.

Payne, P. R., 1972, The nutritive value of Asian dietaries in relation to the protein and energy needs of man, in: *Proceedings of the First Asian Congress of Nutrition* (P. G. Tulpule and K. S. Jaya Rao, eds.), pp. 240–255, Nutrition Society of India, Hyderabad, India.

Pratt, E. L., 1957, Dietary prescription of water, sodium, potassium, calcium and phosphorus for infants and children, *Am. J. Clin. Nutr.* **5**:555.

Rodriguez, M. S., and Irwin, M. I., 1972, A conspectus of research on vitamin A requirements of man, *J. Nutr.* **102**:909.

Rose, H. E., and Mayer, J., 1968, Activity, calorie intake, fat storage, and the energy balance of infants, *Pediatrics* **41**:18.

Sauberlich, H. E., 1964, Human requirements for vitamin B_6, *Vitam. Horm.* **22**:807.

Srikantia, S. G., 1975, Human vitamin A deficiency, *World Rev. Nutr. Diet.* **20**:185.

Srikantia, S. G., and Belavady, B., 1961, Clinical and biochemical observations in phrynoderma, *Indian J. Med. Res.* **49**:109.

Srikantia, S. G., and Pargaonkar, V. U., 1964, Follicular hyperkeratosis, *J. Trop. Med. Hyg.* **67**:295.

Srikantia, S. G., and Reddy, V., 1967, Megaloblastic anaemia of infancy and vitamin B_{12}, *Br. J. Haematol.* **13**:949.

Sullivan, L. W., Luhby, A. L., and Streiff, R. R., 1966, Studies of the daily requirement for folic acid in infants and the etiology of folate deficiency in goat's milk megaloblastic anemia, *Am. J. Clin. Nutr.* **18**:312.

Venkatachalam, P. S., Belavady, B., and Gopalan, C., 1962, Studies on vitamin A nutritional status of mothers and infants in poor communities of India, *J. Pediatr.* **61**:262.

Wadsworth, G. R., 1975, Nutritional factors in anaemia, *World Rev. Nutr. Diet.* **21**:75.

Waterlow, J. C., 1968, Observations on the mechanism of adaptation to low protein intakes, *Lancet* **2**:1091.

WHO, 1950, Expert Committee on Biological Standardization, Report of the Subcommittee on Fat Soluble Vitamins, WHO Tech. Rep. Series No. 3, Geneva.

WHO, 1965, Nutrition in Pregnancy and Lactation, Report of a WHO Expert Committee, WHO Tech. Rep. Series No. 302, Geneva.

Williams, R. D., Mason, H. L., Cusick, P. L., and Wilder, R. M., 1943, Observations on induced riboflavin deficiency and the riboflavin requirement of man, *J. Nutr.* **25**:361.

Woodruff, C. W., 1964, Ascorbic acid, in: *Nutrition: A Comprehensive Treatise*, Vol. II (G. H. Beaton and E. W. McHenry, eds.), pp. 265–295, Academic Press, New York.

Young, V. R., Hussein, M. A., and Scrimshaw, N. S., 1968, Estimate of loss of labile body nitrogen during acute protein deprivation in young adults, *Nature* **218**:568.

Genetic and Nutritional Interactions

Stanley M. Garn, Meinhard Robinow, and
Stephen M. Bailey

Differences in size, proportions, and body composition during growth and into adulthood result from interactions between the genetic material and the nutritional milieu, during both prenatal and postnatal development. So closely are nutrition and genetics related that one of these variables is often confounded with the other, and for many population differences the resolution is not yet fully clear.

There is first a *maternal effect* (or a uterine effect) that not only helps set the size of the conceptus, but may also affect size long after birth, and possibly into the second and third generation. Some large mothers have small babies despite their body size, and some small mothers have large babies, with placental size expressing (if not determining) these differences. From cross-racial marriages there is the suggestion that the smaller birth weights in U.S. blacks may exemplify the "maternal factor."

There is, second, a *nutritional effect* that operates throughout development. This is evident in the neonates of fat mothers, in the overfed infant, and in the obese adolescent. Conversely, the nutritional effect is evident in placental insufficiency, in the undernourished infant or child, and in malabsorption syndromes (themselves often of genetic origin). The rate of growth, the timing of somatic and sexual maturation, and the proportions of the adult are all affected by the level of caloric nutrition—relative to the level of energy expended.

Then there is the *genetic effect,* most evident in such syndromes as cerebral gigantism and constitutional dwarfism, but also within family lines and (to some degree) as population differences in size and proportions, in postnatal ossification timing, and in dental development. However, resemblances between genetically related individuals living together may be inflated

Stanley M. Garn and Stephen M. Bailey • Center for Human Growth and Development, University of Michigan, Ann Arbor, Michigan. *Meinhard Robinow* • School of Medicine, University of Virginia, Charlottesville, Virginia.

by the common environment, and the secular (or generational) trends force us to rethink the meaning of population differences.

These complex interactions of nutrition and genetics have many practical implications in regard to both assessment and diagnosis. Since tall parents tend to have tall children, parent-specific size standards may be appropriate especially for the progeny of the tall and the children of the short. Since body size and maturational timing are both affected by nutritional status, there is need to employ the most recent dimensional and maturational information—especially for developing (or "emerging") national groups. Since the genetically short and the developmentally retarded may be caught disproportionately in the course of nutritional screening, there is need to exclude nonnutritional sources of substandard growth.

Such interactions also bear on the strategy of research. Smoking during pregnancy may be size-reducing, not only for the conceptus but for the infant and preschool child as well. Living with obese parents may generate childhood obesity, even if the parents are adoptive parents and not the biological fathers and mothers. Studies of adopted children, children with mothers in common (but different fathers), and children with fathers in common (but different mothers) may well supplement conventional comparisons of biological relatives living together. Cross-racial adoptions may confirm some differences and deny still others, as in Japanese-American "hybrids" in Tokyo orphanages.

1. Statistics of Growth Parameters

The standard (or customary) growth parameters such as length, weight, proportions (and the indices of maturation) are quantitative traits, and may be treated as such mathematically. They are more or less "normal" in distribution, though when approached closely they are not quite Gaussian in nature. Weight tends to be skewed at many ages, fat notably so, and even body length or stature departs from the Gaussian model when population subgroups are considered, or socioeconomic extremes. The proper statistical treatment then involves calculating age-specific normalized T-scores, or—for strongly age-dependent variables—residuals from segmented regression equations. For small samples, of course, traditional statistical approaches are inevitable, but it would be a great mistake to calculate simple means and standard deviations for fatfolds in poverty-level 17-year-old American girls.

These measures are clearly multifactorial—involving both genetic and nutritional factors. They are also polygenic, for the most part, though the number of genes involved in stature may not be large, if only extremes are considered. However, different populations of the same approximate stature may differ in the number of stature-affecting genes, and this is of nutritional relevance.

How much of a dimension (such as stature) is inherited is commonly calculated by comparing relatives of differing degrees of closeness. Monozygotic (MZ) twins, for example, share all genes in common and dizygotic (DZ) twins fewer, so MZ/DZ comparisons are often used to test for "heritability." Dizygotic twins are essentially just siblings, so DZ sibling comparisons reveal the inflationary effects of very close resemblances in life-style and rearing. From the twin comparisons MZ/DZ, parent–child comparisons, and sibling comparisons, measures of heritability can be calculated that, in theory, reveal how much is genetic and how much not (Table I).

These methods are much used in animal studies, and to very useful effect. They show how much of size can be selected for, and hence the utility of a breed improvement program. Milk production, egg production, and meat production have all benefited by such analyses and their practical utilization. If H^2 (heritability) is high, then the prospects for improvement through selection are good. If H^2 is low, then little improvement can be attained by rigorous selection.

As applied to human beings, the calculations of H^2 are the same, but the implications less certain. Parent–child similarities in stature, for example, are very much the same the world around, once sample size (N) is taken into account. It is a problem if parent–child resemblances are thus as close in the poorly nourished as in the well nourished. Either nutrition has rather little impact on H^2 throughout the caloric range or there are still unaccounted environmental factors that inflate H^2 in human families (Garn and Bailey, 1978).

From our studies of genetically unrelated individuals living together, specifically parents and their adopted children, and on adoptive sibling pairs, we have come to be skeptical of the purely genetic implications of H^2. If parents and adopted children come to resemble each other in fatness and in other parameters to an extent like that of parents and nautral children, then how much genetic meaning in H^2 as commonly calculated for these measures may yet be environmental (Fig. 1)?

Stature is a continuous variable, but perhaps not so Gaussian as long presented, especially when the poor are concerned (or the rich) or the very children viewed at particular nutritional risk. Weight is not so nearly normal in distribution, and the fatness component of weight is definitely nonnormal,

Table I. Approaches to the Genetic Components of Size and Growth

Approach	Comparison	Comment
Correlational	Parent–child	Inflated by living together
Correlational	Sibling	Inflated by living together
Correlational	Monozygotic and dizygotic twins	Partial correction by excess of r_{MZ} over r_{DZ}
"Heritability" estimates	Parent–child	May be inflated if all shared-environment effects not accounted for
"Heritability" estimates	Sibling	As above
"Heritability" estimates	Monozygotic and dizygotic twins	Partial correction by the use of DZ twin pairs $H^2 = (r_{MZ} - r_{DZ})/(1 - r_{DZ})$

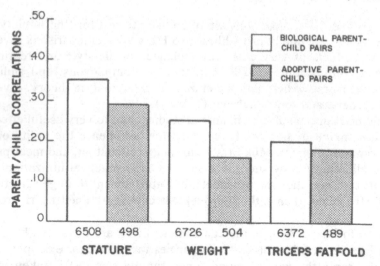

Fig. 1. Parent–child correlations for stature, weight, and the triceps fatfold for biological parent–child pairs (white bars) and adoptive parent–child pairs (solid bars). For all three variables the adoptive pairs are only slightly less similar than the biological pairs, yet remarkably higher than chance expectancy ($r \cong .0$).

nor neatly normalized by a simple logarithmic correction. Again, the simple calculation of heritability (H^2) may not be as simple (in human beings) as the animal studies have led us to think. If stature (for example) can increase 5 cm in a generation, following migration, or 2 cm in the ordinary course of events, or 5–7 cm following "Americanization" of an Oriental diet, then the environmental component of stature may be large, indeed.

Such problems do not negate the use of parent–child, sibling, or twin comparisons in estimating heritability. They do, however, indicate the existence of greater environmental variance than is generally appreciated, leading to the possibility of inflated H^2 values.

Such problems, moreover, do not invalidate populational comparisons, provided environmental differences such as intake and energy expenditure are fully accounted for. In every nation, at every socioeconomic level, tall parents will continue to have tall children, yet absolute size will continue to follow per-capita income and the Gross National Product.

2. Population Differences and National Growth Norms

Given notable population differences in size and the timing of maturation, it is logical to suggest population-specific standards for stature, weight, the ratio of weight to stature, and dental and skeletal maturation. Even the most recent income-matched comparisons in the United States show that blacks are advanced over whites in postnatal ossification timing, and in the age at eruption of the permanent teeth (Garn *et al.*, 1972, 1973b). In addition, these comparisons show that term black neonates are smaller than their white peers, yet black children are taller from the second year until age 14 (Garn *et al.*, 1973a; Garn, 1976; Garn and Clark 1976a). For Mexican-American boys and girls, and adults too, leg length comprises a smaller proportion of total length so that they have a higher weight-to-stature ratio, yet are not necessarily obese (Garn, 1976). Findings such as these, in socioeconomic context, favor the use of population-specific standards for so-called ethnic minorities, even though the genetic basis for the differences is as yet obscure (see Table II).

All standards, however, are complicated by the "secular" or generational trend, which has added 1 cm to adult stature every decade for the last 100 years. Even greater rates of secular gain have been observed in the children of migrants to Westernized countries, and in postwar Japan. In the United States, the secular trend has necessitated repeated updating of anthropometric standards. Recently, however, the secular increase in size and the lowering (advancement) of menarcheal age seem to have leveled off, at least for the middle-income groups, so the "secular trend" may not be a major limitation on growth standards in the United States, the United Kingdom, and Western Europe. Yet body size has so increased in Italy, for example, that Roman schoolchildren now exceed British size standards!

In the past, anthropometric and developmental standards were often based on a sample of the elite, under the assumption that such favored boys and girls were better nourished and enjoyed higher standards of medical care. Indeed, there is evidence that population differences in size and growth are minimal when the elite are compared, at least for the first 10 years of life. In developing countries, where there are large socioeconomic differences in nutritional status and in access to medical care, it makes good sense to develop provisional norms about samples of the elite, as in Nigeria.

In the United States, however, the "New Growth Charts" have been based on a different assumption, involving a National Probability Sample (Hamill *et al.*, 1976). The assumption here is that a National Probability Sample

Table II. Black–White Differences in Size, Maturation, and Hemoglobin Levels in the United States

Variable	Findings	Comment
Body size	Black children smaller at birth to 2 years, larger ages 2–14	Verified in Ten-State, Pre-School, and NINCDS surveys
Ossification	Black children advanced from birth through adolescence	Verified in Ten-State and National Health Examinations
Dental development	Black children advanced in later-emerging deciduous teeth and all permanent teeth	Verified (at constant income levels) in Ten-State and other surveys
Hemoglobins and hematocrits	Black children and adults lower by 0.5–1.0 g/100 ml and 2% in packed-cell volume	Verified in Ten-State and PNS surveys and after exclusion of hemoglobinopathies in Kaiser-Permanente data

best describes the population and that it is a more realistic reference sample than a selection of the elite. In practical terms, the "New Growth Charts" do a disservice to the upper-income segment, just as conventional growth charts of all kinds do a disservice to the children of the tall. Whether the upper-income population is genetically taller is not certain of course, but there are practical and political reasons to employ a National Probability Sample.

Ideally, every child deserves to be assessed against standards that are genetically appropriate, excepting of course those children with genetically determined growth-diminishing defects. The use of population-specific and even family-specific growth charts represents steps in the desired direction, even if it complicates the assessment procedure. In the clinical realm, it becomes vitally important to ascertain whether a child is an example of cerebral giantism, or merely a tall child in a tall family. Conversely, in clinical situations, it is important to employ parent-specific growth charts when evaluating the children of short parents. Moreover, in population context, in nutritional surveys, and in monitoring programs, we may have to incorporate maternal maturational timing as well (Fig. 2).

Even so, updated dimensional data (to cancel out secular trends) and population-specific standards (in mixed population groups) are the minimum requirements for standards we intend to apply.

3. Nongenetic Determinants of the Size of the Newborn

The size and the developmental status of the newborn are both influenced by a large number of factors, and it often becomes difficult to ascertain which of them are purely "genetic." Maternal size is one factor, but maternal size can be fractionated into the fat-free weight (FFW) and the fat weight (FW), both of which are separately influenced by long-term nutritional status. Maternal age is another factor in the size of the newborn (especially in the adolescent mother), parity is a factor (the uterus undergoes "training"), socioeconomic status is a factor, and so is maternal drug usage. Since mothers who smoke during pregnancy have smaller neonates, with all other variables held constant (Garn et al., 1977b,c) the smoking variable can contribute to sibling similarities in size at birth. Race, of course, is a factor, as shown in black–white comparisons at identical income/education/occupation groupings, and it is a complicating fact when we remember that black neonates are simultaneously smaller than, yet skeletally advanced over, their white peers (Table II).

There remains the problem of improving conceptual age estimates using the date of the last menstrual period (LMP). To the traditional confusions of implantation bleeding and lapsed memory have been added more recently those of disrupted or missed menstruation due to miscalculation, or following prolonged use of birth control pills. We are faced, then, with three alternatives: All uncertain LMP reports may be excluded from a given sample, regressions based on a number of anthropometric or neurological indicators may be used

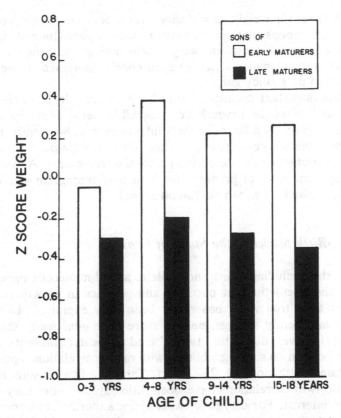

Fig. 2. Effect of maternal maturational timing (age at menarche) on weight of the offspring. As shown, the sons of early maturing mothers (white bars) are consistently heavier than the sons of late maturing mothers (solid bars) even into adolescence. Maternal maturational timing is reflected in the growth rates of the offspring of both sexes.

to verify LMP, or such regression equations may be used as a double check against the reported date of last menstruation. Clearly the first and third methods are most desirable.

Having worked with data in over 50,000 pregnancies, the importance of education, occupation, and income to the pregnancy outcome is obvious. The nonsmoking, educated, above-average-income para produces the very lowest proportion of small-for-term neonates. Viewed in these terms, both gestational prematurity and dimensional inadequacy are so largely preventable as to question their purely "genetic" nature.

In such countries as Ethiopia, economically dictated food limitations may be complicated by self-imposed caloric restrictions during pregnancy. The converse trend, of course, was the nineteenth century English tradition that the pregnant woman "eats for two." If food is withheld during pregnancy, or if eating is encouraged during pregnancy, there may be large population differences in size of the neonate, which then confound genetic meaning. If a

20-lb maximum is set for weight gain during pregnancy, we cannot compare neonatal sizes with groups that set no limit on weight gain. Indeed, initially well-nourished women with minimal weight gain during pregnancy may actually show a poorer pregnancy outcome than poorly nourished women who accept an unlimited pregnancy gain.

All of this is important because size at birth, at least in term singlings, relates to size as late as the seventh year. Small-for-term neonates remain smaller, on the average, than large-for-term infants even at 84 months of age. Although size at 7 years is genetic, in part, and nutritional (in part) we cannot forget the size restrictions imposed during prenatal development. As shown in a variety of ways, the level of prenatal nutrition may reprogram size during the postnatal years, and even into adulthood as well.

4. The Multiple-Birth Model in the Study of Growth

Multiple births, including twins and triplets, afford numerous opportunities to explore the interactions of nutrition and genetics in growth and size attainments, but their true value has rarely been fully exploited. Dizygotic (fraternal) pairs are merely siblings, and no more alike *genetically* than any pair of siblings. However, dizygotic "twins" tend to greater similarity during the growing period than do singling siblings, with many correlations approaching .6 rather than the theoretical .25–.30 (cf. Table I). The reasons for the inflated dizygotic twin correlations have been little explored, but they are of great nutritional interest. For dizygotic twins share a uterine environment in common, though with separate placentation; they share an early feeding regimen in common; they share diseases in common; and they generally share social, educational, and economic opportunities to a greater degree than more widely spaced siblings.

Monozygotic ("identical") twin and triplet sets are genetically the same individual and should approach mathematical identity for the genetically controlled portion of physical growth, maturation, and final size attainment. In addition, they share more intrauterine environment in common than dizygotic twins, and may be treated even more as replicas by their parents, peers, and associates.

It is common practice to subtract dizygotic pair correlations (DZ) from monozygotic pair correlations (MZ) to provide an estimate of heritability, thus canceling out some of the inflationary effects mentioned earlier. Yet monozygotic pairs may still be treated even more alike than dizygotic like-sexed pairs, and they may show—long after birth—the effects of differential nutrition *in utero*. The "runt" of a pair of monozygotic twins or the smallest of a set of monozygotic triplets may long remain the runt, indicating the extent to which the capacity for cellular proliferation may be determined early, and providing a useful if rarely used model in human nutrition research.

Since monozygotic sets are, genetically, the same individual, it follows that the children of monozygotic sets are half-siblings. This is a useful model indeed in experimental mating designs. Not infrequently, a set of monozygotic triplets may have 12–15 children, by three different spouses. The classic animal problems of fleshing and fattening may be so explored, and the genetic half-siblings followed through to adulthood and beyond.

Separated monozygotic sets (twins and triplets) have long been the paramount model for testing heritability. As such, their value in testing the genetic hypothesis for obesity, hyperlipidemia, and numerous other problems of nutritional interest is obvious. Though separated childhood monozygotic sets may be hard to find, separated adult monozygotic sets are the rule. It is impressive to see fatness differences of 2–3 SD (standard deviations) and to discover lipid differences of nearly equal magnitude in separated adult twin sets (Garn, 1962).

Monozygotic and dizygotic sets are accessible through twin registries, twin clubs, and twin contests. One of the best known of the twin studies was carried out entirely on twin pairs involved in a commercial advertising campaign (Osborne and DeGeorge, 1959). In a nutritional context, twins offer the optimal model with which to sort genuine environmental modifications of growth from those of the genome, and as such are well worth the trouble of locating.

5. The Genetics of Fatness and Obesity

Among the more common problems of nutrition, obesity (defined as excessive fatness) seemingly has come closest to fitting the "genetic model." Despite differences in the measure of fatness selected and in the analytic approach, there is much hard evidence that fatness follows family line.

Over 50 years ago Davenport (1925) observed striking family-line similarities in relative weight. Thirty years ago Angel (1949) reported that obese women, interviewed by him, acknowledged a great excess of obese parents and siblings. Much more recently, Withers (1964), analyzing students' reports on their parents and siblings, came to conclusions much like those of Davenport. And there are similar (anecdotal) studies from France and from Japan (Bonnet and Lozet, 1968; Matsuki and Yoda, 1971).

Using fatfolds (triceps, subscapular, iliac, and abdominal) it is easy to demonstrate parent–child correlations and sibling correlations of the orders of magnitude that would agree with a large heritability. If obese parents alone are considered, they do generate an excess of obese children. If both parents are obese, then the children become fatter and fatter through adolescence, and at age 17 their children average three times as fat as the children of the lean (Garn and Clark, 1976b).

However, the familial nature of fatness does not mean that it necessarily has a large genetic component. Husbands and wives come to resemble each other in fatness, yet they are genetically unrelated pairs. Adopted children come to resemble their adoptive "parents" in fatness, whether they are adopted strangers or the children of a spouse by an earlier marriage. Adoptive "siblings" come to resemble each other in fatness, whether they are individually adopted from agencies or orphanages or are the progeny of the separate parents (Garn *et al.*, 1976, 1977a; Hartz *et al.*, 1977) (Fig. 3). Indeed, as Mason has reported (1970), fat people even tend to have obese pets!

These findings, concentrating or genetically unrelated individuals living together, show that if "children" are adopted early, they come to resemble the adoptive parents and their adoptive siblings for reasons other than shared genes (but see Biron *et al.*, 1977).

This discovery has bearing on other "familial" conditions, among them the hyperlipidemias, for there is evidence that cholesterol levels, too, may follow family line. Although there are the essential hypercholesterolemics, high cholesterol levels may also exist within families on a nutritional and dietary basis and not because of specific genes.

6. Importance of Sample Size in Nutrition-Related Growth Research

Although most workers in human growth are familiar with the concept of statistical *significance* and with parametric tests of significance, the implications of sample size have not always been appreciated, or if appreciated put into effect. Not a few accepted "findings" will fall with larger samples, es-

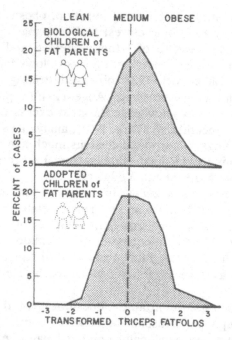

Fig. 3. Fatness distributions of the biological children of fat parents (above) and the adoptive children of fat parents (below). Whether produced by genetic union or adopted, the children of fat "parents" tend to be fat, thereby questioning the genetic hypothesis for fatness and obesity.

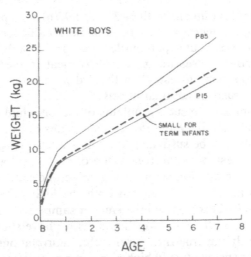

Fig. 4. Effect of birth size upon growth in size during the first 7 years. As shown here for white boys, small-for-term infants tend to remain small, through the early school years at least. Similar results obtained for white girls, black boys, and black girls, and for length and head circumference as well.

pecially after attention to socioeconomic status (SES) as mentioned earlier in this chapter.

With small samples, the null hypothesis is difficult to reject. Differences may yet be there, but with small samples they may fall below the accepted 5% confidence limits, either for the differences between means or for correlations between variables. As a particular example, which we have described in the pediatric literature (Garn *et al.*, 1977c), let us consider the implications of birth size in term neonates. Earlier reports suggested that small-for-term and large-for-term newborn singletons soon "revert to the mean," and that after a few years they are no longer size-delimited. With far larger samples, we now know that there is some canalization of size, and even as late as the seventh year, birth size is still reflected in weight, length, and head circumference as well (Fig. 4).

With relatively small samples ($N < 100$ for age and sex) the percentiles of size embodied in the Stuart–Stevenson "standards" were long accepted and employed in growth assessment the world around. With far larger samples, obtained in the Pre-School Nutrition Survey, the Ten-State Nutrition Survey, and the National Health Examinations, we now know the Stuart and Stuart–Stevenson percentiles to be inadequate. Indeed, the "New Growth Charts" differ from the older norms far more in the percentile values than in the means or medians as such.

Having explored this problem in considerable detail, using successively smaller random subsamples from the same data base, it is clear that samples of conventional size are often not large enough. A sample of 100 boys or 100 girls at a particular age may suffice for the 15th, 50th, and 85th percentiles, but for the outer percentiles (5th and 95th or 3rd and 97th) sex-specific samples of over 1000 each may be necessary for reasonable confidence (Garn *et al.*, 1977d). Remember that these outer percentiles are considerably used in monitoring programs, in growth assessment, and in clinical studies. Surely 1000 boys or girls are not too many to establish the 3rd percentile, if it is to be a clinical standard or the basis of national norms.

Sample size pertains also to such problems of parent–child similarities or similarities between siblings. Since the theoretical values of r are on the order of .2–.3, a sample of 100 may be sufficient to attain statistical significance, but scarcely large enough to establish the true value of an r. Exploring published parent–child similarities in stature, with both genetic and nutritional implications given by various authors, we find that both the high values ($r \cong .5-.7$) and the low values ($r \cong .0-.2$) come from the smaller samples. As N approaches 1000, values of r stabilize within narrower limits, as we have shown (Garn and Bailey, 1978). It is simply not true that parent–child correlations are lower in malnourished groups, or systematically higher in some populations than others, or that r is always larger in sister correlations than in brother correlations. Sample sizes, especially small sample sizes, affect obtained values of r. A great many research reports fail on this simple statistical basis.

There is need, therefore, to consider sample size to a larger degree than hitherto customary, and to repeat previous studies with suitably larger samples to test their validity. Future designs may well embody the calculations we have mentioned, indicating the need for age–specific samples as large as 1600 for many conventional purposes.

7. New Strategies for Genetic Research

For the most part, investigations into the genetics of human growth have relied upon either of two familiar strategies. The more common of these strategies is the population comparison. The less common involves genetically related relatives (Garn and Bailey, 1978).

Population comparisons should, in fact, compare populations of identical SES, in the same climate, and even in the same city in order to fulfill the investigative assumptions. Far too often, however, populations are compared without attention to income, housing, diet, or nutrition. Far too often the comparisons do not even involve the same continent, or (in some egregious examples) the same century! The use of population comparisons to confirm population differences in growth (or to reject them) necessitates truly large samples, as described in the preceding section, and great care in the matching of populations, before making the final genetic claim.

Comparisons of genetically related individuals—parents and their children and brothers and sisters—also demand rather large samples, if only to establish the magnitude of correlations within ±.02–.05. They necessitate concern with paternity, an increasingly common problem, possibly with parity (which may be a complicating factor), and exclusion of emigrants, migrants, and those with malabsorptions and intolerances that may affect the results.

There are, moreover, more novel strategies. One involves attention to dizygotic twins as compared with singleton siblings to test for environmental inflation of similarities (*vide supra*). A second includes the analysis of monozygotic pairings to test for the "runt" effect. Both of these approaches are a necessity in nutrition-oriented growth research. Alternatively, nutritional differences may suggest a third novel strategy—comparing the fatter sibling with the leaner sibling, for example. This third novel strategy is of obvious relevance to lipid research, to research on diabetes, and (in adult siblings) to studies on osteoarthritis.

A fourth novel strategy, which has many useful permutations, is the use of genetically unrelated "relatives" living together. Adopted children may come to resemble their adoptive parents in habits of food and drink, a model of value to obesity research and investigations into alcoholism. Adopted siblings present similar opportunities, especially as they mimic increasing "heritability" with advancing age. Adopted children may be either doubly adopted, i.e., unrelated to either social parent, or singly adopted, i.e., the genetic child of one parent and the adopted child of the other. This situation affords a triad of comparisons, both across and within generations, as pictured in Fig. 5 and as given in Table III.

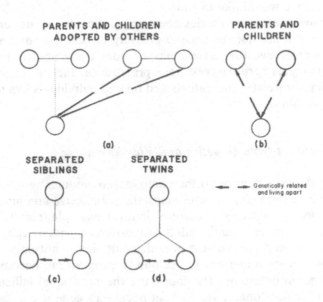

Fig. 5. Novel strategies for research into the interaction of genetic and nongenetic effects. Shown are (a) comparison of adoptive parents and their adopted children as compared with (b) parents and their biological children, (c) separated siblings, and (d) separated twins. Genetically related individuals living apart and genetically unrelated individuals living together help to establish the relative importance of genes held in common and the results of living together.

Table III. Strategies in Separating Nutritional from Genetic Effects

Strategy	Examples
A. Genetically unrelated individuals living together (correct for duration)	1. Husbands and wives 2. Parents and adopted "children" 3. Adoptive "siblings"
B. Genetically related individuals living apart (correct for length of separation)	1. Siblings separated in early life via adoption 2. Siblings separated in later life 3. Separated parent–child pairs
C. Genetically unrelated individuals previously living together and now separated	1. Divorced spouses 2. Separated parents and adopted "children" 3. Separated adoptive "siblings"

Fifth, and last, we come to genetically related individuals living apart, as a result of parental death, adoption, or simply growing up and moving away. Early adoptions provide the most rigid test of the genetic factor, but comparison of adult children and their still older parents, or of adult siblings living apart can be of profound nutritional importance. Parent–child similarities and sibling similarities in the age at death may indicate genetic control of the length of life, but it is also possible that habits learned early may account for familial longevity. A secret we would like to learn.

Lest these novel genetic strategies seem impracticable, let us remember that $\cong 30\%$ of children are not the genetic progeny of both parents, and that over 90% of children leave the nuclear nest. Under these actuarial circumstances, novel strategies here described are practicable. Indeed, in some socioeconomic groups, unrelated individuals and related individuals live together in roughly equal numbers.

8. Separating Nutrition from Genetics and Vice Versa

This chapter directs attention to the complex interactions between nutrition and genetics in determining growth, maturation, final size attainment, and even bodily proportions. Chronic undernutrition, for example, results in diminished adult size and more infantile adult proportions (paedomorphy). With improved nutrition during the growing period, adult size is not only attained earlier, and with greater dimensions, but bodily proportions become more "adultiform" or gerontomorphic. By improving the nutritional milieu, faces as well as bodies become longer, yet without necessary genetic change.

Conventional anthropometric approaches to nutritional assessment dispense with relevant genetic information in the interests of speed, simplicity, and economy. Far too often, programs of surveillance and monitoring oversimplify and thereby fail in part of their purpose. Surely parental size should be taken into account before labeling a small child as at particular nutritional "risk." Surely population differences should be taken into account before labeling a black baby small or a Mexican-American child obese.

In clinical situations, even more information can be even more informative. We would not expect a small-for-term infant to achieve normative values by 2 years of age. We would not expect the child of two small parents to achieve median length at 6 years, or even 12. The runt of a twin pair may long remain size-diminished, without demonstrable feeding impairment. Cleft-palate children and congenital heart disease (CHD) patients may remain small. As with placental insufficiencies, they have been reprogrammed prior to birth.

As shown by the secular or generational trend, many of the differences between populations are not necessarily genetic. With gains of 1 cm per decade, 10 cm in a century, and a lowering of the age of menarche by as much as 4 years, there is great nutritional modification of the capacity to grow and to mature. Under these circumstances, more sophisticated research designs are necessary to separate the energy balance from genetic action. The simple comparison of populations or of parents and their children must give way to still more sophisticated designs and strategies. Adopted children, adoptive sibling "pairs," genetically related individuals living apart, and even genetically unrelated adults living together provide the novel but necessary strategies. As a benefit, though derivative, we may then learn how much of sibling similarity in the length of life and of husband–wife resemblances in the time of death may come from their food and drink and dietary habits in common.

ACKNOWLEDGMENTS. Some of the research mentioned in this chapter was supported by Grant HD 09538 from the National Institutes of Health (Socioeconomic and Genetic Determinants of Obesity), Contract NO1-NS-5-2308 (Development of Physical Growth Measurement Data), and a grant from the Weight Watchers Foundation (Fatness and Lipid Levels). The manuscript was completed by John Humphrey.

9. References

Angel, J. L., 1949, Constitution in female obesity, *Am. J. Phys. Anthropol.* 7:433–472.

Biron, P., Mongeau, J.-G., and Bertrand, D., 1977, Familial resemblances of body weight and weight/height in 374 homes with adopted children, *J. Pediatr.* 91:555–558.

Bonnet, F., and Lozet, H., 1968, Le contexte medico-social de l'obésité chez l'enfant, *Acta Paediatr. Belg.* 22:211–252.

Davenport, C. B., 1925, Body build: Its development and inheritance, in: *Eugenics Record Office Bulletin No. 24,* Carnegie Institute of Washington, Cold Springs Harbor, N.Y.

Garn, S. M., 1962, The genetics of normal human growth, in: *De Genetica Medica* (L. Gedda, ed.), Gregor Mendel Institute, Rome.

Garn, S. M., 1976, The anthropometric assessment of nutritional status, in: *Proceedings of the Third National Nutrition Workshop for Nutritionists from University Affiliated Facilities* (M. A. H. Smith, ed.), Memphis, Tennessee, April 4–9.

Garn, S. M., and Bailey, S. M., 1978, Genetics of maturational processes, in: *Human Growth* (F. Falkner and J. M. Tanner, eds.), Plenum Press, New York.

Garn, S. M., and Clark, D. C., 1976a, Problems in the nutritional assessment of black individuals, *Am. J. Public Health* **66**:262–267.

Garn, S. M., and Clark, D. C., 1976b, Trends in fatness and the origins of obesity, *Pediatrics* **57**:443–456.

Garn, S. M., Wertheimer, F., Sandusky, S. T., and McCann, M. B., 1972, Advanced tooth emergence in Negro individuals, *J. Dent. Res.* **51**:1506.

Garn, S. M., Clark, D. C., and Trowbridge, F. L., 1973a, Tendency toward greater stature in American black children, *Am. J. Dis. Child.* **126**:164–166.

Garn, S. M., Sandusky, S. T., Nagy, J. M., and Trowbridge, F. L., 1973b, Negro–White differences in permanent tooth emergence at a constant income level, *Arch. Oral Biol.* **18**:609–615.

Garn, S. M., Bailey, S. M., and Cole, P. E., 1976, Similarities between parents and their adopted children, *Am. J. Phys. Anthropol.* **45**:539–543.

Garn, S. M., Cole, P. E., Bailey, S. M., and Higgins, I. T. T., 1977a, Effect of parental fatness levels on the fatness of biological and adopted children, *Ecol. Food Nutr.* **7**:91–93.

Garn, S. M., Shaw, H. A., and McCabe, K. D., 1977b, Relative effect of smoking and other variables on the size of the newborn, *Lancet* **2**:667.

Garn, S. M., Shaw, H. A., and McCabe, K. D., 1977c, The effects of socioeconomic status (SES) and race on weight-defined and gestational prematurity in the USA, in: *The Epidemiology of Prematurity*, Urban & Schwarzenberg, Baltimore.

Garn, S.M., Shaw, H. A., Wainright, R. L., and McCabe, K. D., 1977d, The effect of sample size on normative values, *Ecol. Food Nutr.* **6**:1–5.

Hamill, P. V. V., Drizd, T. A., Johnson, C. L., Reed, R. B., and Roche, A. F., 1976, NCHS Growth Charts, 1976, Monthly Vital Statistics Report of Health Examination Survey Data, National Center for Health Statistics, Vol. 25, No. 3, Supplement, Rockville, Maryland

Hartz, A., Giefer, E. and Rimm, A., 1977, Relative importance of the effect of family environment and heredity on obesity, *Ann. Hum. Genet.* **41**:185–193.

Mason, E., 1970, Obesity in pet dogs, *Vet. Rec.* **86**:612–616.

Osborne, R. H., and DeGeorge, F. V., 1959, *Genetic Basis of Morphological Variation: An Evaluation and Application of the Twin Study Method*, Harvard University Press, Cambridge, Mass.

Matsuki, S., and Yoda, R., 1971, Familial occurrence of obesity: An observation about height and weight of college women and their parents, *Keio J. Med.* **20**:135–141.

Withers, R. F. J., 1964, Problems in the genetics of human obesity, *Eugen. Rev.* **56**:81–90.

Nondietary Factors and Nutrition

Paul C. Y. Chen

Growth depends on adequate intake of food which must not only be digested and absorbed but also utilized by the body. Several factors influence and interfere in this process of digestion, absorption, and utilization, the chief among which are infection and metabolic disorders. Food intake itself is influenced not only by infection and loss of appetite but also by the availability of foods, which is partly dependent on socioeconomic factors, catastrophies, and the ecological balance that the human being has achieved with the environment. However, not all available and potentially edible materials are recognized or preferred as foods. Each culture views foods according to its own cultural perspective.

In this chapter, the influence of each of these and other nondietary factors is examined in relation to growth. Although the factors are examined separately, this does not imply that they are not interrelated. For example, poverty is often associated with infection which in turn is frequently accompanied by a loss of appetite and a withdrawal of solid foods or a change in diet as dictated by custom. During this time purgatives as well as other medicines, which interfere with absorption and utilization of nutrients, are often forced upon the growing child. Other social factors, such as the structure of the family, the pattern of meals, the birth interval between the child and his siblings, will in all probability influence his nutrition, the illnesses he suffers, his growth pattern, and even his chances of survival.

1. The Influence of Infection on Growth

The influence of infection on the nutritional status and growth of children has received considerable attention, particularly during the past 20 years or so. It has been observed that infection often precipitates kwashiorkor in chil-

Paul C. Y. Chen • Department of Social and Preventive Medicine, University of Malaya, Kuala Lumpur, Malaysia.

dren suffering from subclinical protein malnutrition. Infection alters not only the absorption, metabolism, and the excretion of various nutrients, but also food intake by its action on the appetite. In addition, it is often customary for solid foods to be withdrawn or to change the diet in some way, with the result that nutrients are often reduced. It may also be customary during illness to prescribe purgatives and other medicines which may interfere with absorption and the utilization of nutrients.

The effects produced by the various infectious diseases differ according to the nature of the agent, the site affected, and the age or physiologic state of the host. Children in the underdeveloped world not only carry a much heavier and more varied burden of infections but are also nutritionally in a more precarious state. Consequently, an infection which would otherwise be relatively innocuous of itself, particularly in the well-nourished populations of the economically favored countries, is often sufficient to precipitate acute malnutrition in children from the poorer areas of the developing world. Morehead *et al.* (1974) noted that two-thirds of moderate to severely malnourished children admitted into a rehabilitation center in Thailand were considered to have a potentially life-threatening infection. Puffer and Serrano (1973), commenting on the close relationship between death rates and nutritional deficiency as a contributory cause, noted that in 10 of the 25 areas studied in the Americas, nutritional deficiency was one of the causes of death in over 60% of the deaths.

Where infection and malnutrition interact for any length of time, growth may be severely retarded. Mata and Behar (1975) studying the role of infection and malnutrition on growth in a Guatemalan village, observed that growth was being retarded by infections particularly during the second year of life. Where infections precipitate kwashiorkor, damage may be serious. Montelli *et al.* (1974) observed that there were significant differences in the electroencephalographic picture and in the neuropsychomotor development of children with kwashiorkor when compared with normal of marasmic children.

Rosenberg *et al.* (1976) note that low birth weight is associated with a high incidence of maternal infection and the likelihood of transmission of infection to the fetus. Mata *et al.* (1971) observed that newborns in a low socioeconomic rural community in Guatemala were not only smaller than newborns from the well-nourished social strata of Guatemala but that a greater proportion had had intrauterine infections as evidenced by the higher levels of IgM in the cord serum

1.1. Bacterial Infections

Bacterial infections of the intestinal tract not only interfere with the absorption of nitrogen but are associated, more importantly, with an increased urinary loss of nitrogen and anorexia which in turn reduces food intake. Consequently, it has been observed that protein–calorie malnutrition is frequently precipitated by an attack of acute diarrheal disease (Jelliffe *et al.*, 1954; Waterlow and Vergara, 1956; Scrimshaw *et al.*, 1968). Jelliffe *et al.*

(1960) have noted that outbreaks of infectious diarrhea which occur in the dry season when houseflies are prevalent, are followed 3-4 weeks later by outbreaks of kwashiorkor. Poskitt (1972) has demonstrated that the incidence of kwashiorkor reaches a peak about a month after the peak incidence of infection due to measles, malaria, lower respiratory tract infection, and diarrhea. Morley (1973) has shown that whooping cough can lead to prolonged poor weight gain and marasmus. Rocha (1975) noted that 67% of the patients below the age of 18 months admitted for salmonellosis in Algeria suffered from protein–calorie malnutrition.

It has been observed that tuberculosis may precipitate the onset of kwashiorkor in children who are chronically malnourished (Pretorius *et al.*, 1956; Morley, 1959; Jayalakshmi and Gopalan, 1958; Bhattacharyya, 1961). Johnston and Maroney (1938) have shown that there is a negative nitrogen balance in children with chronically infected tonsils. It is generally agreed that most bacterial infections produce a negative nitrogen balance, there being increased urinary nitrogen excretion as well as a decreased absorption of nitrogen, and that in a poorly nourished child this may precipitate acute protein–calorie malnutrition.

1.2. Viral and Rickettsial Infections

It has been observed that, of the many communicable diseases that affect children, measles is often the most important from the point of view of nutritional stress, particularly in the economically poorer areas of the developing world. Several workers (Netrasiri and Netrasiri, 1955; Gans, 1961; Morley, 1962, 1969; Avery, 1963; Sai, 1965; Blankson, 1975) have noted that measles precipitates kwashiorkor. It has been demonstrated that, during the febrile phase of the illness, there is an increased urinary excretion of nitrogen, and that measles is accompanied by a drop in the serum albumin level. In addition, measles in poorly nourished children is often accompanied by diarrhea which further impairs intestinal absorption. Morley (1969) emphasizes that not only is measles more severe in the malnourished child but it also makes the nutrition of the child worse. In Nigeria, Morley (1973) observed that 57% of 220 Imesi village children lost 5% or more of their previous body weight during an attack of measles.

In addition to measles other viruses have also been shown to exert a negative influence on the nutritional state of a child. Salomon *et al.* (1968) have observed that not only is kwashiorkor precipitated by an attack of measles but that it is also precipitated in poorly nourished children by German measles, chicken pox, and whooping cough. It has also been observed that children with influenza have an increased nitrogen loss.

The general conclusion is that viral infections have a deleterious effect on the protein nutritional status of poorly nourished children particularly in the economically disadvantaged areas of the developing world.

Rickettsial infections, like all other infections, appear to exert an adverse effect on nitrogen balance. Harrell *et al.* (1946) have noted that patients with

Rocky Mountain spotted fever have low serum protein levels, while Beisel *et al.* (1967) have demonstrated that there is an increased urinary excretion of nitrogen in patients with Q fever.

1.3. Protozoal Infections

Since the beginning of the present century it has been known that patients with an acute febrile attack of malaria suffer from a negative nitrogen balance (Barr and DuBois, 1918). Taylor *et al.* (1949) noted that patients have a lowered serum albumin during febrile attacks while Draper *et al.* (1972) demonstrated that mean hemoglobin levels rose during the period of malaria control but fell during the resurgence of malaria. Oomen (1957) observed that Papuan children with both malaria and malnutrition had larger livers than those who had either of these conditions alone.

Studies on the effect of trypanosomiasis in monkeys and mice (Sheppe and Adams, 1957; Smithers and Terry, 1959; Woodruff, 1959) indicate that, as in malaria, trypanosomiasis produces deleterious effects on nitrogen balance.

1.4. Helminthic Infections

It has been observed that helminthic infections in some way interfere with protein absorption and utilization. This may be due in part to enzymes, such as "ascarase" found in *Ascaris lumbricoides* (Sang, 1938), which inhibit pepsin and trypsin and thus reduce protein absorption, and may be due in part to mechanical damage to the intestinal mucosa. Venkatachalam and Patwardhan (1953) have recorded a decrease in fecal nitrogen following deworming, and Darke (1959) has observed that humans with heavy hookworm infections show a lowered absorption of nitrogen compared with worm-free subjects fed the same diet. Apparently the loss of nitrogen into the gut is more than can be expected from a loss of red cells from the hookworm infection.

There is evidence to suggest that intestinal parasitism is associated with a reduced time period when food remains in the intestines, and that this "intestinal hurry" leads to decreased digestion and absorption. This combined with mechanical blockage, mucosal damage, and a competition for the food by the intestinal parasites, contributes in some degree to affect adversely a child who is heavily infected with helminths. Jelliffe (1953) has demonstrated that antihelminthic treatment of a child with prior growth failure, but without a change in the diet, leads to an increase in body weight.

1.5. Diarrhea and Breast Feeding

Scientific studies have demonstrated that human milk has specific nutritional and antiinfective properties (Jelliffe and Jelliffe, 1971; Mata and Wyatt, 1971; Gothefors and Winbreg, 1975) and that breast-feeding is associated with lower rates of attack and death from infectious diseases. Puffer and Serrano (1973) have shown that in Latin America breast feeding without weaning is beneficial, being accompanied by a smaller proportion of deaths from diarrheal disease when compared with those infants who had not been breast-fed or who had been breast-fed for less than 1 month. Kanaaneh (1972) noted that malnutrition among breast-fed infants is almost absent whereas 30% of bottle-fed infants have been found to be malnourished. It has also been noted (Plank and Milanesi, 1973) that bottle feeding is associated with three times as many deaths as compared to those infants who are wholly breast-fed.

In the economically poorer areas of the developing world, the use of anything other than breast milk often amounts to a death sentence for a young infant. Human milk protects the infant against infection in several ways. Human colostrum is very rich in immunoglobulins which protect the infant against bacterial and viral diseases to which the infant in a developing country is continuously exposed. Further, human milk contains a bifidus factor which promotes the establishment of special microflora of the intestines, thus creating an environment that is unfavorable to pathogenic organisms. It also contains other protective elements such as lysozyme, phagocytes, and antibody-producing cells which can act to contain an infection. Further, breast milk, even though not sterile, contains relatively few bacteria compared with the large doses of organisms that bottle-fed infants are exposed to in the economically poorer areas of the developing world. Finally, as a consequence of better nutrition, the breast-fed infant is better able to withstand repeated infections which in the case of the malnourished infant often lead to death.

1.6. Weanling Diarrhea

In the economically advantaged countries of the developed world, diarrhea in the young child is no longer significant as a cause of morbidity or mortality. Economic advantage as well as associated improvements in nutrition, control of communicable diseases, and improvements in environmental sanitation and in the availability of modern medicine have contributed to reduce diarrheas in the developed world to its present place of relative unimportance. However, in the technically less developed areas of the world, diarrheas continue to claim a heavy toll, in terms of morbidity and mortality, particularly among infants and young children during and after the weaning period. The term "weanling diarrhea" has been used to describe this condition (Gordon *et al.*, 1963; Scrimshaw *et al.*, 1968; Morley, 1973), in which malnutrition and diarrhea exist in synergism. The weanling is a child who is in the course of transition from breast or bottle feeding to the adult type of diet.

Weanling diarrhea is characterized by an acute onset of looseness of bowels. The diarrhea may last for 3–20 days and is often accompanied by a low grade fever, malaise, toxemia, and intestinal cramps. In malnourished children, a chronic recurrent form may develop, with kwashiorkor or other nutritional deficiency disease as a consequence.

In the majority of cases, it has not been possible to incriminate a specific infectious agent. Gordon *et al.* (1964) note that even with the more elaborate laboratory techniques, pathogenic organisms can only be isolated from one-third of diarrheal stools that are examined. Thus this diarrheal syndrome includes a minority of specific disease entities, such as salmonellosis, shigellosis, enteropathogenic *Escherichia coli* infections and amebiasis, a bulk of undifferentiated possibly infectious illnesses, and an uncertain proportion of noninfectious illnesses. In addition to the four specific infections mentioned, infections due to irregularly pathogenic bacteria, enteroviruses, helminths, and protozoa have been implicated.

In most areas of the developing world in which weanling diarrhea is a problem, it has been possible to incriminate salmonellosis, shigellosis, and enteropathogenic *E. coli* infections in less than 20% of cases. Nonetheless, the evidence available indicates that weanling diarrhea is precipitated and maintained by a synergism of infection and poor nutrition. It has been observed in Punjab (Scrimshaw *et al.*, 1968) that the fate of newborns with diarrhea is dependent largely on whether or not they are breast-fed, and that for artificially fed infants the death rate was 950 per 1000 whereas for breast-fed infants the rate was 120.4 per 1000. It was also noted that during the weaning period those that were wholly breast-fed had the lowest rate of diarrheal disease, those who were given additional milk had intermediate rates of diarrheal disease, and those who were on the complex adult type of diet had the highest rate of diarrheal disease. Further it has been noted that the 3-month period after breast-feeding had ceased carried the greatest risk of diarrheal disease.

Not only is the weanling nutritionally disadvantaged compared with the wholly breast-fed child, but he is also exposed to an adult type of diet that is often liberally spiced with peppers, for which his digestive system is poorly prepared. Added to this, environmental conditions tend to expose the child to large doses of organisms present in water and food supplies. Phillips *et al.* (1969), studying the problem of bottle feeding in Uganda, demonstrated that feeding utensils as well as prepared feeds were contaminated with bacteria, especially fecal organisms, revealing that efforts to prepare a clean feed were largely ineffective. As an added problem, during an acute attack of diarrhea, mothers often restrict the food intake of the child and thus aggravate any malnourishment that is already present, thereby precipitating kwashiorkor or other nutritional deficiency diseases.

2. The Influence of Food Practices on Growth

The human being has shown a great deal of adaptability in relation to available foods. So long as certain minimum requirements are provided, he is able to maintain normal growth and health. Consequently, he has been able to subsist on a variety of diets, ranging from a completely carnivorous diet, as among the Eskimos, to a diet that is almost wholly vegetarian, as among some mountain people of New Guinea and some vegetarian Buddhists in Asia. The human's digestive tract is able to accommodate a vast variety of edible materials, yet most cultures, unless forced by famines and other catastrophes, normally restrict themselves to only a limited variety of potentially edible materials, demonstrating their numerous food dislikes and preferences. Today, as a result of population pressures and ecological problems, such food preferences often have a negative effect on nutrient intakes and consequently upon growth itself. Such food preferences are based on the culture's view of what each food item means in relation to other aspects of the culture.

2.1. The Meaning of Food

2.1.1. Cultural Superfoods

In most cultures, one or two foods are elevated to cultural superfoods which thus acquire a semidivine status. In the rice-growing areas of Southeast Asia, rice is not only the staple food but also appears at all major ceremonies concerned with vital events. Thus, among the Malays, rice is ceremonially used at weddings to bless the bridal couple and at traditional healing ceremonies such as the *main puteri,* and is also fed to the newborn infant soon after birth. Nowadays, the practice has largely been reduced to a symbolic touching of rice paste on the infant's lips. In the past, it was usual to supplement breast milk with rice from the neonatal period onward since it was believed that rice was superior to milk. McArthur (1962) noted that if a rural Malay toddler stumbled and fell, the mother often remarked that "he isn't sturdy: that's because he doesn't eat much rice, he wants to eat fish only." Among rural Malays animal foods are served as side dishes and are appreciated not for their direct value but only indirectly since they encourage a greater consumption of rice. Other examples of cultural superfoods include *matoke* (steamed plantain) in Buganda, corn in Central America, and wheat in North India and Northern Europe.

2.1.2. Medicinal Foods

Some foods are consumed not so much for their nutritive value as for their value as medicines. Thus, among Malays, the liver of domesticated animals as well as of fish is normally discarded as waste, and is only prepared for consumption when used as a medicine, e.g., for the treatment of night blindness due to vitamin A deficiency (Chen, 1972).

2.1.3. Prestige Foods

Prestige or status foods are recognized by all cultures. The relative status of each food is often unrelated to its nutritive value. Examples include long-grained, highly polished rice in Southeast Asia, chicken in Africa, the pig in New Guinea, shark's fin soup among Chinese, and steak in the United States. Prestige foods are usually difficult to obtain and are expensive. It is not uncommon, in Malaysia and elsewhere in the developing world, to find that certain brands of expensive milk substitutes are preferred to breast milk, even when families are unable to purchase sufficient supplies for the infant. Consequently, the infant has to be fed on dilute preparations of the milk substitute and suffers as a result. Among the deep-jungle aborigines of Malaysia, the bamboo rat forms an important source of protein. However, as a result of the prejudice of the sophisticated town folk, who do not realize that there are 12 species of the Malaysian forest rat which are as clean as farm animals, the jungle-fringe aborigines avoid this cheap source of protein. Consequently, these aborigines suffer from much more malnutrition than their deep-jungle relatives who have not been shamed into discarding this vital source of animal protein.

2.1.4. Body-Image Foods

Many cultures associate foods with the physiology of the body and ill health. Many of these ideas are systematized into a body of beliefs and practices concerning "hot" and "cold" foods. Such systems exist among the Chinese, Indians, Arabs, Latin Americans, Malays, and other peoples. Molony (1975) shows that in one Mexican community there is a systematic code for determining the valence of a food, and that since women are thought to be in a "cold" condition after birth, they should eat mostly "hot" foods. A similar practice is followed by Malay women in confinement (Chen, 1973).

Malay toddlers are denied many "cold" foods including papaya and fruits for fear of illnesses (Chen, 1970b). Carstairs (1955) observed that in Rajasthan, India, "cold" foods, such as dairy products, are not only preferred but are also the more expensive ones. In Bengal, on the other hand, milk is classified as "hot." Consequently, when a child develops diarrhea, which is classified as a "hot" illness, the village mother stops feeding her child with milk since this would be dangerous even during recovery (Jelliffe, 1969).

A related concept includes the notion that some foods are nutritionally valuable while others are inferior or even poisonous. Among Malays many foods become *bisa* ("poisonous") when one is ill (Wilson, 1971). For example, it is believed that duck, peanuts, eggs, eggplant, chicken, mutton, beef, mackerel, and other foods are harmless to the normal individual but become *bisa* when the individual is ill with sores, scabies, boils, a stomach upset, a cough, or other illnesses. Consequently, when a child is ill, his already meager diet becomes even more severely restricted. Gonzales and Behar (1966) observed that among the traditional Indians, the attitude was that food was something one ate without too much concern so long as one was healthy, but was withheld when one became ill.

Cosminsky (1975) noted that in one Guatemalan community concepts of body-image foods had undergone some change. New concepts of *alimento* (glossed as a highly nutritive substance) and *fresco* (glossed as fresh or cool) had come to accommodate and reinterpret ideas and practices of both Western and traditional medicine and nutrition. Such concepts introduce the idea that certain foods maintain health and prevent illness and are likely to counterbalance the traditional emphasis on the withdrawal of certain foods when a person becomes ill.

2.2. Food Practices during the Period of Growth

2.2.1. Pregnancy

In Vietnam, the pregnant woman often restricts her diet in an attempt to have a small baby and an easy delivery (Coughlin, 1965). Similar practices have been noted among the Burmese (Jelliffe, 1969) and the Bisayan Filipino (Hart, 1965). As a consequence of such restrictions, the fetus receives inadequate stores of nutrients from its mother. In Tamilnad, the pregnant woman avoids practically all types of food, the most important of which are fruits and grain, the reason being a fear of abortion caused by "heating" the body or by inducing uterine hemorrhage (Ferro-Luzzi, 1973). In addition, they may avoid high protein foods in order to have a small baby. The pregnant Kadazan woman of Malaysia will not eat fruits such as the jackfruit as well as sweet or sour foods lest the child be born sickly (Williams, 1965). Among the Orang Asli people of Malaysia, the flesh of a large animal is taboo to women of childbearing age and children. Consequently, men 15–34 years old have a significantly higher plasma albumin level than women of similar age (Bolton, 1972). Mata *et al.* (1971) observed that among low socioeconomic Guatemalan women, the poor nutrition of the mother appeared to be reflected in the low birth weight of their newborns, there being a significant correlation between infant birth weight and the leg circumference of the corresponding mother.

2.2.2. Infancy

The unique qualities of human milk have been noted by numerous researchers, who have emphasized not only the antidiarrheal but also the nutritional importance of human milk as the sole food for infants up to 4–6 months of age and as a protein supplement thereafter during the first 2 years or more of life, particularly in the developing world. The fact that breast milk is a living food with inherent positive antiinfective properties such as the bifidus factor, lysozyme, and immunoglobulins especially in the colostrum, has been detailed by Mata and Wyatt (1971) and Gyorgy (1971) as well as by others.

The irony is that in the developing world, where breast feeding is most needed, it is now rapidly declining. For example, Orwell and Murray (1974) in Ibadan, Nigeria, Knutsson and Mellbin (1969) in Ethiopia, Sousa *et al.* (1975) in Brazil, Dugdale (1970) and Teoh (1975) in Malaysia, and Antrobus (1971) in the West Indies have all noted that breast feeding is on the decline. The high cost of packaged milks, set against the background of low incomes characteristic of the developing world, leads mothers to use dilute preparations far below those recommended by the manufacturers. McKigney (1971) notes that the minimum costs of artificial feeding will practically always be higher than the minimum food costs to support lactation. Antrobus (1971) observed that growth retardation was most marked in the 6- to 18-month-old age group and that the most important nutritional factor was the duration of breast feeding: Children who were breast-fed for longer than 9 months were heavier and grew better than those breast-fed for shorter periods. Gurney *et al.* (1972) noted that a quarter of the 9- to 24-month-old Jamaican children they studied were below 80% of the standard. The peak prevalence of malnutrition could be attributed to the widespread tendency to supplement breast feeding inadequately and early with a bottle and a failure to manage successfully the period between milk-only feeding and the family diet. Haddad and Harfouche (1971) noted that in Lebanese infants of 3 months of age, serum albumin was significantly higher in the completely breast-fed compared with those who were artificially fed.

For those groups who continue to breast-feed their infants, it has been noted that among several cultures, such as the Malays, the colostrum, which is rich in immunoglobulins, is discarded. The loss of this rich source of anti-infective food among the poorer peoples must be a further contributory factor to the burden of infection borne by their infants.

2.2.3. Early Childhood

The weaning of a child can be a much troubled period often associated with a form of diarrhea described previously as weanling diarrhea. The transition from the breast to solid foods is sometimes precipitated by pregnancy and the arrival of another infant, and may be associated with the development of kwashiorkor in the young child. In the Ga language of Accra, Ghana, the term kwashiorkor indicates "the disease of the deposed baby" (Williams and Jelliffe, 1972).

Parents sometimes do not realize that the shift from a predominantly milk diet to an adult diet requires a transitional period during which foods that are easily masticated and digestible should be provided. Often the shift is abrupt and the toddler is offered bulky adult meals that are overspiced, difficult to masticate, and indigestible (Antrobus, 1971; Chen, 1974). Often the transition is from breast to bottle to adult type of diet, and is accompanied by diarrhea. Ashcroft *et al.* (1968) note that the growth of Guyanese infants is retarded from the 4th to the 15th month, a time when mortality is high, and reflects weaning difficulties. These authors note that, during this period, Guyanese mothers reduce breast feeding and introduce bottle feeding, and have usually stopped breast feeding by 9 months. By 15 months, the children are usually on an adult type of diet.

During early childhood and, to a lesser extent, during much of his growing life, the individual is the subject of a variety of food taboos that often deny him essential protein and other foods particularly when he most needs it. For example, in Malaysia, the rural Malay toddler is often denied fish, one of the few sources of protein available to him, in the mistaken belief that it is the cause of ascariasis (Chen, 1970a). The rejection of fish as a human food is common in India among certain Hindu sects as well as in Buddhist areas including Tibet, Mongolia, and the Far East. Simoons (1974) lists over 70 groups who avoid fish as a human food in modern Africa, mainly in north-eastern, eastern, and southern Africa. In both Malaysia (Chen, 1972) and Indonesia (Van Veen, 1971), carotene-rich foods are taboo to young children.

Gonzalez (1964) notes that "strong" foods such as meat, lard, and chili are generally not given to Guatemalan children for fear that they "cannot handle them."

Many foods which would otherwise be available to young children become taboo once the child develops an illness. For example, among rural Malays, certain normally acceptable foods such as fruits and vegetables as well as mutton, beef, some types of fish, duck, prawn, and chicken become taboo during illness (McKay, 1971; Wilson, 1971). Imperator (1969) notes that dietary restrictions imposed on children with measles greatly increase morbidity and mortality. Similarly, the Nyika, like many Africans, believe that the only proper food especially for infants and the ill, is a starchy, protein-deficient porridge made from corn, millet, cassava, or rice, and that meat, fish, eggs, animal milk, and other protein foods are taboo for those who are sick (Gerlach, 1964).

In addition to food restrictions, the weaning period is often associated with the routine use of purgatives such as castor oil and the use of "bush teas." Antrobus (1971) records that in the Caribbean Islands, such teas, even if they are nontoxic, invariably take the place of a milk feed, and because of its lack of nutrients, may aggravate any existing undernutrition. The use of relatively expensive commercial food preparations can be a problem. Ongeri (1975) notes that Kenyan mothers often try to feed the child on one tin of baby food a month under the mistaken belief that tinned foodstuffs contain more protein than fresh foods.

In many parts of the developing world, it is common to have only two meals a day. Consequently, the nutritional needs of young children are difficult to meet unless extra meals are specially prepared for them. Thus, in rural Malaysia, it is not unusual for children to be sent to school with nothing more than a cup of coffee early in the morning. Among the Namu of Micronesia, it is not unusual for lunch to be missed entirely (Pollock, 1974). Hasan (1971) records that in parts of North India meals are served only twice a day. Among the Nasioi of the Solomon Islands, only one meal is cooked each day (Emanuel and Biddulph, 1969). Hautvast (1972) notes that 47% of the Tanzanian schoolchildren he studied did not have a breakfast and 13% did not have lunch. Such long gaps between meals lead to young children temporarily appeasing their hunger with "snacks" which do not provide the balanced meals that the child requires for healthy growth.

The cultural method of food sharing can contribute to undernutrition among young children. In many cultures, men, who are the productive members of the group, have first preference and the greatest share of the food prepared. For example, Selinus (1970) reports that the father, guests, and male youngsters in Ethiopia eat first, followed by the women and children, who get the leftovers. Waldmann (1975) notes that among the Bapedi of South Africa, the men are served meticulously by each wife, but women and children are served haphazardly and from dishes discarded by the men. To compound this problem, some protein foods may be selectively taboo to women and children. Thus, in Malaysia, Bolton (1972) notes that among the aboriginal Orang Asli, if a large deer is trapped and killed, the meat is eaten only by men and elderly women, watched by the younger women and children.

3. The Influence of Socioeconomic and Other Social Factors on Growth

The availability of foods is dependent on the purchasing power of the family as well as on their capacity to produce their own food supplies. This in turn is related not only to the structure of the family but also to their skills and knowledge concerning food production and to the ecological balance that the group has achieved with the environment.

3.1. Socioeconomic Influences

The influence of poverty on the ability to purchase sufficient food supplies and hence to sustain growth in the young in societies dependent on a cash economy is a well-documented fact. Whyte (1974), citing studies in India and Bangladesh, demonstrated that there is a direct correlation between levels of income expenditure and the consumption of foods. He noted that 70% of the rural people of India were at poverty levels and unable to purchase their minimum calorie requirement, and that the lack of protein in the Indian diet was not always the cause of protein deficiency, but rather that protein was diverted to provide energy in a diet deficient in energy sources. Selinus *et al.* (1971) similarly observed that in Ethiopia, the calorie intake of toddlers was below the minimum requirement and that consequently part of the protein had to be consumed for energy production. McKay (1969) demonstrated that the mid-arm circumference and weights of higher income Malay children were greater than those of lower income Malay children. Wray and Aguirre (1969) observed that among Colombian children there was much less malnutrition among those from families with incomes of 500 or more pesos. Belew *et al.* (1972) observed that the serum levels for protein, albumin, and gamma globulin of economically privileged Ethiopian children aged 30–131 months were significantly higher than those of nonprivileged children of the same age.

3.2. Food Production

3.2.1. Overpopulation, Catastrophes, and War

The ability of the group to produce sufficient food supplies is of paramount importance in any society that is not primarily based on a cash economy, since growth is dependent on food consumption. Where population increases have outstripped the ability to produce sufficient food, undernutrition, malnutrition, and even famine can be expected. For example, Pollock (1974) noted that population increases among the Namu of Micronesia had outpaced local food resources and that alternative crops such as rice were urgently required as supplements to breadfruit, the traditional staple. In addition to overpopulation, natural catastrophes due to floods, drought, earthquakes, and the devastation of food crops by pests, can have serious effects on food production and consequently on growth, disease, and death. Unsettled political situations due to wars, revolutions, and civil disturbances lead to similar consequences, as illustrated by the Nigerian/Biafran War (Aall, 1970).

3.2.2. Knowledge of Agriculture and Animal Husbandry

Sociocultural changes brought on by population pressures as well as by political changes may demand new skills in food production. When such skills are absent, groups may be subjected to varying periods when food is scarce. Selinus *et al.* (1971) reported that among the Arsi Galla of Ethiopia, who have had to shift from animal husbandry to agricultural production, the lack of an agricultural tradition and ignorance of dry farming practices kept the yields very low. Consequently the Arsi Galla annually faced "hungry seasons" when milk yields were low and corn was nearly finished.

3.2.3. Ecological Balance

Where the ecological balance with the environment has been upset, food production can be seriously affected. The most common imbalance is due to population increases which have outstripped the ability of the environment to support food production. Not uncommonly, it may be due to poor agricultural techniques or to overgrazing of pasture land leading to soil erosion and a consequent reduction in yield. For example, Cassel (1955), describing the nutritional problems faced by a group of South African Zulus, observed that, as a result of the deep symbolic value of cattle in the Zulu culture, the Zulu had overstocked and overgrazed their lands, which resulted in serious erosion of the soil. Further, as a result of being restricted to their reserves whereas previously they were a roving pastoral people, the Zulus, without changing their now outmoded agricultural practices, had exhausted their plots, oblivious of the need for returning waste products to the soil and rotating their crops.

3.3. Other Social Factors Influencing Growth

3.3.1. Social Structure of Family

The social structure of the family has a direct bearing on the availability of food and hence on growth in the young child. For example, McDowell and Hoorweg (1975) noted that in Uganda, malnourished urban children recovered better if their mothers were married and their fathers were monogamous. On the other hand, they noted that the reverse seemed to hold for the rural family, the reason being that unmarried rural mothers live in an extended family. They conclude that, in Uganda, the nuclear family has advantages in town while the extended family holds advantages in the country. The importance of the extended family in a rural setting is further supported by the observations of

Grewal *et al.* (1973) who noted that in India, children with better nutritional status came from extended families, with higher income and ownership of land, and whose mothers were not working.

3.3.2. Birth Interval

The reduction of the birth interval between children often contributes to the development of malnutrition as illustrated by the "disease of the deposed baby," kwashiorkor. Wray and Aguirre (1969) observed that in Colombian children the rate of malnutrition was high when the birth interval was less than 3 years but that there was a marked decline when the interval was more than 3 years.

3.3.3. Sex of Child

Many cultures show a preference of male children and a tendency to neglect female children with the consequence that growth tends to be impaired in female children. For example, it has been observed (Kimmance, 1972) that there is a marked excess of failure to thrive among Jordanian girls after the age of 6 months, and that there is a greater proportion of malnutrition in Haitian girls than boys (Ballweg, 1972).

3.3.4. Other Social Factors

Other social factors that contribute to low weight gain and its consequences include illiteracy of parents, persistence of traditional health concepts, poor environmental sanitation, and the unavailability of modern health care facilities and services. All these and the various factors elaborated herein interact with one another to form a complex network of impinging forces that synergistically contribute to influence the growth of the child.

4. References

Aall, C., 1970, Relief, nutrition and health problems in the Nigerian/Biafran War, *J. Trop. Pediatr.* **16**:70.

Antrobus, A. C. K., 1971, Child growth and related factors in a rural community in St. Vincent. *J. Trop. Pediatr.* **17**:187.

Ashcroft, M. T., Bell, R., Nicholson, C. C., and Pemberton, S., 1968, Growth of Guyanese infants of African and East Indian racial origins, with some observations on mortality, *Trans. R. Soc. Trop. Med. Hyg.* **62**:607.

Avery, T. L., 1963, An analysis over 8 years of children admitted with measles to a hospital in Sierra Leone, West Africa, *W. Afr. Med. J.* **12**:61.

Ballweg, J. A., 1972, Family characteristics and nutrition problems of preschool children in Fond Parisien, Haiti, *J. Trop. Pediatr.* **18**:230.

Barr, D. P., and DuBois, E. F., 1918, Clinical calorimetry. XXVIII. The metabolism in malarial fever, *Arch. Intern. Med.* **21**:627.

Beisel, W. R., Sawyer, W. D., Ryll, E. D., and Crozier, D., 1967, Metabolic effects of intracellular infections in man, *Ann. Intern. Med.* **67**:744.

Belew, M., Jacobsson, K., Tornell, G., Uppsall, L., Zaar, B., and Vahlquist, B., 1972, Anthropometric, clinical and biochemical studies in children from the five different regions of Ethiopia, *J. Trop. Pediatr.* **18**:246.

Bhattacharyya, A. K., 1961, Aetiological investigations in kwashiorkor and marasmus, *Bull, Calcutta Sch. Trop. Med.* **9**:99.

Blankson, J. M., 1975, Measles and its problems as seen in Ghana, *J. Trop. Pediatr.* **21(1B)**:51.

Bolton, J. M., 1972, Food taboos among the Orang Asli in West Malaysia: A potential nutritional hazard, *Am. J. Clin. Nutr.* **25**:789.

Carstairs, G. M., 1955, Medicine and faith in rural Rajasthan, in: *Health, Culture, and Community* (B. D. Paul, ed.), pp. 107–134, Russell Sage Foundation, New York.

Cassel, J., 1955, A comprehensive health program among South African Zulus, in: *Health, Culture, and Community* (B. D. Paul, ed.), pp. 15–41, Russell Sage Foundation, New York.

Chen, P. C. Y., 1970a, Ascariasis: Beliefs and practices of a rural Malay community, *Med. J. Malaya* **24**:176.

Chen, P. C. Y., 1970b, Indigenous concepts of causation and methods of prevention of childhood diseases in a rural Malay community, *J. Trop. Pediatr.* **16**:33.

Chen, P. C. Y., 1972, Sociocultural influences on vitamin A deficiency in a rural Malay community, *J. Trop. Med. Hyg.* **75**:231.

Chen, P. C. Y., 1973, An analysis of customs related to childbirth in rural Malay culture, *Trop. Geogr. Med.* **25**:197.

Chen, S. T., 1974, Protein calorie malnutrition A major health problem of multiple causation in Malaysia, *Southeast Asian J. Trop. Med. Public Health* **5**:85.

Cosminsky, S., 1975, Changing food and medical beliefs and practices in a Guatemalan community, *Ecol. Food Nutr.* **4**:183.

Coughlin, R. J., 1965, Pregnancy and birth in Vietnam, in: *Southeast Asian Birth Customs* (D. V. Hart, P. A. Rajadhon, and R. J. Coughlin, eds.), Human Relations Area Files Press, New Haven.

Darke, S. J., 1959, Malnutrition in African adults. V. Effects of hookworm infestations on absorption of foodstuffs, *Br. J. Nutr.* **13**:278.

Draper, C. C., Lelijveld, J. L. M., Matola, Y. G., and White, G. B., 1972, Malaria in the Pare area of Tanzania. IV. Malaria in the human population 11 years after the suspension of residual insecticide spraying, with special reference to the serological findings. *Trans. R. Soc. Trop. Med. Hyg.* **66**:905.

Dugdale, A. E., 1970, Breast feeding in a Southeast Asian city, *Far East Med. J.* **6**:8.

Emanual, I., and Biddulph, J., 1969, Pediatric field survey of the Nasioi and Kwaio of the Solomon Islands, *J. Trop. Pediatr.* **15**:56.

Ferro-Luzzi, G. E., 1973, Food avoidances of pregnant women in Tamilnad, *Ecol. Food Nutr.* **2**:259.

Gans, B., 1961, Pediatric problems in Lagos, *W. Afr. Med. J.* **10**:33.

Gerlach, L. P., 1964, Sociocultural factors affecting the diet of the Northeast coastal Bantu, *J. Am. Diet. Assoc.* **45**:420.

Gonzalez, N. L. S., 1964, Beliefs and practices concerning medicine and nutrition among lower-class urban Guatemalans, *Am. J. Public Health* **54**:1726.

Gonzales, N. S., and Behar, M., 1966, Child-rearing practices, nutrition and health status, *Milbank Mem. Fund Q. Bull.* **44**:77.

Gordon, J. E., Chitkara, I. D., and Wyon, J. B., 1963, Weanling diarrhoea, *Am. J. Med. Sci.* **245**:345.

Gordon, J. E., Behar, M., and Scrimshaw, N. S., 1964, Acute diarrhoeal disease in less developed countries. I. An epidemiological basis for control, *Bull. WHO* **31**:1.

Gothefors, L., and Winberg, J., 1975, Host resistance factors, *J. Trop. Pediatr.* **21**:260.

Grewal, T., Gopaldas, T., and Gadre, V. J., 1973, Etiology of malnutrition in rural Indian preschool children (Madhya Pradesh), *J. Trop. Pediatr.* **19**:265.

Gurney, J. M., Fox, H., and Neill, J., 1972, A rapid survey to assess the nutrition of Jamaican infants and young children in 1970, *Trans. R. Soc. Trop. Med. Hyg.* **66:**653.

Gyorgy, P., 1971, Symposium on the uniqueness of human milk: Biochemical aspects, *Am. J. Clin. Nutr.* **24:**970.

Haddad, N. E., and Harfouche, J. K., 1971, Serum protein in Lebanese infants, *J. Trop. Pediatr.* **17:**91.

Harrell, G. T., Wolff, W. A., Venning, W. L., and Reinhart, J. B., 1946, Prevention and control of disturbances of protein metabolism in Rocky Mountain spotted fever; value of forced feedings of high-protein diet and of administration of specific antiserum, *South. Med. J. (Birmingham, Ala.)* **39:**551.

Hart, D. V., 1965, From pregnancy through birth in a Bisayan Filipino village, in: *Southeast Asian Birth Customs* (D. V. Hart, P. A. Rajadhon, and R. J. Coughlin, eds.), Human Relations Area Files Press, New Haven, Conn.

Hasan, K. A., 1971, The Hindu dietary practices and culinary rituals in a North Indian village: An ethnomedical and structural analysis, *Ethnomedizin* **1:**43.

Hautvast, J., 1972, Eating habits in a group of Tanzanian schoolchildren, *J. Trop. Pediatr.* **18:**206.

Imperato, P. J., 1969, Traditional attitudes towards measles in the Republic of Mali, *Trans. R. Soc. Trop. Med. Hyg.* **63:**768.

Jayalakshmi, V. T., and Gopalan, C., 1958, Nutrition and tuberculosis. I. An epidemiological study, *Indian J. Med. Res.* **46:**87.

Jelliffe, D. B., 1953, Clinical notes on kwashiorkor in Western Nigeria, *J. Trop. Med. Hyg.* **56:**104.

Jelliffe, D. B., 1969, *Child Nutrition in Developing Countries,* U. S. Department of State, A.I.D., Washington, D. C.

Jelliffe, D. B., and Jelliffe, E. F. P., 1971, The uniqueness of milk, *WHO Chron.* **25:**537.

Jelliffe, D. B., Bras, G., and Stuart, K. L., 1954, Kwashiorkor and marasmus in Jamaican infants, *W. Indian Med. J.* **3:**43.

Jelliffe, D. B., Symonds, B. E. R., and Jelliffe, E. F. P., 1960, The pattern of malnutrition in early childhood in southern Trinidad, *J. Pediatr.* **57:**922.

Johnston, J. A., and Maroney, J. W., 1938, Focal infection and metabolism; the effect of the removal of tonsils and adenoids on the nitrogen balance and the basal metabolism, *J. Pediatr.* **12:**563.

Kanaaneh, H., 1972, The relationship of bottle feeding to malnutrition and gastroenteritis in a pre-industrial setting, *J. Trop. Pediatr.* **18:**302.

Kimmance, K. J., 1972, Failure to thrive and lactation failure in Jordanian villages in 1970, *J. Trop. Pediatr.* **18:**313.

Knutsson, K. E., and Mellbin, T., 1969, Breast feeding habits and cultural context, *J. Trop. Pediatr.* **15:**40.

Mata, L. J., and Behar, M., 1975, Malnutrition and infection in a typical rural Guatemalan village: Lessons for the planning of preventive measures, *Ecol. Food Nutr.* **4:**41.

Mata, L. J., and Wyatt, R. G., 1971, Symposium on the uniqueness of human milk: Host resistance to infection, *Am. J. Clin. Nutr.* **24:**976.

Mata, L. J., Urrutia, J. J., and Lechtig, A., 1971, Infection and nutrition of children of a low socioeconomic rural community, *Am. J. Clin. Nutr.* **24:**249.

McArthur, A. M., 1962, Assignment Report: Nutrition, WHO Regional Office for the Western Pacific, Manila.

McDowell, I., and Hoorweg, J., 1975, Social environment and outpatient recovery from malnutrition, *Ecol. Food Nutr.* **4:**91.

McKay, D. A., 1969, The arm circumference as a public health index of protein–calorie malnutrition of early childhood, *J. Trop. Pediatr.* **15:**213.

McKay, D. A., 1971, Food illness, and folk medicine: Insights from Ulu Trengganu, West Malaysia, *Ecol. Food Nutr.* **1:**67.

McKigney, J., 1971, Symposium on the uniqueness of human milk: Economic aspects, *Am. J. Clin. Nutr.* **24:**1005.

Molony, C. H., 1975, Systematic valence coding of Mexican "hot"–"cold" food, *Ecol. Food Nutr.* **4:**67.

Montelli, T. B., Ribeiro, V. M., Riberio, R. M., and Ribeiro, M. A. C., 1974, Electroencephalographic changes and mental development in malnourished children, *J. Trop. Pediatr.* **20**:201.

Morehead, C. D., Morehead, M., Allen, D. M., and Olson, R. E., 1974, Bacterial infections in malnourished children, *J. Trop. Pediatr.* **20**:141.

Morley, D. C., 1959, Childhood tuberculosis in a rural area of West Africa, *W. Afr. Med. J.* **8**:225.

Morley, D. C., 1962, Measles in Nigeria, *Am. J. Dis. Child.* **103**:230.

Morley, D., 1969, Severe measles in the tropics—I, *Br. Med. J.* **1**:297.

Morley, D., 1973, *Paediatric Priorities in the Developing World*, Butterworth, London.

Netrasiri, A., and Netrasiri, C., 1955, Kwashiorkor in Bangkok (an analytical study of 54 cases), *J. Trop. Pediatr.* **1**:148.

Ongeri, S. K., 1975, Nutritional problems among Kenyan children, *J. Trop. Pediatr.* **21**(1B):6.

Oomen, H. A. P. C., 1957, The relationship between liver size, malaria and diet in Papuan children, *Doc. Med. Geogr. Trop. (Amsterdam)* **9**:84.

Orwell, S., and Murray, J., 1974, Infant feeding and health in Ibadan, *J. Trop. Pediatr.* **20**:206.

Phillips, I., Lwanga, S. K., Lore, W., and Wasswa, D., 1969, Methods and hygiene of infant feeding in an urban area of Uganda, *J. Trop. Pediatr.* **15**:167.

Plank, S. J., and Milanesi. M. L., 1973, Infant feeding and infant mortality in rural Chile, *Bull. WHO* **48**:203.

Pollock, N. J., 1974, Breadfruit or rice: Dietary choice on a Micronesian atoll, *Ecol. Food Nutr.* **3**:107.

Poskitt, E. M. E., 1972, Seasonal variation in infection and malnutrition at a rural paediatric clinic in Uganda, *Trans. Soc. Trop. Med. Hyg.* **66**:931.

Pretorius, P. J., Davel, J. G. A., and Coetzee, J. N., 1956, Some observations on the development of kwashiorkor. A study of 205 cases, *S. Afr. Med. J.* **30**:396.

Puffer, R. R., and Serrano, C. V., 1973, *Patterns of Mortality in Childhood*, Pan American Health Organization, Washington, D. C., (Scientific Publ. No. 262).

Rocha, E., 1975, Salmonellosis in infants, *J. Trop. Pediatr.* **21**(1B):60.

Rosenberg, I. H., Solomons, N. W., and Levin, D. M., 1976, Interaction of infection and nutrition: Some practical concerns, *Ecol. Food Nutr.* **4**:203.

Sai, F. T., 1965, The challenge of African nutrition, *Isr. J. Med. Sci.* **1**:118.

Salomon, J. B., Mata, L. J., and Gordon, J. E., 1968, Malnutrition and the common communicable diseases of childhood in rural Guatemala, *Am. J. Public Health* **58**:505.

Sang, J. H., 1938, The antiproteolytic enzyme of *Ascaris lumbricoides* var. suis, *Parasitology* **30**:141.

Scrimshaw, N. S., Taylor, C. E., and Gordon, J. E., 1968, *Interactions of Nutrition and Infection*, WHO, Geneva (Monograph Series, No. 57).

Selinus, R., 1970, Home made weaning foods for Ethiopian children, *J. Trop. Pediatr.* **17**:188.

Selinus, R., Gobezie, A., Knutsson, K. E., and Vahlquist, B., 1971, Dietary studies in Ethiopia: Dietary pattern among the Rift Valley Arsi Galla, *Am. J. Clin. Nutr.* **24**:365.

Sheppe, W. A., and Adams, J. R., 1957, The pathogenic effect of *Trypanosoma duttoni* in hosts under stress conditions, *J. Parasitol.* **43**:55.

Simoons, F. J., 1974, Rejection of fish as human food in Africa: A problem in history and ecology, *Ecol. Food Nutr.* **3**:89.

Smithers, S. R., and Terry, R. J., 1959, Changes in the serum protein and leucocyte counts of Rhesus monkeys in the early stages of infection with *Trypanosoma gambiense*, *Trans. R. Soc. Trop. Med. Hyg.* **53**:336.

Sousa, P. L. R., Barros, F. C., Pinheiro, G. N. M., and Gazelle, R. V., 1975, Patterns of weaning in South Brazil, *J. Trop. Pediatr.* **21**:210.

Taylor, H. L., Mickelsen, O., and Keys, A., 1949, The effects of induced malaria, acute starvation and semi-starvation on the electrophoretic diagram of the serum proteins of normal young men, *J. Clin. Invest.* **28**:273.

Teoh, S. K., 1975, Breast feeding in a rural area in Malaysia, *Med. J. Malaysia* **30**:175.

Van Veen, M. S., 1971, Some ecological considerations of nutrition problems on Java, *Ecol. Food Nutr.* **1**:25.

Venkatachalam, P. S., and Patwardhan, V. N., 1953, The role of *Ascaris lumbricoides* in the nutrition of the host effect of ascariasis on digestion of protein, *Trans. R. Soc. Trop. Hyg.* **47**:169.

Waldmann, E., 1975, The ecology of the nutrition of the Bapedi, Sekhukuniland, *Ecol. Food Nutr.* **4**:139.

Waterlow, J., and Vergara, A., 1956, *Protein Malnutrition in Brazil,* F.A.O. Nutritional Studies, No. 14, Rome.

Whyte, R. O., 1974, *Rural Nutrition in Monsoon Asia,* Oxford University Press, Kuala Lumpur.

Williams, C. D., and Jelliffe, D. B., 1972, *Mother and Child Health: Delivering the Services,* Oxford University Press, London.

Williams, T. R., 1965, *The Dusun: A North Borneo Society,* Holt, Rinehart & Winston, New York.

Wilson, C. S., 1971, Food beliefs affect nutritional status of Malay fisherfolk, *J. Nutr. Educ.* **2**:96.

Woodruff, A. W., 1959, Serum protein changes induced by infection and treatment of infection with *Trypanosoma rhodesiense* in monkeys, *Trans. R. Soc. Trop. Med. Hyg.* **53**:327.

Wray, J. D., and Aguirre, A., 1969, Protein–calorie malnutrition in Candelaria, Colombia. I. Prevalence, social and demographic causal factors, *J. Trop. Pediatr.* **15**:76.

Metabolic Anomalies, Nutrition, and Growth

Meinhard Robinow

1. Definitions

Nutrition is the process of assimilating substrate into body tissues and stores. *Nutrients* are the metabolically active constituents of food. *Substrates* are the metabolites required for energy, growth, maintenance, and storage. *Nutritional state* designates success in accomplishing this process in relation to norms. *Growth* is increase in size. Unless otherwise specified, growth will denote skeletal elongation, i.e., increase in stature.

Normal growth is the result of complex interactions of responsive cells with substrate and special regulators. Requirements are adequate intake, digestion, and transport of nutrients, transformation of nutrients to substrate, appropriate supplies of special growth factors, homeostasis of the internal milieu, an intact central nervous system, and a supportive psychosocial enivronment.

Metabolic anomalies can interfere with each of these functions. The "experiments of nature" have allowed us to dissect the complex biochemical processes into discrete steps. Thus they have greatly advanced our knowledge of normal function and our understanding of disease. One can categorize the major dysfunctions which produce growth failure (Table I). Such a scheme provides a logical approach to the understanding of growth disorders. Unfortunately, in many diseases the defective mechanisms are complex and involve more than one category. Even the type of dysfunction may change in the course of a disease. For instance, in cystic fibrosis early failure of nutrition and growth is due to maldigestion and resultant malabsorption, while later in the course pulmonary disease becomes the major determinant (Sproul and Huang, 1964; Lapey *et al.*, 1974).

Meinhard Robinow ● School of Medicine, University of Virginia, Charlottesville, Virginia.

Table 1. Mechanisms of Growth Failure in Metabolic Disease[a]

A.	*Substrate deficiency*
	Disorders of intake, digestion, and absorption
	Disorders of intestinal transport and intermediary metabolism
	Disorders of circulation and oxygen transport
	Disorders of electrolyte conservation
B.	*Disorders of regulation*
	Disorders of hormonal control
	Catabolic states
	Central nervous system disorders
C.	*Defective end-organ response*
	Skeletal dysplasias

[a] Modified from J. Spranger, Mainz, Germany, Some new dwarfing conditions. Presented at 1976 Birth Defects Conference, National Foundation–March of Dimes, Vancouver, June 1976.

2. The Vulnerable Tissues

Next to growing cartilage the tissues most vulnerable to nutritional deprivation are muscle, the blood-forming organs, and bone. In infancy the brain is vulnerable to severe deprivation and, even more, to certain toxic metabolites. Therefore, growth failure in nutrient deprivation and metabolic anomalies is commonly associated with muscle wasting, anemia, osteoporosis and developmental delay. Whether protein–calorie malnutrition in the human can result in permanent brain damage remains a controversial point.

2.1. Substrate Deficiency Disorders

Impaired growth is usually accompanied by poor appetite. It is often impossible to determine whether the poor appetite is the cause or the effect of the growth retardation. Impaired appetite seems to contribute significantly to the growth failure in congenital heart disease (Krieger, 1970; Huse, 1974), renal failure (Broyer *et al.,* 1974; Chantler *et al.,* 1974), and advanced brain disease (*vide infra*).

2.1.1. Celiac Disease

Celiac disease is a familial metabolic disorder characterized by intolerance to a specific food protein, wheat gluten. Ingestion of wheat induces severe atrophy of small-intestinal mucosa with loss of villi. The resulting malabsorption causes diarrhea, weight loss, muscular wasting, osteoporosis, and growth failure. The onset is usually toward the end of the first year. Complete elimination of wheat from the diet permits regeneration of the intestinal mucosa, weight gain, and catch-up growth (Barr *et al.,* 1972). The dietary restriction must be continued through childhood and adolescence, if not for life (Hamilton and McNeill, 1972).

2.1.2. Cystic Fibrosis

Cystic fibrosis is a recessively inherited disorder with a frequency of 1:2500 births in Caucasians; it is uncommon in other races. About 90% of affected newborns have a complete lack of pancreatic enzymes. In affected infants, incomplete digestion and the resultant malabsorption often produce protein–calorie malnutrition and deficiencies of the fat-soluble vitamins A, D, E, and K. Growth failure, emaciation, edema, or bleeding may be the presenting symptoms. The sweat of patients with cystic fibrosis contains abnormally high concentrations of sodium chloride. Profuse sweating in hot weather can cause life-threatening salt losses.

The nutritional deficiencies can often be fully compensated by oral administration of pancreatic extracts, a high protein diet, supplementary water-soluble vitamins, and, when needed, added salt. But sooner or later bronchopulmonary disease, caused by bronchial obstruction from viscid secretions and resulting infections, becomes severe enough to interfere with weight gain and, eventually, with growth (Sproul and Huang, 1964; Lapey *et al.*, 1974). Patients whose pulmonary disease is well controlled may attain normal adult weight and stature (Crozier, 1974) and live into early middle age.

2.1.3. Transport Disorders of Blood

Sickle cell disease and thalassemia may be classified as substrate disorders due to impaired oxygen transport. In both of these genetic anomalies one finds underweight, growth failure, and delayed sexual maturation. Another genetic transport anomaly causing poor growth and nutrition is abetalipoproteinemia (Frederickson *et al.*, 1972).

2.1.4. Disorders of Amino Acid Metabolism

The dysfunctions which usually bring amino acid disorders to medical attention are (a) catastrophic illness in the neonate, (b) failure to thrive in the infant, and (c) mental retardation. Less common findings are disorders of liver, kidney, or skin.

2.1.4a. Severe Metabolic Disorders in the Newborn. Several disorders of amino acid metabolism present as life-threatening diseases in the newborn period. Affected infants, apparently normal at birth, soon develop vomiting, weight loss, muscular hypo- or hypertonia, seizures, hepatic failure, and bleeding disorders. Such catastrophic illnesses occur in some but not all cases of branched-chain ketonuria (maple syrup urine disease), methylmalonic acidemia, the hyperammonemias, and other rare disorders (Nyhan, 1974). Similar catastrophic illnesses may be produced by congenital anomalies of carbohydrate metabolism such as galactosemia (Holzel, 1975) and fructose intolerance (Perheentupa, 1975), though liver disease is usually a more prominent finding in the latter group. Prompt recognition of these disorders is mandatory since most of them respond promptly to appropriate diet therapy, while delay is either fatal or leads to irreversible neurologic damage. These infants are likely to be misdiagnosed as septicemia or intracranial hemorrhage unless the death of a previous child in the family has alerted the physician and the parents to the possibility of a genetic disorder.

2.1.4b. Amino Acid Disorders Causing Failure to Thrive. Some of the disorders which may present as catastrophic illnesses of the neonate may take in other cases a less malignant course and present later in infancy, either as vomiting resembling pyloric stenosis, as failure to thrive, as chronic acidosis, or as liver disease with or without neurologic deterioration (Nyhan, 1974). These disorders include cases of maple syrup urine diseases, proprionic acidemia, tyrosinosis, and the urea cycle disorders. Cystinosis (Schneider and Seegmiller, 1972) is manifested by chronic renal disease (the Fanconi syndrome) with acidosis, rickets, and severe dwarfing. The latter is sometimes aggravated by thyroid deficiency secondary to accumulation of cystine crystals in the thyroid gland.

It should be noted that aminoaciduria, in itself, does not produce nitrogen losses of sufficient magnitude to interfere with health, nutrition, or growth.

In some patients with amino acid disorders failure to thrive has been iatrogenic, caused by therapeutic diets overrestrictive in total protein or some essential amino acid (Rouse, 1966). Providing for nutritional needs while avoiding toxic metabolites requires expert knowledge and careful monitoring, especially during acute infections.

Whereas growth is depressed in most amino acid disorders, tall stature and, especially, increased length of the extremities, suggestive of Marfan's syndrome, are found in homocystinuria (Schimke *et al.*, 1965). The increased growth seems to result from a disturbance of collagen metabolism.

2.1.4c Amino Acid Disorders Causing Primarily Mental Retardation. Mental retardation without significant impairment of general nutrition and growth is produced by several amino acid disorders. Phenylketonuria was the first of these diseases to be recognized (Folling, 1934) and the first to be successfully treated with a restricted diet (Bickel *et al.*, 1953).

2.1.5. Disorders of Carbohydrate Metabolism

2.1.5a. Diabetes Mellitus. Some early observers claimed that children are above average height when their diabetes first becomes manifest (Joslin *et al.*, 1925). Others have been unable to confirm this observation (Craig, 1970), and the matter is still controversial (Drayer, 1974). All authors agree that diabetes tends to retard growth, though the degree of retardation is rather mild (Jiranie and Rayner, 1973). Growth impairment is related more to the duration of the diabetes than to the level of control. Reduced adult stature is largely caused by a poor pubertal growth spurt (Jiranie and Rayner, 1973). The Mauriac syndrome, consisting of hepatomegaly, severe dwarfing, and delayed sexual maturation, was formerly seen in some poorly controlled diabetic children.

This syndrome has become very rare since the introduction of the long-acting insulins (Mandell, 1974).

2.1.5b. Other Diseases of Carbohydrate Metabolism. Galactosemia and fructose intolerance have already been mentioned among the catastrophic metabolic derangements of the newborn. In both diseases normal growth and development can be achieved by strict adherence to diets free of galactose or fructose, respectively.

The glycogen storage disease of liver, types I and III (Howell, 1972), are marked by hepatomegaly and growth retardation, often accompanied by obesity, especially in type I. Patients with type III disease sometimes show spontaneous improvement during adolescence.

2.1.6. Renal Diseases

Renal parenchymal disease is often accompanied by growth failure. The mechanisms are multiple and their relative importance seems to vary, even within well-defined classes of renal disease.

Renal tubular acidosis (Seldin and Wilson, 1972) produces severe failure to thrive and grow, due to acidosis and wastage of base. In the distal tubule type, treatment by base supplementation leads to rapid complete recovery with catch-up growth, while the proximal type is less responsive.

Growth in chronic renal failure varies from mild impairment to virtual standstill. Nutritional status is equally variable. The following factors are implicated in the growth failure accompanying renal disease:

1. Appetite depression (Chantler *et al.*, 1974).
2. The low protein diets generally prescribed for advanced renal disease.
3. Acidosis and the resultant loss of bicarbonate and calcium (West and Smith, 1956; Stickler and Bergen, 1973).
4. Rickets. The kidney is the site of synthesis of dihydroxycholecalciferol, the active metabolite of vitamin D (DeLuca, 1973). Rickets may be the most important cause of growth failure in renal disease (Stickler and Bergen, 1973).
5. Hyperparathyroidism. Renal hyperparathyroidism severe enough to cause roentgenographic skeletal changes is invariably accompanied by virtual cessation of growth.
6. Somatomedin deficiency. The kidney is one of the sites of somatomedin production (Van Wyk *et al.*, 1973). Somatomedin levels in end-stage renal disease are low or unmeasurable (Saenger *et al.*, 1974).

Renal dialysis generally improves nutrition but does not enhance growth (Broyer *et al.*, 1974).

After renal transplantation somatomedin levels may rise and growth may resume (Saenger *et al.*, 1974). The extent of catch-up growth depends on the child's age and on the need for corticosteroids. Alternate-day administration of prednisone may permit resumption of growth (Hoda *et al.*, 1975), but also increases the risk of rejection. Catch-up growth is minimal if transplantation is performed after age 12 (Hoda *et al.*, 1975).

2.1.7. Rickets

Rickets represents faulty ossification of metaphyseal ostoid due to inadequate inorganic substrate. The adult equivalent of rickets is ostomalacia. Growth retardation is a feature of all forms: vitamin D deficiency, vitamin D dependency, and X-linked hypophosphatemic, hepatic, and renal rickets. Inadequacy of inorganic substrate is aggravated by the osteolytic effect of secondary hyperparathyroidism. Vitamin D deficiency responds to ultraviolet light or physiologic doses of vitamin D_3. D-Dependency (Prader *et al.*, 1961) is a metabolic disorder consisting of defective conversion of D_3 to its active metabolite $(OH)_2D_3$. Healing and catch-up growth require huge doses of vitamin D_3 or physiologic doses of the dihydroxyvitamin. The primary metabolic error in X-linked hypophosphatemic rickets is a renal tubular defect of phosphate reabsorption. A secondary anomaly is reduced absorption of calcium from the intestinal tract. Normal or near normal growth can be restored by oral administration of phosphates at frequent intervals combined with modest doses of vitamin D (Glorieux *et al.*, 1972; McEnery *et al.*, 1972). Since rickets, by itself, does not impair general nutrition, the diagnosis is easily missed until dwarfing and bowing of the lower extremities become conspicuous.

2.2. Disorders of Regulation

2.2.1. Thyroid Dysfunction

The thyroid hormones, thyroxin (T_4) and triiodothyronine (T_3), are essential for cerebral growth and development, for skeletal growth, and for skeletal and dental maturation. Thyroid hormone and pituitary growth hormone are both necessary for growth and interact in a complex manner. Congenital hypothyroidism is usually due to dysgenesis of the gland, less often to a genetic block in hormone synthesis. Hypothyroidism developing during childhood is often caused by autoimmune disease. Nutritional hypothyroidism due to iodine deficiency was formerly common in many parts of the world, but has been largely eliminated by the use of iodized salt. In the hypothyroid child, oral or parenteral administration of thyroid hormones produces spectacular growth stimulation. With continued administration normal physical growth is achieved and maintained. Unfortunately, the deleterious effects of thyroid deficiency on infantile brain development become irreversible after a few months of postnatal thyroid deprivation. Mass neonatal blood screening for T_4 and TSH (thyroid stimulating hormone) (Dussault *et al.*, 1975; Klein *et al.*, 1975) should soon eliminate this form of mental retardation.

In hyperthyroidism one finds increased appetite, no weight gain or actual weight loss, often combined with acclerated growth and skeletal maturation. Sexual maturation is not advanced.

2.2.2. Growth Hormone and Somatomedin

Pituitary growth hormone is essential for normal skeletal growth. Acquired growth hormone deficiency, due to destruction of the pituitary or its hypothalamic connections, is often associated with deficiency of other pituitary hormones (ACTH, gonadotrophins, etc.), but isolated growth hormone deficiency is generally due to a genetic defect (Rimoin *et al.*, 1966).

Growth hormone does not stimulate cartilage directly. It induces generation of somatomedins (formerly called sulfation factor), which are the actual stimulators of growth cartilage (Daughaday *et al.*, 1972; Van Wyk *et al.*, 1973). Laron's dwarfism, which clinically mimics pituitary dwarfism, results from a congenital defect of somatomedin production in spite of very high levels of serum growth hormone (Laron *et al.*, 1966).

2.2.3. Psychosocial Dwarfism

Children deprived of parental affection often present with severe growth retardation in spite of ravenous, often bizarre appetites and allegedly ample caloric intakes (Powell *et al.*, 1967a,b). Immediately upon hospital admission fasting growth hormone levels are low and responses to standard growth hormone stimulants may be severely impaired. After removal of the children from their unsuitable homes spectacular weight gain and catch-up growth ensue and normal growth hormone secretion resumes. The role of actual food deprivation during the development of the syndrome remains controversial. This growth disorder is far more common than generally recognized.

2.2.4. Adrenogenital Syndrome

Recessively inherited genetic defects of cortisol production, especially 21- and 11-hydroxylase deficiency, cause feedback stimulation of the adrenals with overproduction of androgens and resultant virilization of females and precocious puberty in males (Zurbrugg, 1975). Androgens are powerful stimulators of growth but, even more, of sexual maturation. Untreated boys are tall for age but their epiphyses close early and they end up as short adults. In the affected neonate inability of the kidneys to conserve salt, with resulting hypochloremic–hypovolumic shock, may be the presenting symptom. Salt losses must be prevented by high oral intake combined with appropriate mineral corticoid administration. Glucocorticoids in physiologic doses suppress the inappropriate androgen secretion by feedback inhibition (Wilkins *et al.*, 1961). Pharmacologic doses of glucocorticoids cause growth retardation and must be avoided by careful monitoring of steroid metabolism.

2.2.5. Glucocorticoid Excess—Cushing's Syndrome

Cortisol and the synthetic glucocorticoids are catabolic agents. In pharmacologic doses they cause protein catabolism, reduce muscle mass, produce a characteristic type of obesity, and inhibit growth by suppressing growth hormone secretion (Frantz and Rabkin, 1964) and by direct action on cartilage (Silberberg *et al.*, 1966). The classical picture of cortisol excess, Cushing's syndrome, results spontaneously from adrenocortical hyperplasia secondary to excessive ACTH production, rarely from adrenal tumors. Far more common is iatrogenic Cushing's syndrome due to corticosteroids prescribed for their anti-inflammatory or immunosuppressive effects.

2.2.6. Growth and Nutrition in Diseases of the Central Nervous System

Emaciation, muscular wasting, and growth arrest are almost universal in advanced stages of metabolic brain diseases, such as Tay-Sachs', Nieman-Pick's, and Hurler's disease, but also occur in severe cerebral palsy. The responsible mechanisms are not yet well understood. One common cause is inadequate food intake which can be due to difficulties in feeding and swallowing or to a disturbed appetite mechanism. Simple intake failure can be overcome by gastrostomy feedings. A rare type of cerebral emaciation is the diencephalic syndrome of infancy (Gareis and Johnson, 1965).

The opposite conditions, obesity and overgrowth, separately or combined, may also be found in cerebral disorders. Obesity, associated with a ravenous appetite and mental retardation, is seen in the Prader-Willi syndrome (Zellweger and Schneider, 1968). A hypothalamic origin is postulated. Cerebral giantism (Sotos *et al.*, 1964) is a genetic disorder characterized by large size at birth, accelerated growth and skeletal maturation, and mental retardation.

2.3. End-Organ Failure

2.3.1. Pseudohypoparathyroidism

The classical form of end-organ failure, manifested as skeletal dysplasia and growth failure, is pseudohypoparathyroidism (Potts, 1972). In this condition the kidney does not respond to parathyroid hormone with increased excretion of cyclic AMP. The clinical picture consists of short stature, a highly variable skeletal dysplasia, obesity, and, often, mental retardation. Pseudopseudohypoparathyroidism appears to be a variant phenotype of this anomaly which is inherited as an X-linked dominant trait.

2.3.2. Hypophosphatasia

Another type of end-organ failure is represented by hypophosphatasia (Sobel *et al.*, 1953), a disorder in which osteoblasts (and various other cell types) are unable to elaborate alkaline phosphatase and thus are unable to calcify osteoid in spite of adequate substrate and regulators.

2.3.3. Lysosomal Storage Diseases

A third variety of end-organ failure is comprised of lysosomal storage diseases, among them the mucopolysaccharidoses and the mucolipidoses. In these disorders lysosomes accumulate material which they are unable to dispose of by enzymatic degradation. As the disease progresses the affected cartilage cells lose their ability to respond to nutrients and regulators. A skeletal dysplasia develops, accompanied by progressive growth failure.

The cause of many skeletal dysplasias remains unknown. The majority of these disorders are inherited as dominant traits. They are more likely anomalies of structural proteins than anomalies of metabolism.

3. References

Barr, D. G. D., Schmerling, D. H., and Prader, A., 1972, Catch-up growth in malnutrition, studied in celiac disease after institution of a gluten-free diet, *Pediatr. Res.* **6**:521.

Bickel, H., Gerrard, J., and Hickmans, E. M., 1953, Influence of phenylalanine intake on phenylketonurics, *Lancet* **2**:812.

Broyer, M., Kleinknecht, M. D., Loirat C., Marty-Henneberg, L., and Roy, M. P., 1974, Growth in children treated with long-term hemodialysis, *J. Pediatr.* **84**:642.

Chantler, C., Lieberman, E., and Holiday, M. A., 1974, A rat model for the study of growth failure in uremia, *Pediatr. Res.* **8**:109.

Craig, J. O., 1970, Growth as a measurement of control in the management of diabetic children, *Postgrad. Med. J.* **46**:607.

Crozier, D. N., 1974, Cystic fibrosis, a not so fatal disease, *Pediatr. Clin. North Am.* **21**:935.

Daughaday, W. H., Hall, K., Raben, M. S., Salmon, W. D., Jr., Van den Brande, J. L., and van Wyk, J. J., 1972, Somatomedin: Proposed designation for sulfation factor, *Nature* **235**:107.

DeLuca, H. F., 1973, Vitamin D in 1973, *Am. J. Med.* **57**:1.

Drayer, N. M., 1974, Height of diabetic children at onset of symptoms, *Arch. Dis. Child.* **49**:616.

Dussault, J. H., Coulomb, P., LaBerge, C., Letarte, J., Guyda, H., and Khoury, K., 1975, Preliminary report on a mass screening program for neonatal hypothyroidism, *J. Pediatr.* **86**:670.

Folling, A., 1934, Uber Ausscheidung von Phenylbrenztraubensaure in den Harn als Stoffwechselanomalie in Verbindung mit Imbezillitat, *Z. Physiol. Chem.* **227**:169.

Frantz, A. G., and Rabkin, M. T., 1964, Human growth hormone. Clinical measurement, response to hypoglycemia and suppression by corticosteroids, *N. Engl. J. Med.* **271**:1375.

Frederickson, D. S., Gotto, A. M., and Levy, R. I., 1972, Abetalipoproteinemia, in: *The Metabolic Basis of Inherited Disease* (J. B. Stanbury, J. B. Wyngaarden, and D. S. Frederickson, eds.), McGraw-Hill, New York.

Gareis, F. J., and Johnson, J. A., 1965, Inanition in infants with diencephalic neoplasms, *Am. J. Dis. Child.* **109**:349.

Glorieux, F. H., Scriver, C. R., Reade, T. M., Goldman, H., and Roseborough, A., 1972, Use of phosphate and vitamin D to prevent dwarfism and rickets in X-linked hypophosphatemic rickets, *N. Engl. J. Med.* **287**:481.

Hamilton, J. R., and McNeill, L. K., 1972, Celiac disease, duration of therapy, *J. Pediatr.* **81**:885.

Hoda, O., Hasinoff, D. J., and Arbus, G. S., 1975, Growth following renal transplantation in children and adolescents, *Clin. Nephrol.* 3:6.

Holzel, A., 1975, Galactosemia (hereditary galactose-1-phosphate uridyl transferase deficiency disease), in: *Endocrine and Genetic Diseases of Childhood* (L. Gardner, ed.), pp. 964–971, Saunders, Philadelphia.

Howell, R., 1972, The glycogen storage diseases, in: *The Metabolic Basis of Inherited Disease* (J. B. Stanbury, J. B. Wyngaarden, and D. S. Frederickson, eds.), pp. 149–173, McGraw-Hill, New York.

Huse, D. M., 1974, Infants with congenital heart disease. Food intake, body weight and energy metabolism, *Am. J. Dis. Child.* 129:65.

Jiranie, S. K. M., and Rayner, P. H. W., 1973, Does control influence the growth of diabetic children? *Arch. Dis. Child.* 48:109.

Joslin, E. P., Root, H. F., and White, P., 1925, Growth, development and prognosis for diabetic children, *J. Am. Med. Assoc.* 85:420.

Klein, A. H., Augstin, A. V., Hopwood, N. J., Penicelli, A., Johnson, L., and Foley, T. P., 1975, Thyrotropin (TSH) screening for congenital hypothyroidism (abstr.), *Pediatr. Res.* 9:291.

Krieger, I., 1970, Growth failure and congenital heart disease: Energy and nitrogen balance in infants, *Am. J. Dis. Child.* 120:497.

Lapey, A., Kattwindel, J., di Sant'Angnese, P., and Laster, L., 1974, Steatorrhea and azotorrhea and their relation to the growth and nutrition in adolescents and young adults with cystic fibrosis, *J. Pediatr.* 84:328.

Laron, Z.., Pertzelan, A., and Mannheimer, S., 1966, Genetic pituitary dwarfism with high serum concentration of growth hormone, *Isr. J. Med. Sci.* 2:152.

Mandell, F., 1974, The Mauriac syndrome, *Am. J. Dis. Child.* 127:900.

McEnery, P. T., Silverman, F. N., and West, C. D., 1972, Acceleration of growth with vitamin D-phosphate therapy of hypophosphatemic-resistant rickets, *J. Pediatr.* 80:763.

Nyhan, W. L., 1974, Patterns of clinical expression and genetic variation in the inborn errors of metabolism, in: *Heritable Disorders of Amino Acid Metabolism* (W. L. Nyhan, ed.), pp. 3–14, Wiley, New York.

Perheentupa, J., 1975, Fructose intolerance, in: *Endocrine and Genetic Disorders of Childhood and Adolescence* (L. Gardner, ed.), pp. 986–1001, Saunders, Philadelphia.

Potts, J. T., 1972, Pseudohypoparathyroidism, in: *The Metabolic Basis of Inherited Disease* (J. B. Stanbury, J. B. Wyngaarden, and D. S. Frederickson, eds.), McGraw-Hill, New York.

Powell, G. F., Brasel, J. A., and Blizzard, R. M., 1967a, Emotional deprivation and growth retardation simulating idiopathic hypopituitarism, *N. Engl. J. Med.* 276:1271.

Powell, G. F., Brasel, J. A., Raiti, S., and Blizzard, R. M., 1967b, Emotional deprivation and growth retardation simulating idiopathic hypopituitarism, *N. Engl. J. Med.* 276:1279.

Prader, A., Illig, R., and Heierli, E., 1961, Eine besondere Form der primaren Vitamin D resistenten Rachitis mit Hypocalcamie und autosomal-dominantem Erbgang. Die hereditare Pseudo-mangel Rachitis, *Helv. Paediatr. Acta* 16:452.

Rimoin, D. L., Merimee, T. J., and McKusick, V. A., 1966, Growth-hormone deficiency in man: An isolated, recessively inherited defect, *Science* 152:1635.

Rouse, B. M., 1966, Phenylalanine deficiency syndrome, *J. Pediatr.* 69:246.

Saenger, P., Hiedemann, E., Schwartz, E., Korth-Schutz, S., Lewy, J. E., Riggio, R. R., Rubin, A. L., Stenzel, K. H., and New, M., 1974, Somatomedin and growth after renal transplantation, *Pediatr. Res.* 8:163.

Schimke, R. N., McKusick, V. A., Huang, T., and Pollack, A. D., 1965, Homocystimuria, *J. Am. Med. Assoc.* 193:711.

Schneider, J. A., and Seegmiller, J. E., 1972, Cystinosis and the Fanconi syndrome, in: *The Metabolic Basis of Inherited Disease*, 3rd ed. (J. B. Stanbury, J. B. Wyngaarden, and D. S. Frederickson, eds.), pp. 1581–1604, McGraw-Hill, New York.

Seldin, D. W., and Wilson, J. D., 1972, Renal tubular acidosis, in: *The Metabolic Basis of Inherited Disease*, 3rd ed. (J. B. Stanbury, J. B. Wyngaarden, and D. S. Frederickson, eds.), McGraw-Hill, New York.

Silberberg, M., Silberberg, R., and Hasler, M., 1966, Fine structure of articular cartilage in mice receiving cortisone acetate, *Arch. Pathol.* 82:569.

Sobel, E. H., Clark, L. C., Fox, R. P., and Robinow, M., 1953, Rickets, deficiency of "alkaline" phosphatase activity and loss of teeth in childhood, *Pediatrics* **11**:309.

Sotos, J. F., Dodge, P. R., Muirhead, D., Crawford, J. D., and Talbot, N. B., 1964, Cerebral giantism in childhood, *N. Engl. J. Med.* **271**:109.

Sproul, A., and Huang, N., 1964, Growth pattern in children with cystic fibrosis, *J. Pediatr.* **65**:964.

Stickler, G. B., and Bergen, B. J., 1973, A review: Short stature in renal disease, *Pediatr. Res.* **7**:978.

Van Wyk, J. J., 1973, The somatomedins, *Am. J. Dis. Child.* **126**:705.

Van Wyk, J. J., Underwood, L. E., Lister, R. C., Marshall, R. N., 1973, The somatomedins, *Am. J. Dis. Child.* **126**:705–711.

West, C. D., and Smith, W. C., 1956, An attempt to elucidate the cause of growth retardation in renal disease, *Am. J. Dis. Child.* **91**:460.

Wilkins, L., Lewis, R. A., Klein, R., Gardner, L. I., Crigler, J. F., Jr., Rosemberg, E., and Migeon, C. J., 1961, Treatment of congenital adrenal hyperplasia with cortisone, *J. Clin. Endocrinol.* **11**:1.

Zellweger, H. U., and Schneider, H. J., 1968, Syndrome of hypotonia-hypomentia-hypogonadism and obesity (HHHO) or Prader-Willi syndrome, *Am. J. Dis. Child.* **115**:588.

Zurbrugg, R. P., 1975, Congenital adrenal hyperplasia, in: *Endocrine and Genetic Diseases of Childhood and Adolescence* (L. Gardner, ed.), pp. 476–500, Saunders, Philiadelphia.

II

Ages of Man (Perspectives)

Maternofetal Nutrition

Aaron Lechtig, Hernan Delgado, Reynaldo Martorell, Charles Yarbrough, and Robert E. Klein

1. Introduction

Growth is a dynamic process that depends on the adequate provision and utilization of nutrients. The human fetus is an adaptive organism and is able to synthesize its own carbohydrates, proteins, and fats from glucose, amino acids, and other blood-borne metabolites transferred from the mother via the placenta (Van Dyne *et al.*, 1962; Page, 1969; Shelley, 1969). The total amount of each nutrient transferred will depend not only on the specific placental transport mechanisms and maternal blood level of each nutrient but also on the amount of placental blood flow and the surface area of the peripheral villi (Wigglesworth, 1964; Robertson *et al.*, 1967; Aherne and Dunnill, 1966; Laga *et al.*, 1972; Lechtig *et al.*, 1975f).

During pregnancy, modifications in the maternal hormonal system help to maintain the availability of nutrients to the fetus relatively independent of alterations in the maternal diet. These hormonal changes cause an increase in nitrogen retention, stimulate the deposition and subsequent mobilization of maternal fat stores, and ensure adequate provision of glucose and amino acids to the fetus (Page, 1969; Nainsmith, 1969; Beaton, 1961; Knopp *et al.*, 1970; Sakurai *et al.*, 1969). However, when maternal dietary deficiencies reach critical low levels, there is a decrease in the effectiveness of these mechanisms in maintaining fetal nutritional homeostasis. When this happens other factors become crucial: present maternal nutrient intake, nutritional status of the mother before pregnancy (availability of maternal body reserves), and the efficiency of conversion of maternal tissue into nutrients available for fetal growth.

Aaron Lechtig, Hernan Delgado, and Robert E. Klein • Institute of Nutrition of Central America and Panama, Guatemala City, Guatemala. *Reynaldo Martorell* • Food Research Institute, Stanford University, Stanford, California. *Charles Yarbrough* • Computers for Marketing Corporation, Kenwood, California.

The importance of maternal nutritional status prior to pregnancy as a determinant of the effect of nutrition during pregnancy on birth weight, is illustrated by the observation that the slope of the relationship between weight gain during pregnancy and birth weight decreases as the prepregnant weight increases (Lechtig *et al.*, 1975e). If the weight of the mother before and during pregnancy reflects her nutritional status, then these results suggest that the influence of improved nutrition during pregnancy on birth weight would be greater in those mothers whose nutritional status before pregnancy was poorer. Also, above a certain level of maternal prenatal nutrition, further improvements in nutrient intake during pregnancy would not result in detectable increases in birth weight. Thus, there may be minimum nutrient requirements which must be provided by the maternal diet and body stores in order to ensure normal fetal growth. Consequently, an effect of improved food intake during pregnancy on birth weight would more likely be detected in populations with severe nutritional deficits than in better nourished populations.

When investigating the relationship between maternal nutrition and birth weight an important consideration is measurement variability (Lechtig *et al.*, 1971). Clearly the finding of a statistical association between two variables depends in part on the precision with which these variables are measured. Also, it is the ratio between the between-individual and within-individual variabilities that is the relevant factor determining the detection of an association, when this exists. Unless this ratio is large enough, it will be impossible to detect an association between maternal nutrition and birth weight. Let us consider as an example food intake during pregnancy. If the intrapersonal variability of food intake measurement is similar to the variability in food intake among the pregnant women of the population, then the probability of finding any statistical relationship between food intake and birth weight will be very low even though such an association may exist (Lechtig *et al.*, 1979b).

Finally, there are problems in inferring cause–effect relationships from statistical associations between nutrition of the mother and birth weight. This is because in the same populations in which maternal malnutrition is widespread, there exist other unfavorable conditions, for example infection, which is itself associated with low birth weight (Lechtig *et al.*, 1976d, 1979a). Thus the challenge is to explore these alternative explanations of an observed association between maternal nutrition and birth weight. One method to prove causality is to determine the consistency of the observed association after a planned modification of the hypothetical causal factor. Consequently, investigations employing experimental or intervention designs, as contrasted to purely observational studies, are most appropriate for exploring the causality of a relationship between maternal nutrition and birth weight. In summary, the probability of demonstrating a causal relationship between maternal nutrition and birth weight increases if (a) a nutritional intervention is involved, (b) the target population consists of malnourished women, and (c) reliable indicators are used to determine the nutritional status of the mother.

In this chapter we review the literature concerning the influence of maternal nutrition on fetal growth and then present data from the INCAP (1969) longitudinal study which incorporates the three considerations just mentioned.

In this study, the influence of maternal nutrition was the main focus of attention. This is an issue of great public health relevance in terms of the possible positive effects of nutritional interventions on lowering the high incidence of low-birth-weight babies and the associated high infant mortality rates in the poorer populations of the world. However, it is equally important to be aware of the effect of maternal nutrition on certain maternal characteristics including lactation performance (Lechtig *et al.*, 1977a). There has been some debate as to whether dietary supplementation of women during pregnancy has any effect on their health or is only beneficial to the growth of the fetus. As part of the INCAP longitudinal study, characteristics relating to physical change during pregnancy, fecundity, and morbidity were investigated and the results are discussed here. Finally, certain implications of the findings presented in this chapter are reviewed and recommendations based on these implications proposed.

2. Influence of Maternal Nutrition on Fetal Growth

2.1. Literature Review

Animal experiments have shown that severe caloric or protein malnutrition during pregnancy delays fetal growth (McCance and Widdowson, 1962), and this growth retardation may be irreversible in those organs in which the nutritional insult has affected the rate of cell division (Winick and Noble, 1966).

In humans, maternal nutrition has been shown to affect birth weight in cases of acute starvation. Birth weights of infants born from pregnancies occurring during famine periods were consistently lower than those of infants born in the same country during times of adequate food supplies (Antonov, 1947; Smith, 1947; Gruenwald and Funakawa, 1967). The influence of moderate levels of chronic maternal malnutrition on fetal growth is less clear.

Maternal nutritional status prior to pregnancy is usually evaluated by anthropometric measurements as there is evidence to suggest that environmental conditions such as nutrition and infection, both closely associated with socioeconomic status, are important determinants of adult size in developing countries (Lechtig *et al.*, 1975b). The anthropometric measurements commonly used are height and weight.

Most of the difference in mean height between adult women from low socioeconomic groups in Guatemala and from a sample of the white population of the United States is already apparent at 7 years of age, suggesting that the difference in height could be accounted for by growth retardation during the first 7 years of life or even earlier. In addition, the heights of 7-year-old children from high socioeconomic groups in developing countries are similar to those of children in the industrialized countries and are greater than the heights of children of low socioeconomic status from the same ethnic group (reviewed by Lechtig *et al.*, 1975b).

Figure 1 presents the results from a prospective study in four rural villages in Guatemala and shows a consistent association between maternal height and the birth weight of her children. A similar consistent relationship has also been found between prepregnancy weight and the birth weight of children among mothers of the same height (Niswander and Gordon, 1972).

Differences in head circumference of adult populations are mainly due to differences in the rate of growth of the head during the first 2 years of life (reviewed by Lechtig *et al.*, 1975b). Decreased head circumference is associated with early severe malnutrition (Mönckeberg, 1968), and within the same ethnic group with lower socioeconomic status (Datta-Banik *et al.*, 1970). Furthermore, it has been found that protein–calorie supplementation during early life improves the rate of growth of head circumference (Lechtig *et al.*, 1976b).

Figure 2 presents results from rural Guatemala which indicate a relationship between maternal head circumference and the proportion of babies with low and high birth weight. This remains significant after controlling for the maternal height and weight. This may well reflect the specific influence of the very early nutritional history of the mother on fetal growth.

Maternal nutritional status during pregnancy is most frequently evaluated by weight gain and dietary intake. Weight gain during pregnancy is strongly correlated with birth weight in both industrialized and developing societies (Niswander and Gordon, 1972; Lechtig *et al.*, 1972b).

In contrast, most studies in industrialized countries have failed to find an association between nutrient intake during pregnancy and birth weight (Dieckman *et al.*, 1944; Thomson, 1959; Eastman and Jackson, 1968). This may be due to the poor reliability of the data on nutrient intake or to the fact that most women in the samples studied were relatively well nourished. In the

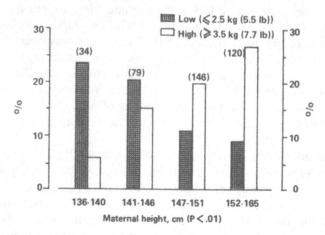

Fig. 1. Relationship between maternal height and birth weight. Number of cases given in parentheses. (From Lechtig *et al.*, 1975b.)

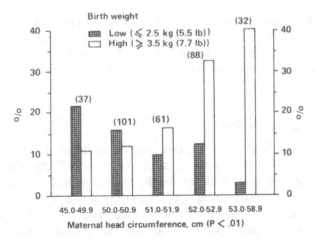

Fig. 2. Relationship between maternal head circumference and birth weight. Number of cases given in parentheses. (From Lechtig *et al.*, 1975b.)

developing countries, several dietary and intervention studies have shown an association between maternal nutritional intake and birth weight. However, other variables which may affect birth weight, such as infectious diseases and medical care, were not explicitly controlled.

Figure 3 presents the results of an analysis of the dietary data of 51 pregnant women from rural villages not receiving food supplements. Average birth weight increased progressively as dietary intake increased, and this re-

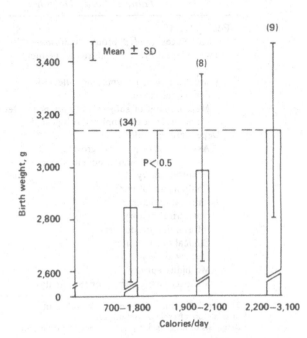

Fig. 3. Relationship between maternal dietary intake and birth weight. Number of cases given in parentheses. (From Lechtig *et al.*, 1972a.)

lationship was still evident when the weight of the newborn was corrected for such influences as height, parity, duration of illness during pregnancy, and sex of the newborn. It should be noted that further analyses with a larger sample size suggested an association in the same direction but this was not statistically significant (Lechtig *et al.*, 1975d).

2.2. The INCAP Longitudinal Study

The INCAP longitudinal study was designed to study the effects of chronic malnutrition on physical growth and mental development (Klein *et al.*, 1973). A nutritional supplement was made available to a chronically malnourished population of women of childbearing age. The study design and the principal examinations made are presented in Table I. Two types of food supplement were provided, *atole* (a gruel) and *fresco* (a refreshing cool drink). Table II presents the nutrient contents of *atole* and *fresco*. It should be stressed that the *fresco* contains no protein and provides only one-third of the calories contained in an equal volume of *atole*. In addition, both preparations contain similar concentrations of the vitamins and minerals which are possibly limiting in the diets of this population.

Two villages received *atole* while the other two received *fresco*. Attendance at the supplementation center was voluntary and this resulted in a wide range of supplement intake during pregnancy. In addition, all villagers received preventive and curative medical care.

Table I. Study Design

Four villages
    Two villages: *Atole*,[a] protein–calorie supplement
    Two villages: *Fresco*,[b] calorie supplement

Maternal and child information collected
    Independent variable:
        Measurement of subject's attendance to feeding center
            and amount of supplement ingested
    Dependent variables:
        Assessment of physical growth
        Assessment of mental development
        Infant mortality
        Components of birth interval
    Additional variables:
        Obstetrical history[c]
        Information on delivery
        Clinical examination
        Dietary survey
        Morbidity survey
        Socioeconomic survey of the family

---

[a] The name of a gruel commonly made with corn.
[b] Spanish for refreshing, cool drink.
[c] Diagnosis of pregnancy by the absence of menstruation.

Table II. Nutrient Content per Cup[a,b]

	Atole	Fresco
Total calories (kcal)	163	59
Protein (g)	11	—
Fats (g)	0.7	—
Carbohydrates (g)	27	15.3
Ascorbic acid (mg)	4.0	4.0
Calcium (g)	0.4	—
Phosphorus (g)	0.3	—
Thiamine (mg)	1.1	1.1
Riboflavin (mg)	1.5	1.5
Niacin (mg)	18.5	18.5
Vitamin A (mg)	1.2	1.2
Iron (mg)	5.4	5.0
Fluor (mg)	0.2	0.2

[a] Review date October 11, 1973; figures rounded to the nearest tenth.
Cup capacity 180 ml.
[b] Taken from Lechtig *et al.* (1975d).

2.2.1. Description of the Population

The study was conducted in four Spanish-speaking subsistence agriculture villages in eastern Guatemala. The villages were chosen on the basis of their relative isolation and homogeneity with respect to language, culture, size, and general economic and social structure. The entire population of the four villages is approximately 3329. The residents are all *Ladinos*,* and are relatively isolated by the lack of paved roads, telecommunication systems, or large stores. Average adult schooling is about 1.5 years and the functional literacy rate is low. The median family income is approximately $200 per year. The typical house is built of adobe and has no sanitary facilities; the drinking water is grossly contaminated.

Before the study began about 15% of the newborns died during the first year of life, a very high figure when compared with the current rates of less than 2% in developed societies. Clinically severe malnutrition (kwashiorkor) is prevalent, and children are severely retarded in physical growth at 7 years of age (Yarbrough *et al.*, 1975). Intrauterine infections are also very common as compared with developed societies (Lechtig *et al.*, 1974). Corn and beans are the principal components of the home diet, with animal protein forming 12% of the total protein intake. Dietary surveys indicate that the average daily intake throughout pregnancy was about 1500 cal and 40 g of proteins. At the time the study began, malnutrition and infectious diseases were endemic in the four villages.

Important characteristics of the maternal population include an average height of 149 cm with a mean weight at the end of the third trimester of

---

* Unlike Indians—descendents and heirs of the Mayan culture—*Ladinos* speak Spanish. Therefore, the two groups are linguistically and culturally different.

pregnancy of 49 kg. The mean weight gain during gestation is 7 kg, about one-half of the norm (Hytten and Leitch, 1971). The median number of previous deliveries is 4 (range, 0–13), and the median age is 26 years (range, 14–46 years).

In a small sample of newborns ($N = 42$) collected at the beginning of this study, the average birth weight was 3000 g, and of the babies with normal gestational ages ($N = 39$), one-third weighed 2500 g or less. Although data are not available from similar villages, it should be noted that in a Mayan Indian rural village in which there was no nutritional intervention, 41% of the newborns weighed 2500 g or less (Mata *et al.*, 1971).

2.2.2. Variables Selected for the Present Analysis

2.2.2a. Maternal Ingestion of Food during Pregnancy. Maternal ingestion of food during pregnancy consisted of two variables: ingestion of the supplement (the experimental treatment) and estimates of daily home diet intake for the last two trimesters of pregnancy.

Maternal home intake was estimated through 24- and 72-hr recall surveys at the end of each trimester of pregnancy. Previous analyses have shown that the home diet intake during the first trimester of pregnancy was significantly lower than the intake during the last two trimesters and that there was no significant difference between the 24- and 72-hr recall surveys (Lechtig *et al.*, 1972a). Both variables, supplement and home diet intake, were expressed in terms of calories because calories were considered to be the main limiting factors in the diet of the population.

2.2.2b. Maternal Anthropometry. Anthropometric variables included height and head circumference as well as weight at the end of the first trimester of pregnancy. First trimester weight may be considered as an indicator of maternal nutritional status at the beginning of pregnancy whereas height and head circumference probably reflect maternal physical growth during infancy and childhood. Previous analyses in this population have shown that the differences in head circumference and height between the adult women of this population and U.S. standards were 3.9 and 16.9 cm, respectively. The differences observed for head circumference at 2 years of age and for height at 7 years of age between the girls of this population and those of the same U.S. standards were 3.0 and 12.1 cm, respectively. These results suggest, therefore, that retardation in maternal height and head circumference reflects physical growth retardation in early life (Lechtig *et al.*, 1975e).

2.2.2c. Obstetrical Variables. Obstetrical variables included parity, gestational age, interval between the current and the previous baby, age of the mother, and duration of lactation during the present pregnancy. Parity was expressed as the number of previous births and was determined by interviewing the mother and reviewing the village civil registry. Gestational age was the time in weeks from the last menstrual period to the birth of the baby. Since all mothers with preschool children were visited every 2 weeks, the onset of pregnancy could be elicited within 15 days of the missed period. Primiparas and those women with postpartum amenorrhea who became pregnant were usually identified somewhat later.

   2.2.2d. Maternal Morbidity during Pregnancy. Maternal morbidity during pregnancy was estimated through the same fortnightly interviews used to monitor menstruation. A composite morbidity indicator was formulated by summing the number of days per months of pregnancy during which the mother was ill with diarrhea or anorexia, or remained in bed due to illness. In previous analyses, these components were shown to be significantly associated with birth weight.

   2.2.2e. Socioeconomic Status of the Family. The family's socioeconomic status was described by a composite scale reflecting the physical conditions of the house, the mother's clothing, and the reported extent of teaching various skills and tasks to pre-school-age children by family members. In previous analyses these three items showed consistent association with birth weight.

   2.2.2f. Fetal Factors. Birth weight (to the nearest 20 g) was determined within the first 24 hr of life. The risk of intrauterine infection was estimated by measuring IgM levels in cord blood (Lechtig *et al.*, 1974). Between June 1970 and February 1973, it was possible to follow 92% of the mothers throughout the entire pregnancy.

2.2.3. Effect of Food Supplementation during Pregnancy on Birth Weight

   The goal of the food supplementation intervention was to increase the total caloric intake during pregnancy with the objective of discerning a relationship between maternal nutrient intake during pregnancy and birth weight. Tables III through VIII and Figs. 4 and 5 show that this goal was reached and the objective achieved.

   Table III shows that the correlation between supplemented calories and home caloric intake was negative, although not significant. It also shows that the home caloric intake for two levels of caloric supplementation was not significantly different. Therefore, these results indicate that there was no strong association between caloric supplementation and home caloric intake and that the supplementation program produced a net increase in the total caloric intake. Finally, after combining the calories from the home diet and the food supplement for each mother, the high supplement group had a total caloric intake significantly higher than that of the low supplement group. The difference between both groups (149 cal/day) indicates that the average net increase in the total caloric intake from supplementation was approximately 26,820 cal.

   Table IV describes the relationship between caloric supplementation during pregnancy and birth weight. It can be seen that the increase in birth weight for the same amount of calories was not significantly different between *fresco* and *atole* villages. Further, in the four villages combined, there was a significant positive correlation between caloric supplementation during pregnancy and birth weight ($p < 0.01$).

   Table V shows that there existed a significant difference ($p < 0.025$)

Table III. Relationship between Supplemented Calories during Pregnancy and Home Caloric Intake[a]

A. For the four villages:
Correlation value(r)   =   −.015
Slope value (b)   =   −3.1 cal home diet/100 supplemented cal
Probability value (p)   >   0.05
Number of cases (N)   =   357

B. For two levels of caloric supplementation during pregnancy:

	Low supplemented group (<20,000 cal)	High supplemented group (≥20,000 cal)	Difference between the means (HSG − LSG)	p
Daily home caloric intake (cal/day)	1,415 ± 443	1,374 ± 364	−41	N.S.
Supplemented calories during pregnancy	7,200 ± 6,221	42,001 ± 19,221	34,822	<.001
Total caloric intake (cal-day)[b]	1,458 ± 446	1,607 ± 380	149[c]	<.001
Birth weight(g)	2,997 ± 471	3,114 ± 476	117	<.05
Number of cases	192	165		

[a] Taken from Lechtig et al. (1975d).
[b] Estimated by adding to the daily home caloric intake the ratio of supplemented calories per 180 days.
[c] Equivalent to 26,820 cal during the last two trimesters of pregnancy.

Table IV. Relationship between Supplemented Calories during Pregnancy and Birth Weight[a]

	Correlation value (r)	Slope value[b] (g birth weight/ 10,000 cal)	Number of cases (N)	Probability value (p<)
Atole villages	.113	23	219	0.10
Fresco villages	.123	30	186	0.10
Four villages (*atole* and *fresco*)	.135	29	405	0.01

[a] Taken from Lechtig *et al.* (1975d).
[b] Slope for *fresco* greater than slope for *atole*: N.S. (test of covariance).

between the mean birth weight for two levels of caloric supplementation during pregnancy (low supplement group $\leq$ 20,000 cal; high supplement group $\geq$ 20,000 cal). Again, this analysis showed no significant difference between the caloric and the protein–calorie supplementation.

Finally, Fig. 4 shows that the percentage of low-birth-weight babies was consistently lower in the high supplement group in both the *fresco* and *atole* populations and that the rate of low-birth-weight babies among low supplement mothers was roughly twice that observed among high supplement mothers.

Table VI presents the correlation values between maternal characteristics and birth weight for those variables that were significantly associated with birth weight. It can be seen that, in addition to caloric supplementation during

Table V. Relationship between Supplemented Calories during Pregnancy and Birth Weight for Two Levels of Caloric Supplementation[a,b]

Supplemented calories during pregnancy	Mean birth weight (g)		
	Atole	Fresco	Total
High $\geq$20,000	3173 (102)	3035 (68)	3105 (170)
Low[c] <20,000	3042 (117)	2948 (118)	2994 (235)
Total	3107 (219)	2992 (186)	3049 ± 469 ($\pm$1 S.D.) (405)

[a] Taken from Lechtig *et al.* (1975d).
[b] Number of cases in parentheses.
[c] Analysis of variance for the differences in birth weight among the low cells:

	Probability value (p) Birth weight (g)
"High" greater than "low"	<0.025
Atole greater than *fresco*	<0.025
Interaction	N.S.

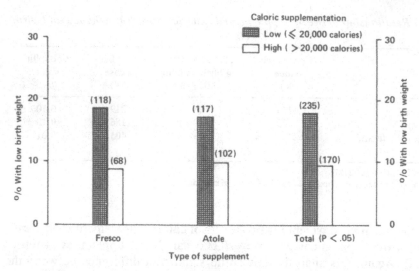

Fig. 4. Relationship between food supplementation and birth weight. Number of cases given in parentheses. (From Lechtig *et al.*, 1975d.)

pregnancy, gestational age, parity, weight at the end of the first trimester of pregnancy, and maternal morbidity were correlated with birth weight.

Table VII shows that the relationship between supplemented calories during pregnancy and birth weight was basically unchanged after statistically controlling for the influence of these maternal variables. The slope values observed on this table (29 or 30 g of birth weight per 10,000 supplemented calories) fall within the range of the expected dose–response relationships (see later).

To determine whether the observed association presented in Table V and Fig. 4 was due to bias produced by missing data, the relationship was analyzed in the population studied from July 1970 through February 1973, where the data coverage was 92%. Table VIII shows that in this population, the relationship between supplemented calories during pregnancy and birth weight was similar in direction and magnitude to that observed in the whole study population. In addition, after controlling for the maternal factors presented in Table VI, the slope value was identical to that of the similarly controlled whole population. Therefore, it is unlikely that the findings are due to bias produced by missing data.

In order to control for consistent maternal factors, either measured or unmeasured, the relationship between caloric supplementation during pregnancy and birth weight within pairs of siblings of the same mother was explored. Figure 5 presents the mean differences in birth weight for 94 pairs of siblings divided into three groups defined by differences in caloric supplementation of the mother between two successive pregnancies. This analysis indicates a positive association between changes in caloric supplementation and changes in birth weight in consecutive pregnancies of the same mother. In

Table VI. Maternal Determinants of Birth Weight in
Four Rural Villages of Guatemala[a,b]

	Correlation coefficient with birth weight	N
At conception:		
Height	.134[c]	399
Head circumference	.284[c]	363
Age	.116[c]	401
Parity	.154[c]	404
Socioeconomic status indicator	.219[c]	363
At the end of the first trimester of pregnancy:		
Weight at the end of first trimester	.277[c]	221
During pregnancy		
Gestational age	.217[c]	395
Morbidity indicator	−.122[d]	240
Caloric supplementation[b]	.135[c]	405

[a] Taken from Lechtig *et al.* (1975d).
[b] Value for the multiple correlation predicting birth weight: $r = 0.410$ ($p < 0.01$).
[c] $p < 0.01$.
[d] $p < 0.05$.

addition, after adjusting the changes in caloric supplementation and in birth weight for the intercorrelations existing between successive pregnancies, the relationship between caloric supplementation and birth weight was roughly similar (slope value 22 g of birth weight per 10,000 cal) to that observed in the entire population.

In summary, the analyses presented in Tables IV–VIII and in Fig. 5 indicate that the relationship of supplemented calories to birth weight was

Table VII. Relationship between Caloric Supplementation during
Pregnancy and Birth Weight (N = 405)[a]

	g birth weight/10,000 cal	
	Slope	SE
Before controlling for suspected confounding factors (simple regression)	29[b]	10.6
After controlling for suspected confounding factors[c] (in multiple correlation)	30[b]	10.6

[a] Taken from Lechtig *et al.* (1975d).
[b] $p < 0.01$.
[c] Height, head circumference, age, parity, socioeconomic status, weight at the end of the first trimester, gestational age, morbidity indicator, and home diet.

*Table VIII. Relationship between Supplemented Calories and Birth Weight
Regarding Different Rates of Coverage on Birth Weight Data[a]*

	Coverage rate (%)	Correlation value (r)	Slope value (g birth weight/10,000 cal)		Number of cases (N)
			Simple regression	Multiple regression[b]	
Total study population (21,812 ± 21,770)	63	.135	29[c]	30[c]	405
Population with high coverage[d] (24,407 ± 22,649)	92	.116	24[e]	30[c]	331

[a] Taken from Lechtig *et al.* (1975d).
[b] After controlling for suspected confounding factors: height, head, age, parity, socioeconomic status, weight at the end of the first trimester, gestational age, morbidity indicator, and home diet. Mean caloric supplementation during pregnancy ± standard deviation in parentheses. Slope for total study population greater than slope for population with high coverage. (Test of covariance: N.S.)
[c] $p < 0.01$.
[d] From July 1970 through February 1973.
[e] $p \leqslant 0.05$.

consistent in the entire population studied, in the population with the highest rate of coverage, and between siblings of the same mother. Therefore it can be concluded that the most suitable interpretation of these results is that caloric supplementation during pregnancy caused an increase in birth weight in this population.

2.2.4. Protein or Calories—Which Is the Most Important Determinant of the Effect of Nutrition on Birth Weight?

Having established a relationship between maternal nutrition and birth weight, a question that logically follows is which nutrient is the most important determinant of this effect? The ensuing discussion concerns only proteins and calories although it could equally well apply to other nutrients.

There are several factors which influence the relative contribution of protein and calories to birth weight. A primary consideration is the diet of the pregnant woman. If the limiting factor is protein, then an increase in the intake of this nutrient would perhaps result in an increment of the average birth weight. However, if calories were the main limiting factor in the diet, the same increase in protein intake would not produce such an effect on birth weight (Lechtig *et al.*, 1975e).

The importance of this consideration is illustrated in the following example using data from the four rural villages in Guatemala. In these villages, the average home caloric intake is very low and provides a relatively small margin for physical activity. On the other hand, protein intake is slightly higher than the average required for maintenance and tissue synthesis. This suggests that the risk of caloric deficiency in this population is higher than that of protein deficiency. At these levels of caloric intake (31 cal/kg daily) no additional increase in the retention of nitrogen would be expected even with an intake of 175–200 mg N/kg a day. Increased protein intake would produce an increment in the nitrogen balance only at caloric intakes higher than 40 cal/kg a day (Oldham and Sheft, 1951; Lechtig *et al.*, 1975e). Conversely, an increment of the caloric intake alone could produce a significant increase in nitrogen retention.

In addition to these dietary considerations there are other factors which may influence the relative effect of protein and calories on birth weight. One of these is the differential capacity of the human placenta to adapt itself to protein deficiency as compared to caloric deficiency. Even with low levels of amino acids in the maternal blood, the placenta is able to maintain an adequate concentration of the needed amino acids in the fetal blood, which is consistently higher than that in the mother's blood (Pearse and Sornson, 1969; Page, 1957). In contrast, the concentration of glucose is always lower in the fetal blood than in the maternal blood (Page, 1957, 1969). Further, experiments with rats and pigs have shown that pregnancy can be sustained on protein-deficient diets because maternal protein is mobilized (Callard and Leathen, 1970; Nainsmith, 1969; Rippel *et al.*, 1965). These data support the theory that the human placenta is better able to buffer deficiencies in protein as compared to calories in the maternal diet.

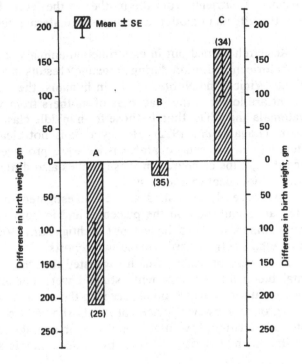

Fig. 5. Relationship between food supplementation and birth weight in consecutive pregnancies. Difference in caloric supplementation: A, from −40,000 to 0 cal; B, from 100 to 20,000 cal; C, from 20,000 to 120,000 cal. Number of pairs of pregnancies given in parentheses. Difference between groups A and C: *p* < 0.01. (From Lechtig *et al.*, 1975d.)

Finally, another factor influencing the relative importance of each nutrient is fetal body composition. The relative contribution of the fetal tissue components to the total fetal weight varies with gestational age and with birth weight. Thus in the range of birth weight between 3.0 and 3.2, the increment in fetal weight would mainly be a function of the synthesis of adipose tissue (Coltart *et al.*, 1969; Knopp *et al.*, 1970). This suggests that for this range of increment in birth weight, the caloric availability to the fetus, either from the maternal diet or stores, may be relatively more important than its protein availability.

In conclusion, the published literature indicates that the relative contribution of calories and protein to an increase in birth weight depends on which is the limiting nutrient in the diet, the functional capacities of the placenta, and the change in birth weight produced in the specific population under study. There are other factors such as physical activity, prevalence of disease, and nutrient availability from the maternal stores before pregnancy which may also be important in determining the relative contribution of calories and protein on birth weight.

2.2.5. Possible Mechanisms of the Effect of Maternal Nutrition on Fetal Growth

Another question which follows from the finding of a relationship between maternal nutrition and birth weight concerns the possible mechanisms of this effect. The postulated mechanism relates to the relative degree of placental transport of nutrients from the mother to the fetus. However, little is known about the effects of moderate changes in maternal nutrition on placental function.

Research carried out in experimental animals has shown that severe protein–calorie malnutrition during pregnancy results in a decrease in the placental weight (Winick and Noble, 1966). In humans, the weight, DNA, and protein content are lower in the placentas of mothers from very poor populations of Guatemala and Chile, than in those from middle class populations in Iowa and Boston (Laga *et al.*, 1972). However, it is not clear whether the observed differences in placental characteristics were produced by maternal nutrition, since the groups compared in these studies also differed in terms of a number of other biological characteristics.

Two recently published studies investigated the influence of moderate maternal malnutrition on the placenta, and in both special care was taken to ensure comparability of the groups (Lechtig *et al.*, 1975f, 1977c). The following paragraphs briefly describe these two reports.

Data were presented which suggested that the concentrations of the placental biochemical components studied were not affected by the moderate protein–calorie malnutrition observed in the mothers of the low socioeconomic status groups. However, placental weight showed a consistent association with maternal nutrition. Placental weight was on average 15% lower in the mothers classified as malnourished (either by socioeconomic status or amount of supplemented calories) than in those defined as well nourished. Because of this difference, the mothers defined as malnourished presented a lower placental content for all chemical compounds studied. Changes in maternal nutritional status were isolated as the main difference between the compared groups. Thus, different factors capable of affecting the relationship between indicators of maternal nutrition and placental weight were controlled, either through the study design or the data analysis. Therefore it was concluded that it is improbable that nonnutritional factors were responsible for the changes observed in the placental weights and that lower placental weights were caused by maternal malnutrition during pregnancy.

The implication that this effect of maternal nutrition on placental weight may be a possible mechanism of the effect of maternal nutrition on fetal growth deserves consideration. Studies in animals have shown that extirpation of part of the placenta during pregnancy results in fetal growth retardation (Hill *et al.*, 1971). In humans, a consistent correlation between the weight of the placenta and size of the newborn has been reported (Hytten and Leitch, 1971). Similar observations have been made in populations of the INCAP longitudinal study (Lechtig *et al.*, 1975f). In addition, in this study a marked reduction in the magnitude of the association between indicators of maternal nutrition and birth weight was observed after the placental weight was entered as a forced variable in a multiple regression to predict birth weight. This result indicates that most of the association between maternal nutrition and birth weight can be explained by placental weight. In the light of animal studies the present data suggest that in humans most of the influence of maternal nutritional status on fetal growth is mediated through the effect on placental weight.

Additional evidence for this hypothesis comes from the observation that placental weight is closely associated with the surface area of peripheral villi (see Fig. 6) (Lechtig *et al.*, 1975f). Since the area of the peripheral villi is a determinant of nutrient transport from the mother to the fetus, this observation suggests a physiological mechanism to explain the statistical association observed between indicators of maternal nutrition, placental weight, and birth weight.

In summary, moderate protein–calorie malnutrition during pregnancy appears to lead to a lower placental weight without significant changes in the concentration of the biochemical components studied. This effect, in turn, may be the mechanism by which maternal malnutrition gives rise to a high prevalence of low-birth-weight (LBW) babies in these populations.

2.2.6. Expected Dose- and Time–Responses of Nutritional Interventions to Improve Birth Weight

Two questions of public health interest relating to the association of maternal nutrition and birth weight concern the dose–response and time–response of nutritional interventions. These two questions are now discussed separately.

2.2.6a. Dose–Response. In order to simplify the discussion and facilitate

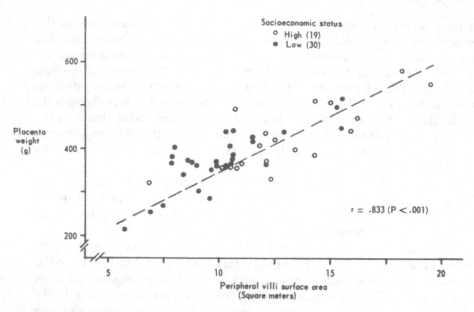

Fig. 6. Urban study. Relationship between peripheral villi surface area and weight of placenta.

the comparison of findings from different sources, caloric intake is used as the measurement of maternal food intake. Slope estimations are utilized to facilitate the computations of the expected effects of maternal nutrition and the comparison of different studies (Lechtig *et al.*, 1975e).

Table IX summarizes the results of analyses made from three different sources:

1. The available data on the relationship between maternal weight gain during pregnancy and birth weight.
2. Data on the relationship between differences in pregnant weight and differences in birth weight between two consecutive pregnancies.
3. Data concerning food intake, body composition, and physical activity.

These three sources provided estimates of the same order of magnitude

Table IX. Expected Dose-Response Relationships between
Caloric Intake during Pregnancy and Birth Weight[a]

Basis for the estimation	Range (g birth weight/10,000 cal)
Weight gain during pregnancy	28–80
Differences in prepregnant weight between two consecutive pregnancies	25–40
Food intake, body composition, and physical activity	36–84

[a] Condensed from Lechtig *et al.* (1975e).

for the relationship between ingested calories and birth weight: from a minimum of 25 g to a maximum of 84 g of birth weight per 10,000 cal during pregnancy.

There are few reports in the literature that present data in such a way that the relationship presented in Table IX can be verified. However, some studies do permit a computation of the dose–response relationship between food intake during pregnancy and birth weight. Table X summarizes the results from five reports in which different approaches were used to estimate changes in food intake. The different approaches were dietary surveys (Burke *et al.*, 1943), acute disease of food availability at a community level (Antonov, 1947; Smith, 1947), and food supplementation under highly controlled conditions (Iyenger, 1975; Lechtig *et al.*, 1975d). Given the heterogeneity and, in some times the data available, it is surprising to find that the computed dose–response relationships (range: 20–75 g of birth weight per 10,000 cal during pregnancy) are of the same order of magnitude as those presented in Table IX.

It should be noted that in all of these computations, the effect of maternal nutrition on birth weight was estimated by means of slope values. There are other ways of looking at the effect of nutritional interventions on birth weight, such as observing the changes produced in the proportion of babies with low birth weight (LBW ≤ 2.5 kg). This approach is useful from a public health point of view because of the known relationship between low birth weight and high risk of infant mortality (Chase, 1969).

The effect of a nutritional intervention on the proportion of LBW babies could be grossly predicted on the basis of the expected increase in the average birth weight and the birth weight distribution of the population before the program is implemented. For example, a slope value of 25 g of birth weight per 10,000 cal with an average supplementation of 40,000 cal during pregnancy could lead to a mean increment of 100 g in the average birth weight. If the proportion of babies with birth weights of 2.5 kg or less is 20% and the proportion of newborns weighing 2.4 kg or less (i.e., 2.5 kg minus 100 g) is

Table X. Estimated Dose–Response Relationships between Caloric Intake during Pregnancy and Birth Weight[a]

Indicator of maternal nutritional status during pregnancy	Estimated slope value (g birth weight/10,000 cal)	Computed from
Home dietary surveys	33	Burke *et al.* (1943)
Decrease in food availability during famine		
Leningrad	33	Antonov (1947)
Holland	20	Smith (1947)
Food supplementation		
In a hospital	75	Iyengar (1975)
In rural villages	30	Lechtig *et al.* (1975a)

[a] Adapted from Lechtig *et al.* (1975e).

8%, the expected reduction in the proportion of LBW babies would be 12% (20% − 8% = 12%).

This reduction would be larger if the proportion of LBW babies is 40% as in some of the rural populations of developing countries (Lechtig *et al.*, 1975a) and very low if the proportion is 7%, as in white populations of the United States (Niswander and Gordon, 1972). Consequently, the same dose–response relationship could have different public health implications depending on the type of population and the net amount of supplemented calories during pregnancy.

2.2.6b. Time–Response. Assertions have been made that food supplementation is most effective during the last trimester of pregnancy. This recommendation is based on the increased nutritional requirements of the fetus during the last month of gestation since the rate of fetal growth increases exponentially up to the 38th week (WHO, 1965). However, one of the functions of weight gain during pregnancy is to provide stores of energy for both the growth of the fetus and lactation. These stores may buffer changes in maternal diet and serve as a protective mechanism, particularly in societies in which pregnant women are engaged in heavy physical activity (Hytten and Leitch, 1971). If the efficiency of storage and subsequent transfer is high, the effect of supplementation during pregnancy would be approximately the same regardless of when the supplementation is given. This is important for food supplementation programs because it would appear that the earlier in pregnancy a woman begins supplementation, the more likely she will consume enough total calories to produce an important increase in birth weight.

Up to this point, we have considered nutrition during pregnancy and its impact on birth weight. In considering different nutritional programs, it would be useful to ask at what ages nutritional interventions can be most useful.

In many populations from low socioeconomic strata, the correlation between attained maternal height or head circumference and birth weight reflects the effect of growth retardation produced during the last 2–7 years of postnatal life (Lechtig *et al.*, 1975b). Therefore, nutritional interventions oriented to avoid physical growth retardation during the first years of life would lead to taller adult women and probably heavier babies. In rural populations of Guatemala, preliminary analyses suggested that the dose–response relationship between improved nutrition in preschool girls and subsequent improved birth weight would be 20 times lower than that predicted for interventions during pregnancy (reviewed by Lechtig *et al.*, 1975e). Interventions that increase maternal prepregnant weight can also increase birth weight (19–20 g of birth weight per 10,000 cal supplemented during adult life before pregnancy) (Lechtig *et al.*, 1975e).

When the data concerning the expected dose–responses of nutritional interventions are compared it can be seen that the nearer the nutritional intervention is to the pregnancy itself, the higher the expected effect on birth weight per unit of supplemented calories.

It should be noted, however, that the expected dose–response relationship, as presented here, is only one of several factors in the discussion of the appropriate time for nutritional intervention. Another important consideration is the possibility that some prepregnancy conditions could limit the effect of nutrition during pregnancy on birth weight. Thus, although the dose–response is greater in mothers with low prepregnant weights, the intercepts are lower, and because of this, mothers with very low prepregnant weight may not deliver babies with satisfactory birth weight even under optimum conditions of food intake during pregnancy [computed from Niswander and Gordon (1972)]. In addition, the lag time between the intervention and the expected benefit, the possible long-term sequelae resulting from malnutrition in early life, and the urgency to take action on malnourished women who are already pregnant, are important considerations in deciding the optimum period for nutritional intervention.

In conclusion, the available data suggest that a nutritional intervention designed to increase birth weight will be more effective the closer the intervention is to the pregnancy itself. Nonetheless, some prepregnancy characteristics, such as low weight or low height, may limit the effect of nutritional intervention programs on birth weight. During pregnancy, the effect of such an intervention may be similar (per unit of calories) regardless of the trimester at which it is given.

3. Influence of Nutrition on Other Maternal Characteristics

Although the relationship between maternal nutrition and fetal growth is very important it is equally important to investigate the effects of nutrition on the pregnant woman herself. Obviously certain conditions affecting the mother have definite implications for the outcome of the pregnancy; toxemia of pregnancy is a prime example of this.

Pregnancy is a time when a woman is exposed to fundamental, physiologically based changes affecting among other things her physique, her fecundity, and her general status of well-being. During the INCAP longitudinal study the effects of food supplementation during pregnancy on several maternal characteristics of public health relevance were explored.

3.1. Maternal Anthropometry during Pregnancy

During pregnancy the physique of the human female undergoes certain changes. Anthropometrical measurements can be made which are both dependent on and independent of these changes. In this section these anthropometric measurements are discussed together with their public health relevance in predicting risk of fetal growth retardation (Lechtig *et al.*, 1978b).

Table XI presents sample sizes, means, and standard deviations for maternal anthropometric variables, both for measurements taken at the end of the third trimester of pregnancy and for monthly changes. These values, particularly those for weight, height, head, and arm circumference, are below those usually reported for well-nourished populations from developed coun-

Table XI. Mean and Standard Deviations of Maternal Anthropometric Measurements during the Last Trimester of Pregnancy (Four Villages)

Measurement	Final[a]			Monthly change[b]		
	Mean	SD	N	Mean	SD	N
Mass and length						
Height (cm)	148.9	5.3	572	N.A.[c]		N.A.
Sitting height (cm)	79.5	3.1	192	N.A.		N.A.
Total arm length (cm)	66.2	3.0	193	N.A.		N.A.
Weight (kg)	53.9	6.6	192	1.20	0.56	137
Circumferences (cm)						
Head	52.8	26.0	193	N.A.	N.A.	N.A.
Chest	79.1	4.6	193	10.96	5.41	136
Arm	22.7	1.9	193	−0.05	0.45	137
Calf	31.5	2.3	192	1.29	2.05	137
Thigh	44.4	3.4	192	0.50	1.23	137
Diameters (cm)						
Biestyloid	4.6	2.9	193	0.11	0.38	137
Bicondylar	8.3	7.1	192	0.05	0.82	136
Skinfolds (mm)						
Bicipital	47.80	18.04	193	0.53	2.39	137
Tricipital	104.37	29.69	193	−0.07	0.46	136
Subscapular	11.7	3.2	193	0.13	0.41	137
Midaxillar	83.55	29.01	193	0.10	0.41	137
Calf	91.91	31.93	192	0.18	0.33	137
Lateral thigh	155.55	58.50	191	0.39	0.81	135
Anterior thigh	154.46	55.93	191	0.46	0.70	136

[a] At the end of the third trimester of pregnancy.
[b] Change per month of pregnancy (see methods in text).
[c] Not applicable.

tries (Hansman, 1970; Hytten and Leitch, 1971). The mean monthly weight gain during the last two trimesters of pregnancy of the study population was 1.2 kg/month. This is approximately half the monthly weight gain observed in well-nourished women from Aberdeen (Hytten and Leitch, 1971).

The relationship between caloric supplementation and maternal anthropometry was similar in the *atole* (protein–calorie supplement) and in the *fresco* (calorie supplement) villages. Thus the relationship appears to be independent of protein in the supplement. Table XII shows the relationship between caloric supplementation during pregnancy and maternal anthropometry for the *fresco* and *atole* villages combined. Most of the correlations between supplemented calories and monthly changes in anthropometric measurements were positive, though small and insignificant. It is of interest that the correlation between supplemented calories and monthly weight gain became stronger after statistically controlling for maternal home diet, head circumference, gestational age, parity, morbidity, socioeconomic status, birth interval, and type of supplement (*atole* and *fresco*). Thus, the correlation between supplemented calories during pregnancy and monthly weight gain became 0.213 (slope value: 294 g weight gain during pregnancy per 104 supplemented calories: $N = 137, p < 0.05$) after controlling for these factors. It should be noted that although the absolute value of the slope observed between supplemented calories and weight gain was higher than that observed between supplemented calories and birth weight (*vide supra*), the relative value expressed as the proportion of the corresponding attained weight was lower (0.5 versus 1.0%) for the mother than for the fetus. This observation suggests that the adaptive processes that protect the mother under conditions of malnutrition, may not be as strong as believed before (Thomson, 1959). Figure 7 shows the relationship between food supplementation and weight gain during pregnancy; the greater the amount of food supplementation, the lower is the proportion of mothers with low weight gain during pregnancy. The proportion of mothers with low weight gain (LWG < 0.5 kg/month) was about seven times higher in the low supplement group compared to the high supplement group. The definition of LWG, less than 0.5 kg/month, during the last two trimesters of pregnancy is equivalent to a total

Table XII. Correlations between Supplemented Calories During Pregnancy and Maternal Anthropometric measurements in Four Rural Villages of Guatemala[a]

Anthropometric measurement	Final[b] (N = 191)[c]	Monthly changes[d] (N = 137)
Mass and length		
Total arm length	0.01	N.A.[e]
Sitting height	0.02	N.A.
Weight	−0.04	0.12
Height	0.01 (N = 572)	N.A.
Weight/height ratio	−0.04	N.A.
Circumferences		
Head	−0.03 (N = 363)	N.A.
Chest	−0.05	0.13
Arm	−0.03	0.15
Calf	−0.01	0.16
Thigh	−0.05	−0.04
Mean change in circumferences	N.A.	0.15
Diameters		
Biestyloid	−0.02	0.06
Bicondylar	−0.11	0.13
Skinfolds		
Bicipital	−0.11	0.07
Tricipital	−0.09	0.04
Subscapular	−0.01	0.08
Midaxillar	−0.04	0.10
Medial calf	−0.18[f]	−0.02
Lateral thigh	−0.07	0.00
Superior thigh	−0.08	0.00
Mean change in skinfold change	N.A.	0.06

[a] Taken from Lechtig *et al.* (1978b).
[b] At the end of the third trimester of pregnancy.
[c] When *N* is different, it is presented in parenthesis.
[d] During last two trimesters of pregnancy.
[e] Not applicable.
[f] $p < 0.01$.

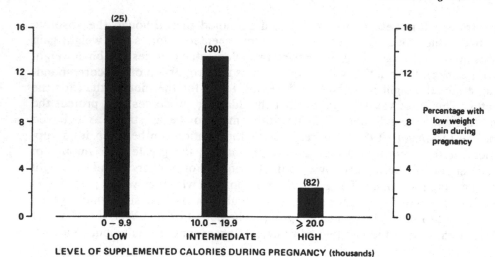

Fig. 7. Effect of supplemented calories on the proportion of mothers with low weight gain (less than 0.5 kg per month of pregnancy) during pregnancy. Number of cases given in parentheses ($p < 0.10$).

weight gain of about 3.0 kg. The risk of delivering low-birth-weight babies in mothers with weight gain of this magnitude is about three times greater than that for mothers of normal weight gain (computed from Niswander and Gordon, 1972).

Table XIII shows correlations between monthly weight gain and other anthropometric changes during pregnancy. It is evident that weight gain is strongly related to changes in almost all circumferences, diameters, and skinfolds measured. These results indicate, therefore, that weight gain reflects generalized changes in almost all body dimensions.

Table XIV shows correlations between maternal anthropometric variables (MAM) and birth weight. Measurements of mass and length, circumferences, and one bone diameter of the mother at the end of pregnancy were associated ($p < 0.05$) with birth weight. Multiple regression analysis indicated that the maternal factors identified in Table VI cannot explain the observed associations with birth weight.

The extent to which the correlations presented in Table XIV are due to the covariation among the various anthropometric variables was also explored. It was found that maternal weight gain by the end of the third trimester was the most powerful predictor of birth weight. Once this variable entered into the regression equation, the association between other maternal anthropometric variables and birth weight was almost totally erased. On the other hand, after controlling for height, head, and arm circumference, most of the initial association between maternal weight gain and birth weight was also erased. These results suggest that height, head, and arm circumference are as good as third trimester weight gain for predicting birth weight.

Since the predictive value of these three variables is almost unaffected by gestational age, they may be used as indicators to select target populations for health and nutritional interventions. Use of these indicators would notably increase the efficiency, coverage, and effectiveness of programs aimed to decrease the incidence of low-birth-weight babies (Lechtig *et al.*, 1976c).

In summary, if caloric supplementation produces an increment of total caloric intake during pregnancy, it can be concluded that caloric supplementation causes an increase in weight gain during pregnancy in this population. The public health implication of this association arises from the fact that low-weight-gain mothers more frequently deliver low-birth-weight babies, who in turn suffer higher rates of infant and neonatal mortality. If caloric supplementation causes a higher level of weight gain during pregnancy, then nutritional intervention programs should have a positive effect on lowering infant mortality rates. The observed associations between maternal anthropometric measurements and birth weight may provide a basis to build indicators predicting risk of delivering low-birth-weight babies. These high risk indicators are an essential element in the improvement of the efficiency of programs aimed at decreasing infant mortality.

3.2. Birth Interval Components

Factors influencing population growth are of definite public health relevance. Changes in population are determined to a large extent by the relative balance of births and deaths. We have mentioned that improving maternal

Table XIII. Correlations between Monthly Anthropometric Changes during Pregnancy and Weight Gain per Month (N = 137)

Anthropometric change per month of pregnancy	r
Circumferences	
Chest	.48[a]
Arm	.18[a]
Calf	.44[a]
Thigh	.07
Mean changes in circumferences	.59[a]
Diameters	
Biestyloid	.02
Bicondylar	.27[a]
Skinfolds	
Bicipital	.36[a]
Tricipital	.34[a]
Subscapular	.37[a]
Midaxillar	.39[a]
Calf	.42[a]
Lateral thigh	.37[a]
Upper thigh	.46[a]
Mean skinfold changes	.51[a]

[a] $p < 0.01$.

Table XIV. Correlations between Maternal Anthropometric Measurements and Birth Weight

Anthropometric measurement	Attained[a] (N = 186)[b]	Monthly changes (N = 135)
Mass and length		
Height	.14[c] (399)	N.A.[d]
Sitting height	.22[c]	N.A.
Total arm length	.08	N.A.
Weight	.35[c]	.21[e]
Weight/height ratio	.34[c]	N.A.
Circumference		
Head	.28[c] (363)	—
Chest	.28[c]	.11
Arm	.27[c]	.02
Calf	.33[c]	.08
Thigh	.26[c]	.06
Mean changes in circumferences	N.A.	.15
Diameters		
Biestyloid	.25[c]	−.03
Bicondylar	.14	.06
Skinfolds		
Bicipital	.10	.05
Tricipital	.12	.01
Subscapular	.14	.04
Midaxillar	.08	.03
Calf	.04	−.05
Lateral thigh	.11	.08
Upper thigh	.10	.01
Mean skinfold changes		.05

[a] At the end of the third trimester of pregnancy.
[b] When N is different it is presented in parentheses.
[c] $p < 0.01$.
[d] Not applicable.
[e] $p < 0.05$.

nutritional status should cause a fall in overall death rate by lowering the infant mortality rate. The question then arises as to what effect improving maternal nutritional status has on fecundity. The relationship between nutrition and fecundity is indeed complex but one aspect of this relationship which is amenable to investigation as part of a long-term nutritional intervention study is the effect of maternal nutrition on birth interval components. The following describes the work by Delgado *et al.* (1978).

A birth interval is defined as the period between one live birth and the next (Potter, 1963). Beginning with a live birth it can be divided into several components: the period of postpartum amenorrhea, the menstruating interval, and the next period of gestation (Henry, 1961; Potter *et al.*, 1965). The length of the birth interval is dependent on the duration of each component with some components having a greater capacity for variability than others. Each component, its capacity for variability, and its vulnerability to a nutritional impact are discussed individually.

3.2.1. Postpartum Amenorrhea

The first component of the birth interval, postpartum amenorrhea, directly follows the previous birth and generally extends to the first menses following that birth. This period is characterized by the absence of ovulation, and although the resumption of menstruation usually precedes that of ovulation, occasionally a woman will conceive on the first ovulation following a birth and menses will not have had a chance to reappear (Pérez *et al.*, 1972). In this case the woman moves directly from amenorrhea of the postpartum period into the amenorrhea of pregnancy.

In the absence of contraception and pregnancy wastage the period of postpartum amenorrhea has the largest possible variability of any of the components of the birth interval (Potter, 1963) after controlling for social factors. Furthermore, the variability is due almost entirely to purely biological rather than sociocultural factors. Thus, a nutritional impact on the birth interval would be expected to operate most strongly through this component.

The duration of postpartum amenorrhea appears to be associated with breast-feeding, pregnancy outcome, the age of the woman, the duration of the previous birth interval, and socioeconomic status (Salber *et al.*, 1966, 1968).

3.2.2. Lactation and Postpartum Amenorrhea

That the duration of breast feeding is related to the length of postpartum amenorrhea has been repeatedly shown in many population groups throughout the world (Ehrenfest, 1915; Peckham, 1934; Booth, 1935; Sharman, 1951; Baxi, 1957; Biswas, 1963; Potter *et al.*, 1965; Salber *et al.*, 1966; Jain *et al.*, 1970; Kamal *et al.*, 1969; Venkatacharya, 1972; Berman *et al.*, 1972; Chen *et al.*, 1974). Results from special field surveys conducted in rural villages in India (Potter *et al.*, 1965), Alaska (Berman *et al.*, 1972), and Bangladesh (Chen *et al.*, 1974) have demonstrated that the median length of amenorrhea for women with a surviving breast-fed child was over 10 months. On the other hand, women who did not nurse their infants experienced significantly shorter durations of amenorrhea, with a median length of 2 months.

It has also been reported that menstruation is absent in practically all women during the early months of lactation and that as lactation proceeds, the proportion of women menstruating increases (Baxi, 1957; McKeown and Gibson, 1954). This suggests that the inhibitory influence of lactation on ovulatory function becomes less powerful as the duration of lactation increases. In addition, lactation could delay pregnancy even after ovulation has been resumed by interfering with the process of implantation.

The length of postpartum amenorrhea is associated not only with the duration of lactation but also with the frequency of suckling as well as supplementary feeding of the child (Sharman, 1951; McKeown and Gibson, 1954; Pérez et al., 1972).

3.2.3. Postpartum Amenorrhea, Pregnancy Outcome, Age of the Mother, and Birth Interval

The length of postpartum amenorrhea is associated with infant survival and pregnancy outcome. Data provided by Potter et al. (1965) indicate that the length of amenorrhea following a miscarriage is less than a month and that it increases to 2½ months following stillbirths and neonatal deaths. Furthermore, the duration of postpartum amenorrhea was about 7 months following infant deaths between 2 and 12 months, and lasted about 12 months when infants survived for more than a year. Lactation probably affects the duration of postpartum amenorrhea in the cases of infant deaths after 2 months. However, in the cases of miscarriages, stillbirths, and neonatal deaths other factors may play a more important role in determining the duration of amenorrhea.

Potter et al. (1965) and Salber et al. (1966) have reported that the duration of postpartum amenorrhea increases with age. Salber et al. (1968) have also shown that birth intervals among women tend to have a characteristic pattern.

3.2.4. Socioeconomic Status and Postpartum Amenorrhea

The duration of postpartum amenorrhea is apparently associated with per capita income; mothers from high socioeconomic groups (Malkani and Michardini, 1960; Salber et al., 1966; McKeown and Gibson, 1954) are amenorrheic for less time than mothers from low socioeconomic groups (Baxi, 1957; Bonte and Van Balen, 1969; Kamal et al., 1969). Table XV presents the results of some selected studies. It shows the probability of being amenorrheic by the end of each of the first 6 months after delivery. Clearly, mothers from low socioeconomic groups present a higher probability of remaining amenorrheic during the first 6 months than mothers from high socioeconomic groups. This comparison must be viewed with caution because of the series of underlying assumptions involved, particularly about the definition of lactation status and the methodology used in each study.

Another way of looking at the reported data consists of comparing the regression slopes between duration of lactation and duration of postpartum amenorrhea in two different socioeconomic groups. Figure 8 shows the regression line computed from Salber's data (Salber et al., 1966) for those women lactating for more than a month and from data reported by Jain et al. (1970), who studied nursing women in Taiwan. Clearly, for the same duration of lactation, mothers from Taiwan (low socioeconomic group) present a longer postpartum amenorrhea than mothers from Boston (middle class). These reports imply that in addition to duration of lactation, factors related to socioeconomic status may affect the duration of postpartum amenorrhea. Nutrition could be one of these factors given that the nutritional demands of pregnancy

Table XV. Probability of Being Amenorrheic in Lactating Women by Postpartum Month[a]

Studies and place	Months						Observations
	1	2	3	4	5	6	
Chen *et al.* Bangladesh	1.00	0.98	0.96	0.95	0.93	0.92	Prospective study of rural populations Low socioeconomic status Full nursing
Potter *et al.* India	1.00	0.94	0.91	0.86	0.83	0.79	Prospective study of rural populations Low socioeconomic status Less than half the infants received only milk by the sixth month after delivery
Berman *et al.* Eskimos	0.97	0.91	0.85	0.75	0.70	0.65	Prospective study of rural population Low socioeconomic status Usually nursing was supplemented when the infant was 6 months old
Pérez *et al.* Chile	1.00	0.99	0.83	0.71	—	—	Prospective study of urban, low middle socioeconomic class. Full nursing
Salber *et al.* Boston	1.00	0.86	0.76	0.71	0.50	0.35	Prospective study of urban, middle socioeconomic status. Computed for those mothers lactating 6 months or more

[a] From Delgado *et al.* (1976).

and lactation are easily coped with in high socioeconomic groups but not in lower class malnourished women (Stix, 1940; Salber *et al.*, 1966; Chen *et al.*, 1974).

3.2.5. Nutrition and Postpartum Amenorrhea

Apart from the variables discussed previously, it has been suggested that the nutritional status of the mother affects the length of postpartum amenorrhea in lactating women (Malkani and Mirchardini, 1960; Solien de González, 1964; Salber *et al.*, 1966; Gopalan and Naidu, 1972; Frisch, 1974). Aside from the implications for the presence of a nutritional impact suggested by socioeconomic comparisons the possibility that nutrition is related to the duration of postpartum amenorrhea is given some support by the increased prevalence of amenorrhea reported during times of severe food shortage and famine (Antonov, 1947; Smith, 1947). Some experimental evidence to date also supports the plausibility of this hypothesis. Chávez and Martínez (1973) compared the postpartum amenorrhea of a small group of lactating women receiving food supplementation through pregnancy and lactation to a control group of lactat-

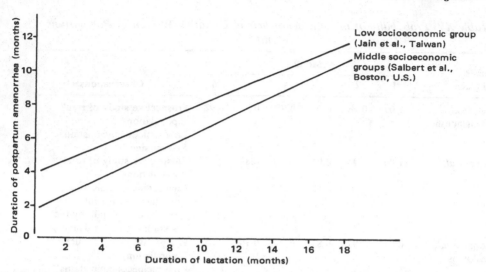

Fig. 8. Relationship between duration of lactation and length of postpartum amenorrhea by two levels of socioeconomic status. (From Delgado *et al.*, 1978.)

ing women receiving no supplementation. Their results suggest that one of the effects of supplementing the diet is to significantly reduce the duration of postpartum amenorrhea. It was reported that postpartum amenorrhea lasts 7.5 months for the supplemented and 14.0 months for the control group although retrospective data show that both groups had similar lengths of postpartum amenorrhea before food supplementation.

3.2.6. Menstruating Interval

Menstruating interval, the second birth interval component, refers to the number of months a woman menstruates after the end of postpartum amenorrhea and before the next conception (Potter *et al.*, 1965). It is generally considered to be equivalent to the period of ovulation.

Potter (1963) has estimated that in noncontracepting populations after adjusting for pregnancy wastage the mean menstruating interval varies within a narrow range. It is affected by a combination of biological and sociocultural factors.

Since the menstruating interval is strongly dependent on the frequency of sexual intercourse it will be affected by all the sociocultural factors that influence this behavior including contraception, absence of the partners from the home, and sexual customs of the population. The physiological factors affecting the menstruating interval include the viability of the sperm and ovum as well as the regularity of ovulation. The apparent length is also prolonged by undetected or unreported pregnancy wastage (Potter *et al.*, 1965). The menstruating interval with all of its physiological and sociocultural factors is closely related to the concept of fecundability defined as the monthly probability of conception.

3.2.7. Anovulatory Cycles

For a long time it was believed that the first menstrual cycles following parturition were anovulatory (Potter, 1963; Holmberg, 1970). Potter (1963) estimated a mean of two anovulatory cycles following postpartum amenorrhea. A recent study by Pérez et al. (1972) indicates that in most cases first menstruation was ovulatory and that the incidence of anovulatory cycles is related to the length of postpartum amenorrhea. Thus, this period would be expected to have relatively little impact on the length of the total birth interval although nutrition theoretically could.

The number of anovulatory cycles throughout the rest of the menstruating interval would also add to the variability of the menstruating interval and could be vulnerable to changes in nutritional status.

3.2.8. Early Pregnancy Wastage

It has been suggested that early pregnancy wastage or intrauterine deaths that go unrecognized as such could prolong the apparent length of the menstruating interval (Potter et al., 1965). Hertig et al. (1959) have estimated that intrauterine mortality decreases rapidly through the first 24 weeks of pregnancy and then stays at a constant low level throughout the rest of the gestation period. According to Hertig et al. (1959), about 50% of the fertilized ova will not survive more than 2 weeks. Since intrauterine deaths in the first 2 weeks are not observable through a disruption of menstruation and subsequent ovulation, additional support for Hertig's findings is difficult to obtain. However, the loss of these fertilized ova would decrease the probability of conception and prolong the menstruating interval.

3.2.9. Menstruating Interval and Fecundability

Because fecundability is usually considered to be the probability of conception after the onset of ovulation findings of fecundability can be used as indicators of possible effects on the menstruating interval.

Barrett (1971) found that fecundability of women declined with age. Fecundability was found to increase from puberty to age 20, remain constant until age 30, and decline progressively afterwards. Potter et al. (1965) suggest that these changes may be due to such age-dependent factors as declining coital frequency, more numerous anovulatory cycles, and rising incidences of unreported fetal wastage.

According to Jain et al. (1970) women of higher socioeconomic status have higher fecundability than those with lower status. Specifically fecundability was found to increase in Taiwanese women with educational level, the number of modern objects owned by the couples, and monthly family income. Jain suggests better health and nutritional status as a possible explanation for these findings.

3.2.10. Gestation Period

The last component of the birth interval is the period of gestation following conception. The duration of this period is determined by the pregnancy outcome since gestation may result in a live birth, a stillbirth, or a spontaneous abortion. Consequently, although the length of most pregnancies ending in a live birth will fall within a narrow range, the duration of pregnancy can be quite variable.

Since the birth interval is defined as the period between live births, pregnancy wastage, while shortening the duration of gestation, will greatly increase the overall birth interval. Potter (1963) has estimated this increase to be as much as 3 months after a miscarriage in the third month in younger Western mothers. The increase could be even greater in older mothers or mothers with low socioeconomic status.

Although nutrition may affect the incidence of premature births and thus pregnancy duration, the greater expected impact would be through changes in spontaneous abortion and stillbirth rates.

3.2.11. Socioeconomic Status, Spontaneous Abortions, and Stillbirths

The higher incidence of fetal death in low socioeconomic groups has been illustrated in India by Soangra *et al.* (1975) and in Columbia by Agualimpia (1969). Moreover, Baird (1966) reports that perinatal mortality is higher in low socioeconomic groups than in high socioeconomic groups. He also found a strong association between maternal height and perinatal mortality within each socioeconomic group. The absence of a significant correlation between current diet and clinical findings led Baird to suggest that the nutritional history of a woman during growth and adolescence may be more important than her dietary intake during pregnancy itself.

3.2.12. Summary of the Literature Review

In summary, an effect of nutrition on postpartum amenorrhea and on the birth interval is not well established at present. Even less is known concerning possible mechanisms involved in such an effect if, as is likely, it does exist. Obviously, there remains a good deal of work to be completed, but it would seem almost certain that as the story of nutrition evolves, the importance of its role in human fecundity will become more evident.

As part of the INCAP longitudinal study the interrelationship among lactation, postpartum amenorrhea, and nutrition was partially explored. In the following paragraphs the preliminary results from this study are discussed.

The data were collected in all those mothers who gave birth between January 1969 and February 1973, and who were followed until January 1975.

Since data were collected prospectively (all families were visited every 14 days), reliable information was obtained in the following areas: (a) total birth interval, (b) length of postpartum amenorrhea, (c) length of menstruating interval (from the first cycle postpartum to conception), (d) duration of lactation, (e) outcome of the previous delivery, and (f) survival of the last child born.

The mean duration of lactation in the study communities was 18 months; the mean duration for postpartum amenorrhea was 14 months. The two variables are highly associated, the correlation coefficient being .62 (377 cases, $p < 0.01$). In these analyses stillbirths and infant deaths were excluded. The median duration of postpartum amenorrhea in nursing women in the study is comparable to those reported in rural populations. Potter *et al.* (1965) in a prospective study in India found a median of 11 months. A study of Eskimo women (Berman *et al.*, 1972) showed a median of 10 months, and Chen *et al.* (1974) in Bangladesh reported a median of 13 months.

As shown in Table XVI data from the longitudinal study reveal a negative association between indicators of past and present nutritional status of the mother, such as height, head circumference, weight, skinfolds, and arm circumference taken during pregnancy and duration of postpartum amenorrhea.

Negative associations were found between the length of postpartum amenorrhea and both home caloric intake and caloric supplementation ingested during pregnancy. A negative association was also found between total caloric intake during pregnancy (sum of home caloric intake and caloric supplementation) and the duration of postpartum amenorrhea in both *atole* and *fresco* villages. However, in *atole* villages this correlation was significant ($r = -.178$,

Table XVI. Relationship between Length of Postpartum Amenorrhea and Past and Present Indicators of Maternal Nutritional Status Taken during Pregnancy[a]

Variable	Correlation coefficient	Number of cases
Maternal height	−.10	395
Head circumference	−.09	298
Weight	−.14	160
Skinfolds		
Biceps	−.07	160
Triceps	−.08	160
Subscapular	−.07	160
Midaxillary	−.05	160
Arm circumference	−.13	160
Home caloric intake	−.09	339
Caloric supplementation	−.07	398
Total caloric intake	−.12	340

[a] From Delgado *et al.* (1978).

$p < 0.05$, $N = 186$), whereas in *fresco* villages it was not. Significant differences in the duration of postpartum amenorrhea between *atole* and *fresco* villages were not found. The reduction in the duration of postpartum amenorrhea for the same number of supplemented calories, or slope value, was not significantly different between *fresco* and *atole* villages. However, the slope value for the *fresco* villages was very low ($b = -0.8$ months/10,000 cal supplemented during pregnancy) as compared with the slope for *atole* villages ($b = -2$ months/10,000 cal supplemented during pregnancy).

All these data suggest that the better the nutritional status of the mother during pregnancy, the shorter is the length of postpartum amenorrhea. Further indications can be obtained by comparing the duration of postpartum amenorrhea in three groups (terciles) with increasing total caloric intake within several categories of lactation duration. As shown in Table XVII, within all lactation categories the length of postpartum amenorrhea was shorter in the groups with high calorie intake than in the groups with low calorie intake.

Furthermore, analyses showed that the weight gain of the infant during the first 9 months after birth is positively related to the duration of postpartum amenorrhea ($r = .15$, $N = 301$) and that caloric supplementation ingested by the lactating infant during the first 9 months is negatively related to the length of postpartum amenorrhea ($r = -.14$, $N = 401$). These results suggest that the frequency of suckling and the amount of supplemental feeding of the infant during the first year of life may be important determinants of the duration of postpartum amenorrhea. Similar results have been reported previously (Pérez *et al.*, 1972; McKeown and Gibson, 1954).

Finally, other variables, such as parity, length of previous birth interval, and age of the mother, were found to be positively correlated with the duration of postpartum amenorrhea. The last two associations have been reported previously in the literature (Salber *et al.*, 1966, 1968). The association between parity and duration of postpartum amenorrhea may be explained by the high association between parity and age in these mothers.

All these results give partial support to the hypothesis that improvement of maternal nutrition is associated with a decrease in the duration of postpartum amenorrhea and that this may increase the probability of a shorter birth interval.

Table XVII. *Length of Postpartum Amenorrhea in Three Different Total Caloric Intake[a] Groups within Categories of Duration of Lactation[b]*

Total caloric intake (cal/day)	Categories of lactation (months)				
	0–6[c]	7–12	13–18	19–24	≥25
Low (<1308)		6.54 (13)[d]	12.03 (35)	16.44 (45)	19.93 (14)
Middle (1309–1630)		6.31 (13)	11.64 (33)	15.47 (57)	18.90 (10)
High (≥1631)		5.47 (14)	11.13 (30)	14.60 (47)	19.64 (11)

[a] Total caloric intake = sum of home caloric intake plus caloric supplementation.
[b] From Delgado *et al.* (1978).
[c] No cases.
[d] Number of cases.

One further aspect of maternal nutritional supplementation that can be considered here is the effect on the quality of the milk produced during lactation. It has been found in the INCAP study population where calories are the main limiting factor in the diet of lactating mothers and their infants, that an increment of caloric intake in the lactating mothers will allow enough protein to be spared to increase the breast milk protein supplied to the infant and the proportion of ingested protein available for growth (Lechtig *et al.*, 1976e).

3.3. Blood Pressure, Edema, and Proteinuria during Pregnancy

Toxemia of pregnancy is a potentially lethal condition of unknown etiology that affects women throughout the world. An essential part of antenatal care is to screen women for early signs of pregnancy toxemia (preeclampsia). The earliest sign is a sudden weight gain usually after the 28th week of pregnancy which cannot be explained on the basis of caloric intake alone. This rapid weight gain is due to the accumulation of extracellular fluid in excess of that of normal pregnancy. When the interstitial fluid has doubled in amount, generalized edema becomes clinically detectable. This generalized edema must be distinguished from postural edema which is localized to the lower extremities and may have no pathological significance. Further clinical signs of preeclampsia are elevated blood pressure and proteinuria.

Pregnant women from low socioeconomic strata groups in which protein-calorie malnutrition (PCM) is highly prevalent present a high incidence of toxemia of pregnancy (Davis, 1971; Neutra, 1973). In these same populations a high incidence of elevated diastolic blood pressure and a low frequency of postural edema (PE) have also been reported (Hytten and Leitch, 1971; Davis, 1971).

This section explores the hypothesis that PCM is a determinant of high DBP and low incidence of PE in low socioeconomic populations and briefly discusses the possible relationship between PCM and toxemia of pregnancy. In the analyses the dependent variables are both diastolic and systolic blood pressure (DBP and SBP, respectively), PE, and proteinuria. The independent variable is ingestion of the supplement during pregnancy.

3.3.1. Diastolic Blood Pressure

Table XVIII presents the relationship between supplemented calories during pregnancy (SCDP) and DBP. A significant association was observed with the dose–response relationship being −0.4 mm Hg/10,000 SCDP. The other maternal characteristics measured, either alone or combined, cannot explain the originally observed association. In addition, the correlation between SCDP and change in DBP from the first to the third trimester of pregnancy was −.176 ($N = 107$; $b = -0.7$ mm Hg/10,000 SCDP: $p < 0.10$). Given the small sample size it was not possible to select "pathological" cutoff points for DBP and SBP. Therefore, it was decided to choose as breaking points those closest to

Table XVIII. Relationship between Supplemented Calories during Pregnancy (SCDP)
and Diastolic Blood Pressure (DBP) at the Third Trimester of Pregnancy

A. Continuous variable analysis (N = 392)	Correlation value (r)	Slope value (mm Hg blood pressure/ 10,000 cal)	Probability value (p∠)
First-order correlation	−.104	−0.4	0.05
Multiple correlation[a]	−.124	−0.4	0.05

B. Discrete variable analysis	Level of SCDP			High minus low (t-test) p∠
	Low	Middle	High	
1. Percentage with DBP ≥70 mm Hg (N = 392)	30.3	19.6	19.7	0.05

	Supplemented calories during latter pregnancy were			
	Lower	Similar	Higher	
2. Siblings subsample. Percentage with decrement of 20 mm Hg or greater in latter pregnancy as compared with preceding pregnancy (N = 135 pairs)	0.0	10.3	13.9	0.01

[a] Values obtained after controlling for height, head circumference, age, parity, socioeconomic status, weight at the end of the first trimester, gestational age, morbidity indicator, and home diet within the population with highest coverage (from July 1970 through February 1973). Slope for total study population greater than slope for population with high coverage: N. S. (test of covariance). Slope for *fresco* greater than slope for *atole*: N.S. (test of covariance).

the upper tercile of the population distribution. These values corresponded to 70 and 110 monthly for DBP and SBP, respectively.

As Table XVIII indicates, the percentage of cases with DBP higher than 70 mm Hg was lower in the high than in the low supplemented group. When caloric supplementation during the latter pregnancy was lower than during the preceding pregnancy none of the women showed decrement in DBP in the latter pregnancy when compared with the preceding one. When the caloric supplementation during the latter pregnancy was more than 10,000 cal higher than during the preceding pregnancy, the proportion of women with this decrement in DBP was 13.9%. These analyses appear to indicate a consistently inverse association between caloric supplementation and changes in DBP in the entire population studied, within the same pregnancy, and between pregnancies of the same mother.

3.3.2. Systolic Blood Pressure

Table XIX describes the relationship between caloric supplementation during pregnancy and systolic blood pressure. Although the slope value shows a trend to negative association (-0.2 mm Hg/10,000 cal), the correlations were not statistically significant. The analysis with discrete variables and with siblings did not show significant associations. A similar result was observed when analyzing changes in SBP from the first to the third trimester ($r = -.022$; $b = -0.1$ mm Hg/10,000 SCDP, $N = 107$; $p > 0.10$). However, all seven cases with SBP ≥ 130 mm Hg fell within the low supplemented group (t-test: $p < 0.05$).

3.3.3. Postural Edema

Table XX shows the discrete variable analyses with three levels of supplemented calories and shows a trend in the total sample to a higher proportion of pregnant women with PE in the high supplemented than in the low supplemented groups. This trend was also observed within pairs of siblings. The proportion of women with increased PE during the latter pregnancy as compared with the preceding one was three times higher in those who increased than in those who decreased the amount of supplemented calories during their latter pregnancy. Moreover, a similar trend was observed in sibling analysis within the *fresco* and *atole* groups (sign test, two independent comparisons, $p = 0.02$). On the basis of these analyses it was inferred that an association

Table XIX. *Relationship between Supplemented Calories during Pregnancy (SCDP) and Systolic Blood Pressure (SBP) at the Third Trimester of Pregnancy*

A. Continuous variable analysis ($N = 392$)	Correlation value (r)	Slope value (mm Hg blood pressure/ 10,000 cal)		Probability value (p)
First-order correlation	.003	0.0		N.S.
Multiple correlation[a]	$-.044$	-0.2		N.S.

B. Discrete variable analysis	Level of SDCP			Probability value (t-test; high minus low $p\angle$)
	Low	Middle	High	
1. Percentage with SBP $\geqslant 110$ mm Hg ($N = 392$)	35.9	37.0	34.8	N.S.
	Lower	Similar	Higher	
2. Sibling analysis. Percentage with decrement of 20 mm Hg or greater in latter pregnancy ($N = 135$ pairs)	5.3	7.7	12.7	N.S.

[a] Values obtained after controlling for height, head circumference, age, parity, socioeconomic status, weight at the end of the first trimester, gestational age, morbidity indicator, and home diet within the population with highest coverage (from July 1970 through February 1973). Slope for total study population greater than slope for population with high coverage: N. S. (test of covariance). Slope for *fresco* greater than slope for *atole*: N.S. (test of covariance).

Table XX. Relationship between Supplemented Calories during Pregnancy (SCDP) and Presence of Postural Edema (PE) at the Third Trimester of Pregnancy

	Level of SCDP			High minus low (*t*-test)
Discrete variable analysis[a]	Low	Middle	High	$p \angle$
1. Percentage with PE (*N* = 353)	7.2	15.4	13.1	0.10
	SC during latter pregnancy were			
	Lower	Similar	Higher	
2. Sibling analysis. Percentage not having PE in preceding pregnancy and having PE in latter pregnancy (*N* = 106 pairs)	7.7	14.7	22.0	N.S.

[a] Values obtained after controlling for height, head circumference, age, parity, socioeconomic status, weight at the end of the first trimester, gestational age, morbidity indicator, and home diet within the population with highest coverage (from July 1970 through February 1973). Slope for total study population greater than slope for population with high coverage: N. S. (test of covariance). Slope for *fresco* greater than slope for *atole*: N.S. (test of covariance).

existed between increased caloric supplementation and higher frequency of postural edema in the pregnant women of this population.

3.3.4. Proteinuria

Finally, the analyses presented in Table XXI, exploring the relationship between SCDP and proteinuria, are contradictory and hence difficult to interpret. The analyses across three levels of SCDP showed a weak trend to inverse association. Given the small sample size involved it was not possible to perform sibling analyses to further explore this relationship. Therefore, no clear association was detected between SCDP and proteinuria.

3.3.5. Discussion

These results may be summarized as follows: SCDP is inversely related to DBP and directly related to PE; no clear association was detected between SCDP and either SBP or proteinuria. Given the fact that caloric supplementation produced a biologically significant increment in the total nutrient intake during pregnancy (Lechtig *et al.*, 1975d), the most suitable interpretation of these results is that caloric supplementation during pregnancy caused a decrease of DBP and an increase of the presence of PE in this population.

Several mechanisms may be proposed for these effects. There is increasing evidence that most of the effect of improved maternal nutrition may be explained by increased hormone secretion, particularly estrogens. Both maternal nutrition and birth weight are directly associated with urinary estriol excretion (Iyengar, 1975), and it is known that increased estrogen production may lead to vasodilatation and consequently decreased peripheral resistance. This in turn may lead to decreased blood pressure, particularly in the diastolic component (Hytten and Leitch, 1971). Increased estrogen production also causes changes in the nature and reactions of the ground substance of connective tissue. As a result, there is an increased hygroscopic capacity of the mucopolysaccharide, with consequent water and salt retention in connective tissues (Hytten and Leitch, 1971; Gersh and Catchpole, 1960; Langgard and Hvidberg, 1969) which is clinically detectable in the lower limbs of pregnant women. Thus, postural edema would appear to have physiological rather than pathological significance in pregnancy. In turn, the increment of extracellular fluid may decrease the activity of the renin–angiotensin–aldosterone system and by this means, lead to decreased blood pressure. Higher estrogen production may also lead to increased activity of plasma aminopeptidases which may increase the rate of destruction of polypeptides with vasopressin activity in plasma, and therefore produce a decrease of arterial blood pressure (reviewed by Hytten and Leitch, 1971). These mechanisms are not mutually exclusive and not necessarily the only ones that may be operating. Given the present state of knowledge, however, they represent useful models to orient further research in the area. It is evident that the results presented have important implications for nutrition and public health policies, provided they are replicated by further studies performed on different populations.

The observation that postural edema is produced by improved nutrition during pregnancy is an indication that it is a normal phenomenon observed in pregnant women with good nutritional and health status and that its prevalence may be a simple indicator of good nutritional status during pregnancy. In the study population, the proportion of women presenting with PE during the third trimester of pregnancy was 10.2%, a very low figure when compared to 40% in well-nourished normotensive pregnant women of Aberdeen (Thomson *et al.*, 1967). In addition, the presence of PE may be a gross indicator of adequate fetal growth.

The observation that improved nutrition during pregnancy appears to lower DBP may indicate that PCM may be an important determinant of the reported high incidence of toxemia in populations of low socioeconomic status. This may be another pathway through which maternal PCM affects fetal prognosis, since both the fetus and the placenta of women with pregnancy toxemia show important alterations in growth and development when compared with the products of conception from normal women (Alvarez *et al.*, 1972; Hen-

Table XXI. Relationship between Supplemented Calories during Pregnancy (SCDP) and Proteinuria at the Third Trimester of Pregnancy

| | Level of SCDP | | | High minus low |
Discrete variable analysis	Low	Middle	High	(*t*-test) $p \angle$
Percentage with proteinuria (*N* = 199)	28.6	17.1	22.1	N.S.

dricks and Brenner, 1971; Fox, 1970; Tominaga and Page, 1966; Friedman *et al.*, 1962).

4. Implications and Recommendations

It has been estimated that on the order of 22 million low-birth-weight (LBW ≤ 2.5 kg) babies were born in the world during 1975 (Lechtig *et al.*, 1977b, 1978a,c). Throughout the world LBW is associated with higher infant mortality rates than normal-birth-weight babies. This is especially so in the developing countries which account for 94% of all LBW babies and where fetal growth retardation resulting in small-for-dates babies is the main associated factor.

LBW babies are less likely to survive during the first year of life than normal-birth-weight babies (Chase, 1969; Mata *et al.*, 1971; Lechtig *et al.*, 1976a). The higher mortality entails decreased return to the familial and societal investments made during pregnancy and early postnatal years.

The implications of LBW babies and infant mortality for population dynamics are poorly understood and it cannot be predicted exactly what effect changes in infant mortality rates will have on fertility levels. Consequently, it is suggested that programs designed to decrease the incidence of LBW babies be integrated with programs in family planning.

Those LBW babies who survive present a high risk of physical growth retardation and impaired mental development (Wiener *et al.*, 1968). In poor populations growth retardation is a useful risk indicator of higher morbidity which entails increased costs of medical care services and possibly unfavorable effects on productivity. Decreased ability to cope with school situations with a higher proportion of children repeating the first school years, suboptimal learning, and therefore lower return to education expenditures can also be expected. Further, the association between low birth weight and suboptimal learning may contribute to inequality in employment opportunity, lower productivity, lower income, and a poor quality of life which may in turn affect future generations. In addition, as development proceeds in any country, skills such as initiative and receptivity to and understanding of technological innovations become critical determinants of productivity in both urban and rural populations. In conclusion, it can be appreciated that a high incidence of LBW is undesirable in that it may entail a heavy economic burden and present a serious obstacle to development in many countries.

From this, it can be inferred that investment in programs oriented to decrease the incidence of LBW babies may have an economic return important enough to emphasize policies designed to stimulate social and economic development (Lechtig *et al.*, 1974). This investment is justified not only because the ultimate goal of development is to improve the quality of human existence but also because human quality is a key to development.

There is increased recognition that present health planning methodologies are usually inadequate for the efficient utilization of the limited health care resources available, particularly in poor populations (Ahumada *et al.*, 1965; Hilleboe *et al.*, 1972). Recently, attempts have been made to develop simplified models in the area of nutrition and health. Stickney *et al.* (1976) have developed a model based on results of two long-term prospective studies conducted by INCAP in rural Guatemala (Mata *et al.*, 1971; Klein *et al.*, 1973; Lechtig *et al.*, 1975d). With similar models and a minimum amount of pertinent data a planning group may be able to compare, for example, the benefits of a nutritional program for pregnant women that reduces the risk of LBW by 50% with a program for infants oriented to reduce by 50% the risk of becoming malnourished between 0 and 24 months of age. Further, meaningful evaluative comparisons of interventions may be made in terms of cost-effectiveness. It is believed that the development of simple models will lead to a more effective communication between research groups and planners by indicating the most important factors influencing effectiveness and efficiency. The development and testing of models also provides useful feedback to guide research efforts by indicating which type of information is most essential to planning. Emphasis should be placed on the careful assessment of these models when applied to real situations beginning with the selection of interventions and continuing through the final program evaluation.

An important problem in program design is that most of the scales currently used to identify women with greater risk of delivering LBW babies are impractical. Maternal height, head and arm circumference, and a simple description of the family house appear to be appropriate as simple risk indicators for routine use to identify women who would most benefit from a nutritional intervention program (Lechtig *et al.*, 1976c).

Figure 9 indicates that a significant reduction in the proportion of LBW babies occurred in the high risk categories (defined by maternal head circum-

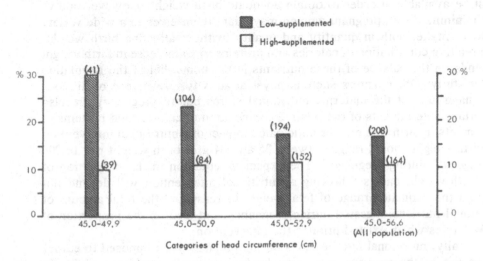

Fig. 9. Proportion of low-birth-weight babies per risk category in the low and high supplemented groups.

ference) of the well-supplemented group in contrast to the same high risk categories of the low supplemented group. Similar results were observed when maternal height or family house was analyzed. Therefore, the indicators proposed discriminate groups of mothers whose probability of delivering a LBW baby will decrease significantly if they are adequately covered by nutritional interventions.

For illustrative purposes, Table XXII presents the high risk category selected for each variable, as well as its expected impact on the effectiveness, cost, and efficiency of a program focused on those mothers identified to be at high risk. It can be seen that the efficiency of the program would be notably increased if any of these risk factors were used as selection criteria.

In conclusion, from the literature reviewed and INCAP data, it is believed that the high incidence of LBW babies in many poor communities probably comprises one of the major public health problems in many countries. Development of simple intervention and assessment models will help to select and evaluate the most convenient programs to decrease the incidence of LBW babies. The use of simple risk indicators may notably improve the coverage, efficiency, and effectiveness of these intervention programs. The improvements in maternal nutritional status will result in improved fetal growth with the birth of healthier, more viable babies. Such an effect will be a most positive contribution to the evolution and development of nations.

5. Summary

Review of the literature and data from the INCAP longitudinal study suggest that maternal nutrition, both before and during pregnancy, has an effect on birth weight. There appears to be a minimum level of nutrients which must be available in order to obtain adequate birth weight. However, above this minimum level, pregnant women may adapt themselves to a wide variety of food intake, both in quantity and quality, without affecting birth weight. The relative contribution of calories and proteins to an increase in birth weight depends on the balance of these nutrients in the home diet of the population under study. Other factors such as physical activity, prevalence of disease, and magnitude of the maternal nutritional stores before pregnancy are also important determinants of the relative contribution of calories and proteins to increments in birth weight. The anticipated impact of a nutritional intervention on birth weight should range between 25 and 84 g of birth weight per 10,000 cal ingested during pregnancy. The expected reduction in the proportion of low-birth-weight babies following a nutritional intervention will depend not only on the estimated range of fetal weight increase and the total amount of additional nutrients ingested during pregnancy, but also on the proportion of LBW babies which existed prior to the intervention.

Finally, nutritional interventions during pregnancy, as opposed to earlier in the life of the mother, may have the higher impact on birth weight. In

Table XXII. Use of Dichotomous Single Variables to Define Groups of Women at High Risk of Delivering Low-Birth-Weight Babies[a]

Variables used	Criteria to define the high risk group	Sensitivity (%)	Specificity (%)	Total cost of the program[b,c] (dollars)	Effectiveness[c] (decrement of infant mortality rate)	Efficiency[c] (cost of one infant death prevented) (dollars)	Increase in the efficiency of the program[d] (%)
Head circumference	∠51.00 cm	68.6	53.8	5,050	10.6	476(B)	102
Height	∠152.0 cm	79.3	30.5	7,120	9.4	758(B)	27
House score	∠4	69.7	59.9	4,490	9.8	458(B)	110
Coverage of the entire population (no use of high risk indicators)	—	100.0	—	10,000	10.4	962(A)	—

[a] From Lechtig *et al.* (1976c) (*Am. J. Obstet. Gynecol.*).
[b] Computations made for a target group of 1000 women (cost $10 per woman).
[c] Assuming coverage of the high risk group only.
[d] Computed as follows: $[(A - B)/B] \times 100$.

consequence, interventions during pregnancy are recommended as a priority measure.

Regarding the effect of nutrition on other maternal characteristics, data from the INCAP longitudinal study suggest that caloric supplementation during pregnancy reduces the risk of low weight gain during pregnancy. Several maternal anthropometric measurements, mainly height, head, and arm circumference, were associated with birth weight. Since their predictive value is unaffected by gestational age, these measurements may be used as simple indicators to select target populations for health and nutritional interventions. Results from the literature and from the INCAP data suggest that maternal nutrition is associated with a decrease in the duration of postpartum amenorrhea and that this may increase the probability of a shorter birth interval. However, there was no detectable evidence of a nutritional effect on the probability of conception. Food supplementation during pregnancy also appeared to decrease diastolic blood pressure and increase the prevalence of lower limb edema.

The public health relevance of these findings is discussed and recommendations are made to decrease the high worldwide incidence of fetal growth retardation.

6. References

Agualimpia, C., 1969, Demographic facts of Colombia, the national investigation of morbidity, *Millbank Mem. Q.* **47**:255.

Aherne, W., and Dunnill, M. S., 1966, Quantitative aspects of placental structure, *J. Pathol. Bacteriol.* **91**:123.

Ahumada, J., Arreaza-Guzmán, A., Durán, H., Pizzi, M., Sarue, E., and Testa, M., 1965, *Health Planning: Problems of Concept and Method*, Pan American Health Organization, Scientific Publication No. 111, Washington, D.C.

Alvarez, H., Vera-Medrano, C., Sala, M. A., and Benedetti, L. R., 1972, Trophoblast development gradient and its relationship to placental hemodynamics. II. Study of fetal cotyledons from the toxemic placenta, *Am. J. Obstet. Gynecol.* **114**:873.

Antonov, A. N., 1947, Children born during the siege of Leningrad in 1942, *J. Pediatr.* **30**:250.

Baird, D., 1966, Variations in fertility associated with changes in health status, in: *Public Health and Population Change: Current Research Issues* (M. C. Sheps and J. C. Ridley, eds.), pp. 353–376, University of Pittsburgh Press, Pittsburgh.

Barrett, J. C., 1971, Fecundability and coital frequency, *Pop. Studies* **25**:309.

Baxi, P., 1957, A natural history of childbearing in the hospital class of women in Bombay, *J. Obstet. Gynecol, India* **8**:26.

Beaton, G. H., 1961, Nutritional and physiological adaptations in pregnancy, *Fed. Proc.* **20**:196.

Berman, M. L., Hanson, K., and Hellman, I., 1972, Effect of breast-feeding on postpartum menstruation, ovulation and pregnancy in Alaskan Eskimos, *Am. J. Obstet. Gynecol.* **114**:524–534.

Biswas, S., 1963, A study of amenorrhea after childbirth and its relationship to lactation period, *Indian J. Public Health* **7**:9.

Bonte, M., and Van Balen, H., 1969, Prolonged lactation and family spacing in Rwanda, *J. Biosoc. Sci.* **1**:97.

Booth, M., 1935, The time of reappearance and the character of the menstrual cycle following gestation, *Yale J. Biol. Med.* **8**:215.

Burke, B. S., Beal, V. A., Kinkwood, B., and Stuart, H. C., 1943, The influence of nutrition during pregnancy upon the condition of the infant at birth, *J. Nutr.* **26**:569.

Callard, I. P., and Leathen, J. H., 1970, Pregnancy maintenance in protein-deficient rats, *Acta Endocrinol.* **63**:539.

Chase, H. C., 1969, Infant mortality and weight at birth: 1960 United States birth cohort, *Am. J. Public Health* **59**:1618.

Chávez, A., and Martínez, C., 1973, Nutrition and development of infants from poor rural areas. III. Maternal nutrition and its consequences on fertility, *Nutr. Rep. Int.* **7**:1.

Chen, L. C., Ahmed, S., Gesche, M., and Mosley, W. H., 1974, A prospective study of birth dynamics in rural Bangladesh, *Pop. Studies* **28**:277.

Coltart, T. M., Beard, R. W., Turner, R. C., and Oakley, N. W., 1969, Blood glucose and insulin relationships in the human mother and fetus before onset of labour, *Br. Med. J.* **4**:17.

Datta-Banik, N. D., Krishna, R., Mane, S. I. S., and Raj, L., 1970, Longitudinal growth pattern of children during preschool age and its relationship with different socio-economic classes, *Indian J. Pediatr.* **37**:438.

Davis, A. M., 1971, Geographical epidemiology of the toxemias of pregnancy, *Isr. J. Med. Sci.* **7**:751.

Delgado, H., Lechtig, A., Martorell, R., Brineman, E., and Klein, R. E., 1978, Nutrition, lactation and postpartum amenorrhea, *Am. J. Clin. Nutr.* **31**(2):322.

Dieckman, W. J., Adain, F. L., Michael, H., Kiamen, S., Dunkle, F., Arthur, B., Costin, M., Campbell, A., Winsley, A. C., and Lorgang, E., 1944, Calcium, phosphorus, iron and nitrogen balance in pregnant women, *Am. J. Obstet. Gynecol.* **47**:357.

Eastman, N. J., and Jackson, E., 1968, Weight relationships in pregnancy. I. The bearing of maternal weight gain in pre-pregnancy weight on birth weight in full term pregnancies, *Obstet. Gynecol. Survey* **23**:1003.

Ehrenfest, H., 1915, The reappearance of menstruation after childbirth, *Am. J. Obstet. Dis.* **72**:577.

Fox, H., 1970, Effect of hypoxia on trophoblast in organ culture. A morphologic and autoradiographic study, *Am. J. Obstet. Gynecol.* **107**:1058.

Friedman, E. A., Little, W. A., and Sachtleben, M. R., 1962, Placental oxygen consumption in vitro. II. Total uptake as an index of placental function, *Am. J. Obstet. Gynecol.* **84**:561.

Frisch, R. E., 1974, Demographic implications of the biological determinants of female fecundity, Paper presented at the Population Association of America Annual Meeting, New York (April).

Gersh, I., and Catchpole, H. R., 1960, The nature of ground substance of connective tissue, *Perspect. Biol. Med.* **3**:282.

Gopalan, C., and Naidu, A. N., 1972, Nutrition and fertility, *Lancet* **2**:1077.

Gruenwald, P., and Funakawa, I. L., 1967, Influence of environmental factors on foetal growth in man, *Lancet* **1**:1026.

Hansman, C., 1970, Anthropometry and related data, in: *Human Growth and Development* (R. W. McCammon, ed.), pp. 101–154, Thomas, Springfield, Ill.

Hendricks, C. H., and Brenner, W. E., 1971, Toxemia of pregnancy: Relationship between fetal weight, fetal survival, and the maternal state, *Am. J. Obstet. Gynecol.* **109**:225.

Henry, L., 1961, Some data on natural fertility, *Eugenics Q.* **8**:81.

Hertig, A. T., Rock, J., Adams, E. C., and Menkin, M. C., 1959, Thirty-four fertilized human ova, good bad and indifferent recovered from 210 women of known fertility; a study of biological wastage in early human pregnancy, *Pediatrics* **23**:202.

Hill, D. E., Myers, R. E., Holt, A. B., Scott, R. E., and Cheek, C. B., 1971, Fetal growth retardation produced by experimental placental insufficiency in the rhesus monkey. II. Chemical composition of brain, liver, muscle and carcass, *Biol. Neonate* **403**:1.

Hilleboe, H. E., Barkhuus, A., and Thomas, W. C., 1972, *Approaches to National Health Planning,* World Health Organization, Public Health Papers No. 46.

Holmberg, I., 1970, *Fecundity, Fertility and Family Planning,* Report 10, Demographic Institute, University of Gotersburg, Almquist & Wiksell, Stockholm.

Hytten, F. E., and Leitch, I., 1971, *The Physiology of Human Pregnancy,* 2nd ed., Blackwell, London.

INCAP, 1969, *Evaluación Nutricional de la Población de Centro América y Panamá,* Instituto de Nutrición de Centro América y Panamá, (INCAP), Oficina de Investigaciones de los Institutos Nacionales de Salud (EE.UU.); Ministerio de Salud Pública, Guatemala, Publicaciones INCAP-V-25-V-28.

Iyengar, L., 1975, Influence of diet on the outcome of pregnancy in Indian women, in: *Nutrition* (A. Chávez, H., Bourges, and S. Basta, eds.), Karger, Basel.

Iyengar, L., and Rajalakshmi, K., 1975, Effect of folic acid supplementation on birth weights of infants, *Am. J. Obstet. Gynecol.* **122:**332.

Jain, A. K., Hau, T. C., Freedman, R., and Chang, M., 1970, Demographic aspects of lactation and post-partum amenorrhea, *Demography* **7:**255.

Kamal, I., Hefnawi, F., Ghoneim, T., Tallat, M., Younis, N., Taqui, A., and Abdalla, M., 1969, Clinical, biochemical and experimental studies on lactation, *Am. J. Obstet. Gynecol.* **105:**314.

Klein, R. E., Habicht, J.-P., and Yarbrough, C., 1973, Some methodological problems in field studies of nutrition and intelligence, in: *Nutrition, Development and Social Behavior* (D. J. Kallen, ed.) pp. 61–75, U.S. Govt. Printing Office, DHEW Publication No. (NIH) 73-242, Washington, D.C.

Knopp, R. H., Herrera, E., and Freinkel, N., 1970, Carbohydrate metabolism in pregnancy from fed and fasted pregnant rats during late gestation, *J. Clin. Invest.* **49:**1438.

Laga, E. M., Driscoll, S. G., and Munro, H. N., 1972, Comparison of placentas from two socioeconomic groups. I. Morphometry, *Pediatrics* **50:**24.

Langgard, H., and Hvidberg, E., 1969, The composition of oedema fluid provoked by oestradiol and of acute inflammatory oedema fluid, *J. Reprod. Fertil. Suppl.* **9:**37.

Lechtig, A., Arroyave, G., Habicht, J.-P., and Béhar, M., 1971, Nutrición materna y crecimiento fetal (Revisión). *Arch. Latinoam. Nutr.* **21:**505.

Lechtig, A., Habicht, J.-P., de León, E., Guzmán, G., and Flores, M., 1972a, Influencia de la nutrición materna sobre el crecimiento fetal en poblaciones rurales de Guatemala. I. Aspectos dietéticos, *Arch. Latinoam. Nutr.* **22:**101.

Lechtig, A., Habicht, J.-P., Guzmán, G., and Girón, E. M., 1972b, Influencia de las características maternas sobre el crecimiento fetal en poblaciones rurales de Guatemala, *Arch. Latinoam. Nutr.* **22:**255.

Lechtig, A., Mata, L. J., Habicht, J.-P., Urrutia, J. J., Klein, R. E. Guzmán, G., Cáceres, A., and Alford, C., 1974, Levels of immunoglobulin M (IgM) in cord blood of Latin American newborns of low socioeconomic status, *Ecol. Food Nutr.* **3:**171.

Lechtig, A., Delgado, H., Lasky, R. E., Klein, R. E., Engle, P. L., Yarbrough, C., and Habicht, J.-P., 1975a, Maternal nutrition and fetal growth in developing societies. Socieconomic factors, *Am. J. Dis. Child.* **129:**434.

Lechtig, A., Delgado, H., Lasky, R., Yarbrough, C., Klein, R. E. , Habicht, J.-P., and Béhar, M., 1975b, Maternal nutrition and fetal growth in developing countries, *Am. J. Dis. Child.* **129:**553.

Lechtig, A., Delgado, H., Lasky, R., Yarbrough, C., Martorell, R., Habicht, J.-P., and Klein, R. E., 1975c, Effect of improved nutrition during pregnancy and lactation on developmental retardation and infant mortality, in: *Proceedings of the Western Hemisphere Nutrition Congress IV, 1974* (P. L. White and N. Selvey, eds.), pp. 117–125, Publishing Science Group Inc., Acton, Mass.

Lechtig, A., Habicht, J.-P., Delgado, H., Klein, R. E., Yarbrough, C., and Martorell, R., 1975d, Effect of food supplementation during pregnancy on birth weight, *Pediatrics* **56:**508.

Lechtig, A., Yarbrough, C., Delgado, H., Habicht, J.-P., Martorell, R., and Klein, R. E., 1975e, Influence of maternal nutrition on birth weight, *Am. J. Clin. Nutr.* **28:**1223.

Lechtig, A., Yarbrough, C., Delgado, H., Martorell, R., Klein, R. E., and Béhar, M., 1975f, Effect of moderate maternal malnutrition on the placenta, *Am. J. Obstet. Gynecol.* **123:**191.

Lechtig, A., Delgado, H., Martorell, R., Richardson, D., Yarbrough, C., and Klein, R. E., 1976a, Socioeconomic factors, maternal nutrition and infant mortality in developing countries, Paper presented at the VIII Congress of Gynecology and Obstetrics, México, D. F., October 17–23, 1976.

Lechtig, A., Delgado, H., Martorell, R., Yarbrough, C., and Klein, R. E., 1976b, Effect of maternal nutrition on infant growth and mortality in a developing country, in: *Perinatal Medicine. Fifth European Congress of Perinatal Medicine, Uppsala, Sweden, 9–12 June, 1976* (G. Rooth and L.-E. Bratteby, eds.), pp. 208–220, Almqvist & Wiksell International, Stockholm, Sweden.

Lechtig, A., Delgado, H., Yarbrough, C., Habicht, J.-P., Martorell, R., and Klein, R. E., 1976c, A simple assessment of the risk of low birth weight to select women for nutritional intervention, *Am. J. Obstet. Gynecol.* **125**:25.

Lechtig, A., Martorell, R., Delgado, H., Yarbrough, C., and Klein, R. E., 1976d, Effect of morbidity during pregnancy on birth weight in a rural Guatemalan population, *Ecol. Food Nutr.* **5**:225.

Lechtig, A., Martorell, R., Yarbrough, C., Delgado, H., and Klein, R. E., 1976e, The urea/creatinine ratio: Is it useful for field studies? *J. Trop. Pediatr.* **22**:121.

Lechtig, A., Habicht, J.-P., Wilson, P., Arroyave, G., Guzmán, G., Delgado, H., Martorell, R., Yarbrough, C., and Klein, R. E., 1977a, Maternal nutrition, breast milk composition and infant nutrition in a rural population of Guatemala (abstract), presented at Research Forum, Western Hemisphere Nutrition Congress V, Quebec, Canada, August 15–18, 1977.

Lechtig, A., Margen, S., Farrell, T., Delgado, H., Yarbrough, C., Martorell, R., and Klein, R. E., 1977b, Low birth weight babies: Worldwide incidence, economic cost and program needs, in: *Perinatal Care in Developing Countries* (G. Rooth and L. Engström, eds.), Chapter II, pp. 17–30, University of Uppsala, Uppsala, Sweden.

Lechtig, A., Rosso, P., Delgado, H., Bassi, J., Martorell, R., Yarbrough, C., Winick, M., and Klein, R. E., 1977c, Effect of moderate maternal malnutrition on the levels of alkaline ribonuclease activity of the human placenta, *Ecol. Food Nutr.* **6**(2):83.

Lechtig, A., Margen, S., Farrell, T., Delgado, H., Martorell, R., and Klein, R. E., 1978a, Birthweight and society. The societal cost of low birth-weight, in: *SAREC Report. Birth-Weight Distribution—An Indicator of Social Development. Report from a SAREC/WHO Workshop* (G. Sterky and L. Mellander, eds.), pp. 55–58, No. R2 1978, Swedish Agency for Research Corporation with Developing Countries, Uppsala, Sweden.

Lechtig, A., Martorell, R., Delgado, H., Yarbrough, C., and Klein, R. E., 1978b, Food supplementation during pregnancy, maternal anthropometry and birth weight in a Guatemalan rural population, *J. Trop. Pediatr. Environ. Child Health* **24**:217.

Lechtig, A., Sterky, G., and Tafari, N., 1978c, Conclusions, in: *SAREC Report. Birth-Weight Distribution—An Indicator of Social Development. Report from a SAREC/WHO Workshop* (G. Sterky and L. Mellander, eds.), pp. 87–90, No. R2, 1978, Swedish Agency for Research Cooperation with Developing Countries, Uppsala, Sweden.

Lechtig, A., Delgado, H., Irwin, M., Klein, R. E., Martorell, R., and Yarbrough, C., 1979a, Intrauterine infection, fetal growth and mental development, *J. Trop. Pediatr. Environ. Child Health*, in press.

Lechtig, A., Martorell, R., Yarbrough, C., Delgado, H., and Klein, R. E., 1979b, Influence of food supplementation on the urinary urea/creatinine (U/C) ratio of the child, *J. Trop. Pediatr. Environ. Child Health*, in press.

Malkani, P. K., and Michardini, J. J., 1960, Menstruation during lactation. A clinical study, *J. Obstet. Gynecol., India* **11**:22.

Mata, L. J., Urrutia, J. J., and Lechtig, A., 1971, Infection and nutrition of children of a low socioeconomic rural community, *Am. J. Clin. Nutr.* **24**:249.

McCance, R. A., and Widdowson, E. M., 1962, Nutrition and growth, *Proc. Soc. Biol.* **156**:326.

McKeown, T., and Gibson, J. R., 1954, A note on menstruation and conception during lactation, *J. Obstet. Gynecol Br. Emp.* **61**:824.

Mönckeberg, F., 1968, Effect of early marasmic malnutrition on subsequent physical and psychological development, in: *Malnutrition, Learning and Behavior* (N. S. Scrimshaw and J. E. Gordon, eds.), p. 269, MIT Press, Cambridge, Mass.

Nainsmith, D. J., 1969, The foetus as a parasite, *Proc. Nutr. Soc.* **28**:25.

Neutra, R., 1973, A case-control study for estimating the risk of eclampsia in Cali, Colombia, *Am. J. Obstet. Gynecol.* **117**:894.

Niswander, K. R., and Gordon, M., 1972, *The Women and Their Pregnancies. The Collaborative*

Perinatal Study of the National Institute of Neurological Disease and Stroke, DHEW Publication No. (NIH) 73-379, U. S. Department of Health, Education and Welfare, Washington, D.C.

Oldham, H., and Sheft, B. B., 1951, Effect of caloric intake on nitrogen utilization during pregnancy, *J. Am. Diet. Assoc.* **27**:847.

Page, E. W., 1957, Transfer of materials across the human placenta, *Am. J. Obstet. Gynecol.* **74**:705.

Page, E. W., 1969, Human fetal nutrition and growth, *Am. J. Obstet. Gynecol.* **104**:378.

Pearse, W. H., and Sornson, H., 1969, Free amino acids of normal and abnormal human placenta, *Am. J. Obstet. Gynecol.* **105**:696.

Peckham, C. H., 1934, An investigation of some effects of pregnancy noted six weeks and one year after delivery, *Bull. Johns Hopkins Hosp.* **54**:186.

Pérez, A., Vela, P., Masnick, G. S., and Potter, R. G., 1972, First ovulation after childbirth: The effect of breast-feeding, *Am. J. Obstet. Gynecol.* **114**:1041.

Potter, R. G., 1963, Birth intervals: Structure and change, *Pop. Studies* **17**:155.

Potter, R. G., New, M., Wyon, J., and Gordon, J., 1965, Applications of field studies to research on the physiology of human reproduction, *J. Chronic Dis.* **18**:1125.

Rippel, R. H., Harmon, B. G., Jensen, A. H., Norton, H. W., and Becker, D. E., 1965, Response of the gravid gilt to levels of protein as determined by nitrogen balance, *J. Anim. Sci.* **24**:209.

Robertson, W. B., Brosens I., and Dixon, H. G., 1967, The pathological response of the vessels of the placental bed to hypertensive pregnancy, *J. Pathol. Bacteriol.* **93**:581.

Sakurai, T., Takagi, H., and Hasaya, N., 1969, Metabolic pathways of glucose in human placenta. Changes with gestation and with added 17-beta-estrodiol, *Am. J. Obstet. Gynecol.* **105**:1044.

Salber, E. J., Feinleib, M., and MacMahon, B., 1966, The duration of postpartum amenorrhea, *Am. J. Epidemiol.* **82**:347.

Salber, E. J., Feldman, J. J., and Hannigan, M., 1968, Duration of postpartum amenorrhea in successive pregnancies, *Am. J. Obstet. Gynecol.* **100**:24.

Sharman, A., 1951, Menstruation after child birth, *J. Obstet. Gynaecol. Br. Emp.* **58**:440.

Shelley, H. J., 1969, Carbohydrate metabolism in the foetus and the newly borns, *Proc. Nutr. Soc.* **28**:42.

Smith, C. A., 1947, Effects of maternal undernutrition upon the newborn infant in Holland (1944–45), *J. Pediatr.* **30**:229.

Soangra, M. R., Joahi, C. K., and Sharma, R., 1975, Socio-economic and environmental factors affecting stillbirth, *Indian J. Med. Sci.* **29**:5.

Solien de Gonzalez, N., 1964, Lactation and pregnancy: A hypothesis, *Am. Anthropol.* **68**:873.

Stickney, R. E., Beghin, I. D., Urrutia, J. J., Mata, L. J., Arenales, P., Habicht, J.-P., Lechtig, A., and Yarbrough, C., 1976, Systems analysis in nutrition and health planning: Approximate model relating birth weight and age to risk of deficient growth, *Arch. Latinoam. Nutr.* **26**:177.

Stix, R. K., 1940, Factors underlying individual and group differences in uncontrolled fertility, *Milbank Mem. Fund. Q.* **18**:239.

Thomson, A. M., 1959, Diet in pregnancy. 2. Assessment of the nutritive value of diets, specifically in relation to differences between social classes, *Br. J. Nutr.* **13**:190.

Thomson, A. M., Hytten, F. E., and Billewicz, W. Z., 1967, The epidemiology of oedema during pregnancy, *J. Obstet. Gynaecol. Br. Commonw.* **74**:1.

Tominaga, T., and Page, E. W., 1966, Accommodation of the human placenta to hypoxia, *Am. J. Obstet. Gynecol.* **94**:679.

Van Dyne, C. M., Havel, R. J., and Felts, J. M., 1962, Placental transfer of palmitic acid 1-C14 in rabbits, *Am. J. Obstet. Gynecol.* **84**:1069.

Venkatacharya, K., 1972, Demographic effects of variations of amenorrhea and fetal wastage, *Indian J. Public Health* **16**:39.

Wiener, G., Rider, R. V., Oppel, W. C., and Harper, P. A., 1968, Correlates of low birth weight: Psychological status at eight to ten years of age, *Pediatr. Res.* **2**:110.

Wigglesworth, J. S., 1964, Experimental growth retardation of the fetal rat, *J. Pathol. Bacteriol.* **88**:1.

Winick, M., and Noble, A., 1966, Cellular response in rats during malnutrition at various ages, *J. Nutr.* **89**:800.

World Health Organization, 1965, Nutrition in Pregnancy and Lactation, WHO Expert Committee Report, Technical Series No. 302, Geneva.

Yarbrough, C., Habicht, J.-P., Malina, R., Lechtig, A., and Klein, R. E., 1975, Length and weight in rural Guatemalan Ladino children birth to seven years of age, *Am. J. Phys. Anthropol.* **42**:439.

6

The Newborn

Gordon B. Avery

1. Nutrition of the Full-Term Newborn

Feeding the vigorous, term newborn is normally relatively straightforward. Milk is the staple food of the newborn, and breast milk, the natural and preferred feeding, contains virtually all the baby's requirements for several months. The mother's milk supply adapts itself to the baby's needs when both are healthy and when frequent nursing is encouraged in the early weeks. The physical intimacy of breast feeding enhances mother–infant bonding; the immunological properties of human milk protect against certain kinds of infection; early contact with protein antigens from a foreign species is avoided; and the hazards of formula preservation are absent. The advantages of human milk feeding are even more pronounced in developing countries, where refrigeration may not be available, where formulas are expensive and frequently watered down, and where infectious diarrhea exacts a heavy toll in the artificially fed newborn.

The balance of considerations may be somewhat different in technically advanced societies. Commercially prepared formulas are safe and readily available, and refrigeration is nearly universal. Working mothers or those with other young children may have difficulty giving the time which successful and sustained breast feeding requires. Under these circumstances, a valid choice between breast and bottle feeding is available to the new mother. Unfortunately, what should be a free and unencumbered choice is frequently complicated by strong social factors. On one hand, breast feeding is sometimes pushed aggressively, as though infant formulas could not be used effectively. On the other hand, the inexperienced mother desiring to breast-feed may be hampered by rigid hospital routines, unsympathetic physicians and nurses, and by a lack of suitable helping persons who are knowledgeable in the techniques of breast feeding.

Gordon B. Avery • Division of Neonatology, Children's Hospital, National Medical Center, Washington, D.C.

1.1. Breast Feeding

A number of publications have described the technique of breast feeding (Applebaum, 1975; Haire, 1969). In addition, volunteer groups such as the La Leche League offer supportive services and advice.

The breast is prepared for lactation by the hormonal environment in late gestation. Critical to successful breast feeding is the letdown reflex, in response to sucking by the baby, which is mediated by oxytocin and releases milk which otherwise accumulates and causes serious engorgement of the breast. Many failures of lactation are attributable to inadequate letdown reflex and failure to empty the breast at frequent intervals as the milk is starting to come in. Good nutrition of the mother, a relaxed and supportive environment, and frequent suckling favor development of milk production. Many authorities feel that nursing should be on demand during the first few days, and that the baby should be put to the breast in the first few hours of life (Klaus and Kennell, 1976). If the breast is allowed to become engorged, the baby may have difficulty getting enough of the nipple into his mouth, and nursing may become quite uncomfortable.

The content of human milk varies from individual to individual, from early to late lactation, and between the first milk (foremilk) and last milk (hindmilk) passed during a single feeding. Colostrum, the milk secretion of the first few days before lactation is well established, contains abundant viable lymphocytes, macrophages, and surface immunoglobulins (IgA). It is rich in protein, but relatively low in fat and limited in amount. The protein content in mature human milk may vary from 0.8 to 1.3%, and the calorie content per ounce may vary between 67 and 75/100 ml (Fomon and Filer, 1974). As lactation continues beyond the first few weeks, there is some decline in protein and calorie content. At a single feeding, the early milk is watery, relatively low in calories, and has little fat. As the feeding progresses, in the presence of an adequate letdown reflex, a fat-rich milk of high caloric value becomes available. A mother who is herself considerably undernourished may produce milk of somewhat lower total volume and decreased vitamin content (Fomon and Filer, 1974).

Under normal circumstances these distinctions are not critical, and the baby who nurses until satisfied by enough calories will adjust his intake to these variations and will receive adequate nutrition. In the event that human milk is used for a sick or premature infant, variations in milk composition may become critical, and it is desirable to know the concentration of key nutrients, especially protein.

Human milk is so good a food that under normal circumstances infants receiving only breast milk for the first year of life obtain adequate protein, essential fatty acids, electrolytes, vitamins, and minerals provided calories are adequate. Exceptions may sometimes be iron, fluoride, and vitamin D. Supplementation with these substances is still officially recommended in breast-feeding infants, although new evidence suggests that this may not be necessary (see later).

Concern is frequently expressed about the effects on the baby of drugs administered to the nursing mother. Indeed, many substances taken by the mother are secreted in breast milk. A detailed tabulation of the concentrations of various substances passing into breast milk has been published by Knowles (1965). In general, the safest assumption is that some level of excretion will occur, unless information to the contrary is at hand. Nevertheless, the effect on the infant is in most instances minimal, and is often not grounds for discontinuing *needed* medications. Exceptions include antithyroid drugs such as propylthiouracil and antimetabolites given for cancer therapy. Occasionally, excessive use of alcohol or nicotine by the mother will affect a nursing infant. Basically, therapeutic drugs should be used in lactating women only when absolutely indicated, for the least period of time, in the lowest effective dosage, and using the least toxic remedy.

1.2. Formula Feeding

Unmodified cow's milk is not a suitable feeding for human newborns. The protein content is too high and predisposes to azotemia and acidosis. Without heat treatment, cow milk protein is poorly digested and may cause gastrointestinal blood loss through sensitization. Ash content is excessive, and iron, fluoride, and vitamins A, C, and D are deficient.

For these reasons, cow's milk is commonly modified for infant feeding by heat treatment, dilution, carbohydrate enrichment, and vitamin supplementation, yielding the traditional "formula." Commercially available formulas have been modified to conform to the main components of human milk, and certainly are nutritionally adequate for the growth and development of human infants. Most commercial formulas contain adequate vitamins when 600–1000 ml/day is taken. A comparison of the composition of human milk, cow milk, and a representative commercial formula is given in Table I. In addition, satisfactory infant formulas can be prepared by dilution of evaporated milk, addition of carbohydrate, and separate vitamin supplementation. Virtually all routine infant formulas are adjusted to be 67 cal/100 ml (20 cal/oz). Unless specified, formulas ordinarily do not contain adequate iron and fluoride, and these elements may require supplementation (see later).

1.3. Vitamins

Recommended vitamin allowances for the newborn and their content in human and unmodified cow's milk are listed in Table II. Routine commercial formulas contain adequate vitamins, provided intake is adequate, and supplementation is ordinarily unnecessary. Unmodified cow's milk should not be fed in the newborn period, but evaporated milk formulas may be deficient, and should be supplemented with a preparation containing vitamins A, C, and D. Multivitamin preparations also containing B vitamins are satisfactory. It is still officially recommended that babies receiving breast milk should receive vitamin D, 400 IU/day, although recent findings of high levels of water-soluble

Table 1. Composition of Mature Breast Milk, Cow's Milk, and a Routine Infant Formula[a,b]

Composition per 100 ml	Mature breast milk	Cow's milk	Routine formula with iron[c]
Calories	75.0	69.0	67.0
Protein (g)	1.1	3.5	1.5
Lactalbumin (%)	60.0	18.0	
Casein (%)	40.0	82.0	
Water (ml)	87.1	87.3	
Fat (g)	4.0	3.5	3.7
CHO (g)	9.5	4.9	7.0
Ash (g)	0.21	0.72	0.34
Minerals (mg)			
Na	16.0	50.0	25.0
K	51.0	144.0	74.0
Ca	33.0	118.0	55.0
P	14.0	93.0	43.0
Mg	4.0	12.0	9.0
Fe	0.1	Tr.	1.2
Zn	0.15	0.1	
Vitamins			
A (IU)	240.0	140.0	158.6
C (mg)	5.0	1.0	5.3
D (IU)			42.3
E (IU)			0.83
Thiamine (mg)	0.01	0.03	0.04
Riboflavin (mg)	0.04	0.17	0.06
Niacin (mg)	0.2	0.1	0.7
Curd size	Soft, flocculent	Firm, large	Mod. firm, mod. large
pH	Alkaline	Acid	Acid
Anti-infective properties	+	−	−
Bacterial content	Sterile	Nonsterile	Sterile
Emptying time	More rapid		

[a] From Avery and Fletcher (1975), reprinted with permission.
[b] Composite of a number of sources.
[c] Enfamil. Trademark, Mead Johnson.

vitamin D in human milk may make this unnecessary (Lakdawala and Widdowson, 1977).

1.4. Iron

The iron content of human milk is officially considered to be inadequate to provide the intake of 1 mg/kg a day for term infants or 2 mg/kg a day for prematures, up to a maximum of 15 mg, recommended by the Committee on Nutrition of the American Academy of Pediatrics (1976). This is certainly the case with unfortified commercial formulas. Iron deficiency is the most common nutritional disease seen in the United States, and is so widespread that routine iron prophylaxis has been urged. Iron stores at birth are sufficient for 2–3 months in the premature and 5–6 months in the term baby, so iron deficiency is most commonly encountered in the second 6 months of life. However, since most of the iron stores are in the large hemoglobin mass of the newborn, the infant who has bled or is of extremely low birth weight is particularly at risk.

Once solid feedings have been initiated, they will provide additional dietary iron. The iron content of some representative infant foods is listed in Table III. However, patterns of intake are highly variable, and iron absorption may be quite poor. Rios *et al.* (1975) presented evidence that iron absorption from infant cereal may be less than 1%. More recent preparations have resulted in about 4% uptake. Other iron sources may achieve 2–5% absorption. The ability of four exclusively breast-fed infants to avoid iron deficiency, recently reported by McMillan *et al.* (1976), may be the result of excellent absorption (about 50%) and lack of the gastrointestinal blood loss engendered by cow's milk protein. Iron uptake is greater in the deficient than in the replete individual, and conversely a number of dietary factors bind iron and decrease its absorption. Nevertheless, as iron deficiency is both common and preventable, the current official recommendation is that infants receiving human milk or routine formulas be given supplementary iron either as medicinal drops or in food sources, in the dosage indicated earlier.

1.5. Fluoride

Abundant evidence indicates that an adequate fluoride intake during tooth formation deposits in enamel and decreases dental caries. On the other hand, in localities in which fluoride content in the water exceeds 4–5 ppm, fluorosis

Table II. Vitamin Allowances for the Newborn[a]

Nutrient	Recommended daily intake		Content/liter of milk			
			Cow		Human	
Vitamin D	400	IU	13–31	IU (fortified— 400 IU)	21	IU
Vitamin A	1400	IU	1000–1600	IU	1900–2800	IU
Vitamin K	5	μg	60	μg	15	μg
Vitamin E	5	IU (0.5 mg/kg)	1.0	mg	6.6	mg
Ascorbic acid	35–50	mg	12–25	mg	52	mg
Thiamine	0.3	mg	0.32	mg	0.15	mg
Riboflavin	0.4	mg	1.7	mg	0.2–0.4	mg
Niacin	5	mg	0.9	mg	1.5	mg
Pyridoxine	200–400	μg	250–600	μg	100	μg
Pantothenic acid	405	mg	3	mg	1.8	mg
Folacin	50	μg	30–100	μg	20	μg
Vitamin B$_{12}$	0.3	μg	4	μg	0.3	μg
Biotin	150–300	μg	44	μg		

[a] From Avery and Fletcher (1975), reprinted with permission. (Data of Fomon, Mitchell *et al.*, and RDA.)

Table III. Iron Content of Commercially Prepared Strained and Junior Foods[a]

Food	Mean iron (mg/100 g)	Range (mg/100 g)
Dry cereals	71.6–80.1	50–100.2
Mixed 1:6 with milk	10.8–11.4	7.1–14.3
Cereal with fruit, strained		
High protein	5.3	—
Rice	0.18	—
Junior cereal with fruit		
Oatmeal	3.91	—
Strained		
Juices	0.46	0.1–0.95
Fruits	0.36	0.1–0.9
Plain vegetables	0.66	0.29–1.32
Creamed vegetables	0.56	0.24–0.88
Meats	1.57	0.7–4.43
Egg yolks	3.12	3.02–3.2
High meat dinners	0.76	0.47–1.71
Soups and dinners	0.51	0.12–2.13
Desserts	0.30	0.10–0.71
Junior[b]		
Meat sticks	1.49	0.9–2.0

[a] From Avery and Fletcher (1975), reprinted with permission.
[b] Other foods similar to strained [Fomon, (1974) Tables 16-1 and 16-2].

with tooth discoloration is common. Little fluoride is present in breast milk and most commercial infant formulas. Thus the infant's intake of fluoride is largely determined by the fluoride content of the local water supply and by how much water is used in formula preparation. Powdered formulas will use twice the water employed to dilute concentrated liquids, and ready-to-use formulas will require no local water. It is currently recommended that infants receive a total of 0.5 mg/day of fluoride from all sources (Wei, 1974). According to the fluoride content of the local water and the amount consumed, this will require a variable amount of supplementation (Table IV).

1.6. Minerals

The mineral content of human and cow's milk and the recommended daily intake for a term newborn are depicted in Table V. Requirements for some of the trace elements have not yet been quantified. The deficient content of iron and fluoride has already been noted. In addition, milk alone provides insufficient copper. However, clinically significant deficiencies of these latter elements have rarely been seen in otherwise healthy infants. An exception may be hyponatremia encountered in rapidly growing prematures. Although cow's milk contains abundant calcium and phosphate, absorption may be poor because the high phosphate content encourages formation of insoluble complexes in the gut lumen. Many commercial formulas have modified calcium–phos-

Table IV. Recommended Fluoride Supplementation during the First Year of Life[a]

Milk or formula	Desirable fluoride supplementation (mg/day) when fluoride concentration of water supply (ppm) is			
	<0.3	0.3–0.7	0.8–1.1	>1.1
Human milk	0.5	0.5	0.5	0
Cow's milk	0.5	0.5	0.5	0
Commercially prepared formula				
Ready-to-use	0.5	0.5	0.5	0
Concentrated liquid	0.5	0.25	0	0
Powder	0.5	0	0	0[b]
Evaporated milk formula	0.5	0.25	0	0

[a] Source: From Wei (1974), reprinted with permission. Data from Fomon, S. J. [Personal communication (1973)]. Recommendations are aimed at providing approximately 0.5 mg of fluoride daily.
[b] In reconstituting commercially prepared powdered formulas, it is desirable to avoid use of water containing more than 1.1 ppm of fluoride.

phorus ratios to more closely approximate human milk, with the result that calcium uptake is better and hypocalcemia less common.

1.7. Solid Foods

With the exception of the vitamin and mineral constituents already mentioned, milk is an excellent infant food and by itself is sufficient for the first

Table V. Minerals and Trace Elements[a]

Nutrient	Recommended daily intake	Content/liter of milk	
		Cow	Human
Calcium	360.0 mg	1,250.0 mg	340.0 mg
Phosphorus	240.0 mg	960.0 mg	140.0 mg
Sodium	8.0 meq (184 mg)	25.0 meq	7.0 meq
Potassium	8.0 meq (320 mg)	35.0 meq	13.0 meq
Chloride	8.0 meq (280 mg)	29.0 meq	11.0 meq
Iron	10.0 mg	1.0 mg	0.3 mg
Iodine	35.0 μg	47.0 μg	30.0 μg
Magnesium	60.0 mg	120.0 mg	40.0 mg
Zinc	3.0 mg	3.8 mg	1.2 mg
Copper	0.3 mg	0.03 mg	0.03 mg
Manganese	0.8 mg		
Fluoride	0.5 mg		
Chromium			
Cobalt			
Molybdenum			
Selenium			

[a] Source: From Avery and Fletcher (1975), reprinted with permission. Data of Fomon, Mitchell *et al.*, and RDA.

year of life. Solid or, more correctly, semisolid food is therefore introduced for other reasons, some valid and some invalid. Among the valid indications for nonmilk food are the following: provision of nutrients that are low in milk, such as iron and vitamin D; development of ability to handle a variety of tastes and textures of food and development of skills of mastication and swallowing necessary for further dietary advancement; and adaptation to foods of higher caloric density to avoid overstretching stomach capacity as size and caloric intake grow. Invalid reasons include a competitive regard for dietary advancement as a sign of development; expectation that solids will "hold" the baby longer; and the desire to substitute cheaper foods, often carbohydrates, for commercial formula. This last consideration results in serious malnutrition in less developed areas of the world, where milk or animal products may severely strain the resources of the family and the substitute is often an inferior carbohydrate "gruel" that lacks many components of milk, but most especially protein.

These considerations suggest some of the advice that might be given regarding solids in the first year of life. There is no urgency to starting solids, and the process is considerably easier after 3 months, when the muscular coordination needed for handling feeding from a spoon is better developed, and the interfering dextrusor reflex has subsided. Second, solids can be introduced cautiously, one at a time, making certain that each is well tolerated before adding others. Those most likely to cause allergic sensitization, notably animal proteins, are delayed. Thus cereal, fruits, and vegetables normally precede meats, and eggs are ordinarily not given until after 6 months. Probably, overall, solids are rushed more than is optimal in this country. Table VI demonstrates the decreasing percentage of calories derived from milk during the first year of life, according to a survey by Fomon (1975).

Once solids have been started, they should not be allowed to upset the balance between nutrients inherent to human milk or formula feedings. The caloric density and distribution between protein, carbohydrate, and fat of a

Table VI. Estimated Total Caloric Intake and Intake from Milk or Formula by Male Infants[a]

Age (months)	Total caloric intake (kcal/day) by percentile			Calories from milk or formula (% of total)[b]
	10th	50th	90th	
0–1	275	400	580	93
1–2	465	565	680	85
2–3	505	625	795	70
3–4	550	640	785	67
4–5	550	675	885	64
5–6	615	740	960	60
6–9	710	820	1020	45
9–12	795	925	1230	34

[a] Source: From Fomon (1975), reprinted with permission. Data of Beal (1970) as summarized by Fomon (1975).
[b] Estimates based on reports from various sources.

Table VII. Caloric Distribution of Strained and Junior Foods[a]

Category	Number of products	Energy (kcal/100 g)	Percentage of calories		
			Protein	Fat	Carbohydrate
		Strained			
Juices	32	65[b]	2	2	96
		(45–98)	(0–5)	(0–7)	(89–100)
Fruits	33	85	2	2	96
		(79–125)	(0–10)	(1–7)	(87–99)
Vegetables					
Plain	24	45	14	6	80
		(27–78)	(5–32)	(1–19)	(62–92)
Creamed	6	63	13	13	74
		(42–94)	(5–26)	(3–25)	(57–90)
Meats	25	106	53	46	1
		(86–194)	(20–72)	(28–80)	(0–8)
Egg yolks	4	192	21	76	3
		(184–199)	(20–23)	(69–80)	(0–8)
High meat dinners	15	84	29	47	29
		(63–106)	(20–42)	(20–57)	(19–40)
Soups and dinners	42	58	16	28	56
		(39–94)	(7–33)	(2–54)	(34–86)
Desserts	37	96	4	7	89
		(71–136)	(0–18)	(0–31)	(51–99)
		Junior			
Fruits	30	85	2	2	96
		(69–116)	(0–11)	(0–5)	(86–98)
Vegetables					
Plain	16	46	12	7	81
		(27–71)	(6–24)	(1–19)	(65–92)
Creamed	5	64	13	17	70
		(45–72)	(5–24)	(3–38)	(51–90)
Meats	17	103	56	43	1
		(88–135)	(43–64)	(36–55)	(0–4)
Meat sticks	6	168	32	63	5
		(112–204)	(27–38)	(52–70)	(1–12)
High meat dinners	15	85	30	42	28
		(64–110)	(22–39)	(17–57)	(16–44)
Soups and dinners	45	61	15	27	58
		(39–100)	(6–21)	(2–55)	(34–86)
Desserts	32	93	4	6	90
		(73–112)	(0–14)	(0–25)	(68–99)

[a] From Anderson and Fomon (1974), reprinted with permission.
[b] Mean with range in parentheses.

number of infant foods is given in Table VII. In some instances there is great disparity between nutritional value and cost, or between actual composition and what is implied by the label. The manufacturer must list ingredients in the order of their concentrations, and when sugar or wheat filler is listed first, this is the largest constituent, even in a "high-meat dinner." Another hazard of

Table VIII. Average Sodium Content
(meq/100 kcal) of Infant Foods[a]

Milk	
Human	0.9
Cow	3.3
Strained foods	
Juices	0.1
Fruits	0.9
Vegetables	
Plain	11.3
Creamed	7.4
Meats	7.0
High meat dinners	7.6
Soups and dinners	10.8
Desserts	2.5

[a] From Anderson and Fomon (1974), reprinted with permission.

commercial baby foods may be their high sugar and/or salt contents. The average sodium content of a number of strained foods is shown in Table VIII; it is probably excessive in view of the possible contribution of salt intake to adult hypertension. The potential ill effects of needless additiives is also a present-day concern. Parents who prepare their own baby foods by blenderizing table foods probably realize both a nutritional and a cost advantage, provided they avoid fatty, allergenic, and highly seasoned foods. For a complete discussion of solid feedings in early infancy, the reader is referred to the excellent chapter by Anderson and Fomon (1974).

2. Nutrition of the Premature

2.1. Special Problems of the Premature

The premature is at a nutritional disadvantage compared with the full-term infant in a number of respects. Were he not prematurely delivered, he would obtain nutrients in abundance from his mother across a placenta that, in some instances, actively transfers substances in his direction, against a concentration gradient, even under conditions of maternal deprivation. In this way, he would receive 28 g of calcium in the last trimester, and similarly large amounts of magnesium, phosphorus, and iron (Widdowson, 1968). Even the infant of a moderately iron-deficient mother will ordinarily be born with a normal hemoglobin mass and total iron content. Neither breast milk nor usual infant formulas provide intakes of calcium equivalent to those acquired transplacentally during the last trimester.

For his part, the premature is seldom able to take enough feeding by mouth to supply his needs during the first 1–3 weeks of life. Sucking and swallowing are poorly developed until 33–34 weeks of gestation, and tube feedings are often required. The premature's stomach capacity may be only a few milliliters, and distension and gastric residua will result from exceeding this amount. Because of a relaxed esophageal sphincter, short esophagus, and recumbent posture, aspiration is a frequent consequence of overfeeding. Gastric emptying is variable, but distension will usually result when feeding is more frequent than every 2–3 hr. Respiratory distress is very common in the smaller premature, and this both increases the baby's metabolic needs and decreases his tolerance of a full stomach which impedes motion of the diaphragm.

The premature has a high metabolic rate, is engaged in active growth, and yet has very minimal caloric stores. The relatively small amount of body fat present at 28 weeks of gestation and the calorically expensive fat synthesis occurring during the last 12 weeks of gestation are illustrated in Fig. 1. Even brief periods of partial starvation thus result in serious protein catabolism and are poorly treated. Up to 20% of ingested calories may be lost through fat malabsorption when animal fat (butterfat) makes up a major ingredient of the feeding. The relatively large surface-to-volume ratio of the premature makes him particularly vulnerable to heat loss to his environment. Our practice of caring for prematures nude facilitates observation but maximizes radiant heat loss.

As a result of these and other nutritional disadvantages of the premature, intrauterine growth rates are rarely sustained in the first few weeks after premature birth. The normal, almost uninterrupted, prenatal and postnatal

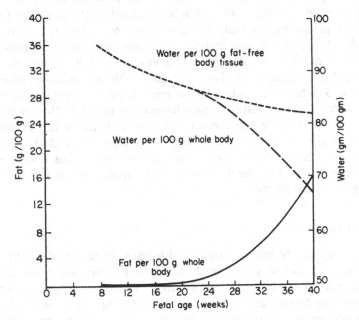

Fig. 1. Changes in body water and fat are plotted against fetal age in weeks. (Courtesy of Widdowson, 1968.)

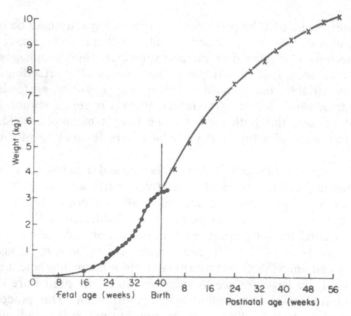

Fig. 2. Growth of the fetus and newborn with weight plotted against fetal and postnatal age. (Courtesy of Widdowson, 1968.)

growth curve is pictured in Fig. 2. This represents the ideal rate of gain for the premature, in that it carries out his genetic potential and nutritionally supports age-appropriate differentiation. However, in actual practice the attempt to achieve intrauterine growth rates may result in overloading the premature with metabolites which cannot be handled, or it may simply be impossible to provide the full calculated intake. Accordingly, a judicious compromise must be reached between ideal intake for growth and the amount which will be tolerated and can be feasibly administered. How far short this has usually fallen is illustrated by Fig. 3, which shows families of growth curves for prematures of differing birth weights. These rates of gain have been taken as "average" or "satisfactory," although they represent considerable delay in the establishment of growth in the smallest weight groups. Although it might be argued that many of these infants later experience catch-up growth, this reassurance is hardly adequate in that the smallest group of prematures still suffers from unexplained neurological damage.

2.2. Strategies of Intake

Because of the need to initiate nutritional support of the small premature immediately, and because this can rarely be done by the oral route alone, many neonatologists have become opportunists. In infants weighing under 1500 g, it is customary to start an intravenous with 10% glucose upon admission to the nursery as a minimal source of calories to which other intake will be added. When the baby's condition permits, small gavages of milk will be initiated after demonstrating the patency of the upper gastrointestinal tract with a sterile water feeding. Over the next 1–3 weeks an overlapping regimen will be followed, decreasing intravenous intake as oral feeding is tolerated. An example of such an overlapping intravenous and oral regimen is given in Fig. 4. Should the infant prove ill or unable to progress rapidly with oral feedings, intravenous alimentation is given (see later), providing enough calories and nutrients for growth during the baby's recovery.

Needless to say, the need for intravenous feedings will be minimal in the larger premature who is not ill, and indeed robust prematures near term can often feed from a bottle or even nurse at the breast.

Other feeding regimens for the small or sick premature are basically modifications of the one just described. Parental infusion may be via an umbilical artery catheter, peripheral vein, or central vein. It may consist of glucose alone or of a complete feeding including amino acids, vitamins, electrolytes, minerals, and perhaps fat in the form of Intralipid. The more elaborate parenteral infusion will be appropriate when oral feedings fall far short of the infant's requirements for a significant period of time. The site and timing of the gastrointestinal feedings will also be determined by the infant's condition. When nipple or periodic intragastric gavage is unsuccessful, other methods may be chosen, as shown in Table IX.

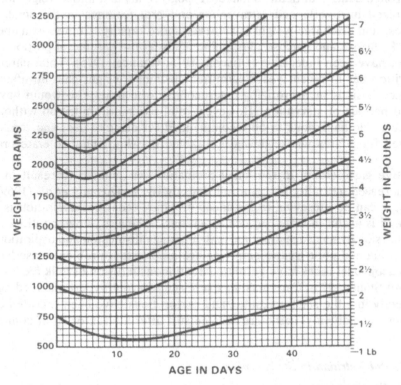

Fig. 3. Postnatal growth curves typical of prematures of varying birth weights. (Courtesy of Dancis *et al.*, 1948.)

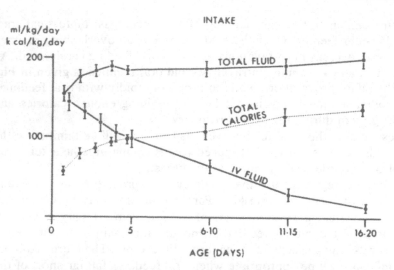

Fig. 4. Fluid and caloric intake during the first 20 days of life, in a group of prematures receiving overlapping intravenous and oral feedings. (Courtesy of Chance, 1975.)

Comparison between these various feeding techniques, on the basis of well-controlled data, is difficult to make. Hypoglycemia and initial weight loss are minimized by overlapped intravenous and oral feedings in the smaller prematures, and discharge weight is achieved more rapidly. Comparisons between 10% glucose and glucose plus amino acids as the parenteral portion of the intake, have been indecisive (Driscoll, 1975; Chance, 1975). Total parenteral nutrition has in recent years permitted sustained weight gain in infants with surgical lesions or necrotizing enterocolitis, who otherwise would have protracted periods of starvation. However, total parenteral nutrition without oral intake is too hazardous to be recommended for elective use, and should be reserved for infants whose bowel is disabled or who cannot tolerate oral feedings.

Elective gastrostomy in the small premature was shown to result in a slightly increased mortality in a controlled study by Vengusamy *et al.* (1969). However, it can be lifesaving in instances in which recurrent aspiration is a problem and is useful in many surgical situations. Likewise, slow, continuous nutrient infusion by a nasogastric tube may minimize distension and aspiration. Where residua and poor gastric emptying are a persistent problem, nasoduodenal or nasojejunal tubes have permitted continuous infusion of milk feedings (Cheek and Staub, 1973). Care must be exercised to avoid hypertonic feedings given directly into the small intestine, and perforations have been reported as polyethylene or polyvinyl catheters became stiff after a week or so of continuous usage.

2.3. Parenteral Nutrition in the Newborn

The indications for total and partial parenteral nutrition have already been summarized. In general, the techniques and side effects are similar to those in

Table IX. Strategies of Intake

Nipple
Gavage
Intragastric drip
Gastrostomy
Nasojejunal feedings
Essential fatty acids transcutaneously

other age groups, and have been extensively reviewed elsewhere (Winters and Hasselmeyer, 1975). The daily allowances for newborns are, of course, different; these are indicated in Table X. When nutrition is given by vein, adequate weight gain has been seen on as little as 90 cal/kg a day. Fluid intake may be raised to as high as 200 ml/kg a day in an effort to permit more dilute infusions, or when radiant warmers increase evaporative water loss, or when other abnormal losses are present. Some ingredients, such as phosphorus and essential fatty acids, can be deferred during brief parenteral nutrition but their omission will result in serious deficiencies over the course of 6–8 weeks. In

Table X.
A. Daily Requirements of Total Parenteral Nutrition[a]

Protein	2–4 g/kg
(5% protein hydrolysate)	(recommended 3 g/kg/24 hr)
Calories	100–120.0 kg^{-1}
H_2O	125.0 ml/kg
Na	3–4.0 meq/kg
K	2–3.0 meq/kg
Ca	2–2.5 meq/kg
P	2.5 mmol/day[b]
Mg	1 meq/day
Multivitamins (Multivit)[c]	1 ml/day

B. Requirements Given at Intervals

Vitamin K	1–5 mg	Given weekly or as needed Monitored by prothrombin levels	i.m. or i.v.
B_{12}	<1 μg/day required	given weekly	i.m. or i.v.
Folic acid	5–50 μg/day	Given daily or weekly	i.v.
Trace minerals and essential fatty acids		Twice weekly infusions of whole blood or plasma 10–20 ml/kg	
Iron (Imferon)[d]		As needed to maintain Hbg in normal range	

[a] From Avery and Fletcher (1975), reprinted with permission.
[b] Blood levels may be watched at intervals and P as K_2PO_4 given as needed.
[c] Trademark, U. S. Vitamin and Pharmaceutical Corporation.
[d] Trademark, Lakeside Lab.

contrast to older individuals, a substantial portion of the newborn's intake is required for growth (see later). Hence allowances must be adjusted to activity, disease state, and whether growth is occurring.

Complications associated with parenteral nutrition are summarized in Table XI. Sepsis is perhaps the most serious of the common problems. Its frequency varies tremendously with the care in asepsis practiced in a given institution, but is rarely less than 5%. Hyperglycemia and urine sugar spill are useful signals of sepsis when similar glucose levels were previously tolerated. Because peripheral veins are often used for relatively hypertonic infusions, serious sloughs from infiltration are common, particularly when used with pumps. Because of these many considerations, close monitoring of the infant is required (Table XII).

Small prematures during the first 2 weeks of life may be intolerant to the *glucose* infusions of 10% and greater concentration which are needed for adequate caloric intake. Thus blood glucose concentrations may rise to 500–600 mg/dl, with serious hyperosmolarity, and glycosuria may be sufficient to cause osmotic diuresis with rapid loss of fluid and electrolytes. Stress hormones—adrenal steroids and catecholamines—may play a role, as may immaturity of hepatic glucose homeostasis. Insulin response to a rise in blood glucose is blunted in many prematures, but may be better in response to amino acids, an argument for including amino acids in intravenous regimens when high glucose concentrations are needed (Chance, 1975).

Until recently, inclusion of *lipid* in intravenous regimens has not been feasible. One result was the gradual development of essential fatty acid deficiency when total parenteral nutrition was prolonged. Gross manifestations, such as skin breakdown and poor wound healing, are usually delayed for 6–8 weeks, but biochemical evidence of essential fatty acid deficiency may be noted within days (Friedman *et al.,* 1976a). A fall in serum arachidonic acid and a rise in 5,8,11-eicosatrienoic acid have been considered useful indicators. While essential fatty acids can apparently be absorbed transcutaneously fol-

Table XI. Known Complications in Parenteral Nutrition[a]

Catheter	Metabolic
Sepsis	Persistent glycosuria
Extravasation of fluid	Acidosis
Local skin infections	Radiographic bone changes
Plugging or dislodgement	Hyperammonemia
Thrombosis of major vessel	Dehydration
Improper placement	Hepatic damage
	Postinfusion hypoglycemia
	Hypocalcemia, hypophosphatemia
	Hypokalemia
	Skin rash
	Essential fatty acid deficiency
	Copper deficiency

[a] From Avery and Fletcher (1975), reprinted with permission.

Table XII. Variables to Be Monitored in Infants Receiving Total Parenteral Nutrition[a]

Variable	Frequency
Metabolic (blood)	
Electrolytes	Daily, then 3 times weekly
Ca, P	Twice weekly
Glucose	As needed
BUN	Daily, ×2, then weekly
Osmolarity	As needed
Serum protein	Weekly
SGOT, SGPT	Baseline, then weekly
Ammonia	Weekly, particularly in prematures
Hemoglobin	Twice weekly
Acid–base	As needed, or weekly
Urine—glucose, s.g.,	Each voiding
Protein	
Growth	Wt. daily, ht. and head circumference weekly
Intake–output	Daily
Detection of infection	As needed
(activity, CBC's, cultures,	
temperature)	

[a] From Avery and Fletcher (1975), reprinted with permission.

lowing applications of sunflower seed oil (Friedman *et al.*, 1976b), the provision of linoleic acid equal to at least 2% of total calories is highly desirable. Another advantage of including lipid in intravenous nutrition is its high caloric density and low osmolarity, permitting the use of peripheral veins and decreasing the concentration of glucose required to provide adequate calories.

Since the availability of Intralipid, total and partial intravenous alimentation have been greatly facilitated. A mixture of soybean oil, egg yolk phosphatides, and glycerol, Intralipid is about 50% linoleic acid and is generally well tolerated provided a number of specific precautions are observed. Detailed discussions of Intralipid usage in the newborn have been published elsewhere (Coran, 1975). In infants it is customary to give a small test dose to exclude hypersensitivity. Then infusions are gradually increased over several days to a final infusion of 4 g/kg a day, given as a 10% solution which provides 44 cal/kg a day. The serum is checked for turbidity to ensure adequate fat clearance. Since the lipid emulsion is unstable when mixed with many other intravenous solutions, the Intralipid is given by a separate intravenous line or is added to the regular intravenous line close to the patient. In any event, a slow rate of infusion, spanning as nearly as possible the entire 24 hr, will minimize the tendency for hyperlipidemia.

Significant complications of Intralipid therapy include hyperlipidemia, liver damage, interference with pulmonary gas exchange, and the potential of uncoupling albumin–bilirubin binding with kernicterus at low bilirubin levels. Kernicterus has not actually been reported, but poses a threat that suggests that Intralipid not be used when indirect bilirubin is significantly elevated or during the first few days of life when physiologic jaundice in prematures is expected. Each individual infant has a different capacity to clear lipids, and hence frequent checks of serum turbidity should be made, with occasional measurement of free fatty acids or total lipids. With slow infusions lasting nearly 24 hr, most infants will clear the fat without significant serum elevations. However, clearance may be less optimal in small-for-dates babies. Elevated transaminases and mixed jaundice are not uncommon occurrences with Intralipid therapy, and indicate liver toxicity. However, their manifestations usually clear when therapy is discontinued or the dose reduced. The demonstration of interference with pulmonary gas exchange has been with bolus infusion and has been temporary only. On the other hand, no evidence suggests that modest doses of lipid, infused slowly, have a significant effect on lung function. The issue is important, for prematures with chronic lung disease are prime candidates for intravenous alimentation to prevent inanition during their long convalescence. At present our policy is not to withhold Intralipid during the convalescent period, when nutrition may become a critical issue.

2.4. Nutritional Monitoring

The sick newborn, and particularly the premature, presents a complex nutritional problem. Dietary allowances are customarily computed and given, orally or parenterally, on a per-kilogram basis. Figures such as 120 cal/kg and 3 g protein/kg, together with allowances such as are contained in Tables I–V, are useful rules of thumb. However, variable rates of growth, activity, body composition, absorption, enzyme maturation, presence of stress hormones, and so on, vastly modify the infant's response to a given intake. An imaginary nutritional balance, pictured in Fig. 5, suggests the complexity of the factors involved. Add to this the immaturity of renal and hepatic function that minimize the infant's ability to achieve homeostasis during periods of nutritional overload or deprivation.

A few examples will serve to illustrate this complexity. A marasmic infant with previous deficiencies has essentially no inactive depot fat and a large active cell mass per kilogram of body weight. Such an infant may require 150 or even 180 cal/kg a day to achieve adequate growth. In the premature, a relatively large share of the day's intake goes into growth of new tissue (see Table XIII). Thus, if growth fails to occur, the protein load may be excessive, resulting in metabolic acidosis, azotemia, and an increased osmolar load to be excreted by the kidney. Variable intestinal absorption, especially of dietary fat, makes uncertain the intake actually available to the infant. Selective malabsorption of vitamin E and folic acid occurs in the infant of less than 32 weeks of gestation.

In the presence of so many variables, *precise dietary allowances cannot be computed.* The solution is to provide an intake calculated by rule of thumb to be approximately correct, and then titrate the intake against the response, using anthropomorphic indices of growth and laboratory indices of body chemistry as a guide. An example of such a monitoring regimen is given in Table

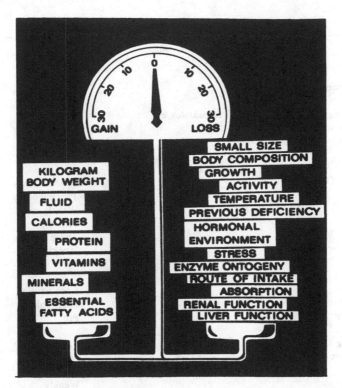

Fig. 5. A hypothetical nutritional balance, showing the influence of both intake per kilogram of body weight, and also physiologic variables in the infant which modify the growth which will occur on a given intake. From Avery (1976).

XIV. Protein overload would be signaled by azotemia, increased urinary urea, metabolic acidosis, and elevation of certain plasma amino acids such as methionine, leucine, isoleucine, and valine. Protein lack will have the opposite chemical stigmata, together with decreased serum albumin and elevation of plasma glycine, serine, glutamine, and taurine. Carbohydrate overload will be signaled by glycosuria, and inadequate fluid intake by elevated urine specific

Table XIII. Requirements for Growth[a]

	Wt. = 1000 g: 15 g/day	Wt. = 3000 g: 30 g/day
Calories	20%	39%
Protein	69%	52%
Water	10%	4%
Calcium	99%	99%
Na	33%	17%
K	24%	17%
mOsm	42%	33%

[a] From Avery (1976). Calculated from data of Fomon *et al.* (1974) as a percentage of total requirement needed to achieve indicated rates of growth.

Table XIV. Nutritional Monitoring[a]

I. Anthropomorphic	
Weight	Daily
Length	Weekly
Head circumference	Weekly
II. Hematologic	
Hemoglobin and hematocrit	Twice weekly
Blood smear	Weekly
III. Serum	
Na, K, Cl, CO_2	Twice weekly
BUN	Twice weekly
Protein	Weekly
Albumin	Prn
Ca, P	Twice weekly
Alkaline phosphatase	Weekly
IV. Urine	
Volume	Each voiding
Specific gravity	Each voiding
Osmolarity	Prn
Glucose	
Qualitative	Each voiding
Quantitative	Prn
Urea	Weekly
Na, K, P	Weekly
V. Other	
Bone films	Monthly
Amino acids	Prn

[a] From Avery (1976).

gravity and/or osmolarity. Failure of weight gain may be the result of inadequate calories, infection, chronic metabolic acidosis, or stress with its accompanying metabolic demand and adverse hormonal environment. It is difficult to judge sodium intake in a sick premature without adjusting to measured serum electrolytes. Macroelectrolytes can be adjusted by serum values, but insidious deficiencies of vitamins and trace elements may occasionally develop.

In summary, the sick premature is in an "unphysiologic" and uniquely vulnerable position nutritionally. Even with the most careful management, growth is almost always less than ideal. No set routine will always be successful, and both opportunism of mode of feeding and constant adjustment of intake are necessary.

2.5. Necrotizing Enterocolitis

Necrotizing enterocolitis (NEC) is a severe disease involving large and small intestine, which is particularly characteristic of small prematures and which conditions the approach to oral feedings in current nursery practice. The predilection of this disease for prematures is illustrated in Table XV. It has been estimated to account for 2% of admissions to nursery intensive care units (Barlow *et al.*, 1974). Its causation is complex and probably includes inadequate perfusion of the intestine during periods of perinatal shock and hypoxia, overgrowth of intestinal flora after feedings are initiated, and invasion of the damaged intestinal wall by gas-producing organisms. Symptoms include paralytic ileus, distension, blood in the stools, apneic spells, and in severe cases perforation, shock, and septicemia.

Of significance to this discussion are the onset of symptoms following the start of oral feedings, the increased risk when feedings are significantly hypertonic, and the early symptoms of distension and increased gastric residua. So fearful have been some of the outbreaks of necrotizing enterocolitis that in their wake, oral feedings are started very slowly in prematures with perinatal distress, and reliance on early intravenous alimentation has increased. In the past, small volumes of concentrated feedings have been used to offset the small size of the premature's stomach capacity. This practice, and the use of elemental formulas, has decreased because of fear of NEC. Nursery staffs have become justifiably gun-shy about minor degrees of distension and gastric residua, and at the first sign of trouble will place the infant on nothing by mouth and gastric suction. Undoubtedly these practices have resulted in aborting numerous cases of NEC and have minimized the danger of gastric overload and aspiration. However, they have also increased reliance on alternative nutritional strategies and have sometimes delayed weight gain. On balance, an increased respect for the delicate and sometimes sluggish intestinal tract of the premature seems beneficial.

2.6. Human Milk in the Premature

A resurgence of interest in the use of human milk for prematures has resulted from consideration of its possible benefit in preventing necrotizing enterocolitis. The initial observation that NEC was less common in nurseries

Table XV. Necrotizing Enterocolitis Incidence[a]

Weight (g)	Inborn (%)	Outborn (%)
<1500	8	11
1500–2500	3	7
>2500	0.5	4

[a] Incidence of NEC by weight group at McMaster, 1973–1975 (J. C. Sinclair, personal communication).

in which breast-feeding predominated has not proved consistent, but is still suggestive. On the other hand, animal investigation showed a clear protection by maternal milk against experimental NEC in the newborn rat (Gryboski, 1975). The immune benefits of fresh human milk, and its protective effect against infection by certain enteric bacteria and viruses are indisputable (Pitt, 1976). Unfortunately, the precise contribution of infection to NEC has not been delineated at this time. The initiation of feedings in the perinatally stressed premature with small amounts of colostrum from his own mother presents important theoretical advantages not yet established clinically. Many of these advantages are lost if pooled human milk is substituted, for freezing or pasteurizing destroys several of the immune factors.

Meanwhile, argument continues about the nutritional appropriateness of human milk for the smallest premature. Historically, it was concern about the marginal protein content of human milk, in the particular circumstances of the premature, which resulted in a switch to the formula feeding of prematures in this country during the past few decades. Räiha *et al.* (1976) have demonstrated good growth in healthy prematures of less than 30 weeks of gestation who were able to take 170 ml human milk/kg by the end of the first week of life. Lower BUN values and little trouble with metabolic acidosis were features of this regimen. However, plasma proteins declined slightly during the study period in the smallest prematures fed breast milk. Many infants less than 1500 g are unable to tolerate the large volumes of milk given in this study. S. J. Fomon, E. E. Ziegler, and H. D. Vasquez [1978 (unpublished data)] have presented data suggesting that human milk is deficient in protein, calcium, and sodium when compared with the needs of rapidly growing prematures of the smallest weight groups (Table XVI).

The development of "new look" breast milk banks are intended to recapture the immunological and physiological benefits of colostrum and human milk in the feeding of small prematures. Over and against these considerations are the logistical problems of obtaining fresh breast milk, sterility control, the loss of immune benefit when pasteurization or freezing is employed, and possibly the variable protein and caloric content of human milk. A possible solution might be the feeding of milk donated by the mother, which could be reinforced with added protein, calcium, electrolytes, vitamins, and iron, and then given by gavage.

Table XVI. Marginal Substances in Human Milk[a]

	Protein (g/100 kcal)	Calcium (mg/100 kcal)	Sodium (meq/100 kcal)
Estimated requirements[b]	2.84	184	1.1
Banked human milk	1.50	43	0.8

[a] From S. J. Fomon, E. E. Ziegler, and H. D. Vazquez, unpublished data.
[b] Assumed body weight 1200 g, weight gain 20 g/day, energy intake 140 kcal/kg a day.

2.7. The Small-for-Date Infant

The foregoing discussions of the nutritional needs of the premature have referred to the infant born before term after a period of normal intrauterine growth, and therefore appropriate in weight for gestational age. About one-third of infants weighing less than 2500 g are of a relatively more advanced gestational age but underweight because of poor intrauterine growth: the small-for-date baby. These latter infants may have intrinsic defects (rubella syndrome, trisomy, etc.), may be the victims of poor uteroplacental function (maternal toxemia, hypertension), or even be the result of maternal malnutrition. The majority of cases are, however, unexplained. The period of poor intrauterine growth may be chronic, resulting in an infant with reduced height, weight, and head circumference, or acute, resulting in an infant with reduced weight, normal length and head circumference, and loose skin suggesting loss of subcutaneous fat. The diagnosis of intrauterine growth failure is made by comparing the baby's measurements at birth and his estimated gestational age with normal curves from the literature.

Nutrition and growth in the small-for-dates infant are different from those of the appropriately grown premature. The baby with short-term intrauterine malnutrition is relatively mature and tolerates rapidly advancing oral feedings. He is prone to hypoglycemia in the first days of life, but frequently experiences catch-up growth and achieves discharge well before a true premature of the same birth weight. The chronically and severely undergrown fetus may have suffered lasting damage and may grow slowly both before and after discharge from the nursery, never achieving full genetic potential for stature. The infant with intrinsic disease such as chronic intrauterine infection or chromosomal anomaly will usually grow poorly both before and after birth, regardless of how nutrients are provided.

2.8. Summary

The nutritional management of the premature is complex and taxing. Interruption of growth, metabolic derangements, and unrecognized deficiencies may contribute to the sequelae of markedly premature birth. Our present approach is a patchwork quilt of improvisations in no way comparable to continuous placental nutrition. Thus, research in this area is still greatly needed.

3. References

Anderson, T. A., and Fomon, S. J., 1974, Beikost, in: *Infant Nutrition* (S. J. Fomon, ed.), p. 425, Saunders, Philadelphia.

Applebaum, R. M., 1975, The obstetrician's approach to the breasts and breast feeding, *J. Reprod. Med.* **14**:98.

Avery, G. B., 1976, Analysis of growth failure in sick prematures, in: *Intensive Care in the Newborn* (L. Stern, B. Friis-Hansen, and P. Kildeberg, eds.), Masson, New York.

Avery, G. B., and Fletcher, A. B., 1975, Nutrition, in: *Neonatology* (G. B. Avery, ed.), Lippincott, Philadelphia.

Barlow, B., Santulli, T. V., Heird, W., Pitt, J., Blanc, W., and Schullinger, J., 1974, An experimental study of acute necrotizing enterocolitis—The importance of breast milk, *J. Pediatr. Surg.* 9:587.

Beal, V. A., 1970, Nutritional intake, in: *Human Growth and Development* (R. W. McCammon, ed.), p. 63, Charles C Thomas, Springfield.

Chance, G. W., 1975, Results in very low birth weight infants (1300 gram birth weight), in: *Intravenous Nutrition in the High Risk Infant* (R. W. Winters and E. G. Hasselmeyer, eds.), Wiley, New York.

Cheek, J. A., and Staub, G. F., 1973, Nasojejunal alimentation for premature and full-term newborn infants, *J. Pediatr.* 82:955.

Committee on Nutrition, American Academy of Pediatrics, 1976, Iron supplementation for infants, *Pediatrics* 58:765.

Coran, A., 1975, Intravenous use of fat for the total parenteral nutrition of the infant, in: *Intravenous Nutrition of the High Risk Infant* (R. W. Winters and E. G. Hasselmeyer, eds.), p. 343, Wiley, New York.

Dancis, J., O'Connell, J. R., and Holt, L. E., 1948, Grid for recording the weight of premature infants, *J. Pediatr.* 33:570.

Driscoll, J. M., Jr., 1975, A preliminary study of total intravenous alimentation in low birth weight infants, in: *Intravenous Nutrition of the High Risk Infant* (R. W. Winters and E. G. Hasselmeyer, eds.), Wiley, New York.

Fomon, S. J., 1975, What are infants fed in the United States? *Pediatrics* 56:350.

Fomon, S., J., and Filer, L. J., Jr., 1974, Milk and formulas, in: *Infant Nutrition* (S. J. Fomon, ed.), p. 361, Saunders, Philadelphia.

Fomon, S. J., Ziegler, E. E., and O'Donnell, A. M., 1974, Infant feeding in health and disease, in: *Infant Nutrition* (S. J. Fomon, ed.), 2nd ed., Saunders, Philadelphia.

Friedman, Z., Danon, A., Stohlman, M., and Oates, J., 1976a, Rapid onset of essential fatty acid deficiency in the newborn, *Pediatrics* 58:640.

Friedman, Z., Schochat, S. J., Maisels, J., Marks, K. H., and Lambeth, E. L., Jr., 1976b, Correction of essential fatty acid deficiency in newborn infants by cutaneous application of sunflower-seed oil, *Pediatrics* 58:650.

Gryboski, J., 1975, *Gastrointestinal Problems in the Infant*, p. 284, Saunders, Philadelphia.

Haire, D., 1969, *Instructions for Nursing Your Baby*, International Childbirth Association, Seattle.

Klaus, M. H., and Kennell, J. H., 1976, *Maternal–Infant Bonding*, pp. 59–62, Mosby, St. Louis, Mo.

Knowles, J. A., 1965, Excretion of drugs in milk—A review, *J. Pediatr.* 66:1070.

Lakdawala, D., and Widdowson, E., 1977, Vitamin D in human milk, *Lancet* 1 (8004):167.

McMillan, J. A., Landau, S. A., and Oski, F. A., 1976, Iron sufficiency in breast-fed infants and the availability of iron from human milk, *Pediatrics* 58:686.

Pitt, J., 1976, Breast milk leukocytes, *Pediatrics* 58:769.

Räiha, N. C. R., Heinonen, L., Rassin, D. K., and Gaull, G. E., 1976, Milk protein quantity and quality in low-birth-weight infants: I. Metabolic responses and effects on growth, *Pediatrics* 57:659.

Rios, E., Hunter, R. E., Cook, J. D., Smith, N. J., and Finch, C. A., 1975, The absorption of iron as supplements in infant cereal and infant formulas, *Pediatrics* 55:686.

Vengusamy, S., Pildes, R. S., Raffensperger, J., Levine, H. D., and Cornblath, M., 1969, A controlled study of feeding gastrostomy in low-birth-weight infants, *Pediatrics* 43:815.

Wei, S. H. Y., 1974, Nutritional aspects of dental caries, in: *Infant Nutrition* (S. J. Fomon, ed.), Saunders, Philadelphia.

Widdowson, E. M., 1968, Growth and composition of the fetus and newborn, in: *Biology of Gestation*, Vol. II (N. S. Assoc., ed.), Academic Press, New York.

Winters, R. W., and Hasselmeyer, E. G. (eds.), 1975, *Intravenous Nutrition of the High Risk Infant*, Wiley, New York.

The Young Child: Normal

Bo Vahlquist

1. Preschool Age—Definitions of Subgroups

In many contexts there is a need for a more precise definition of age groups in the "preschool period." Admittedly only half of the children in the world today have access to schools and thus even the term preschool may not always be relevant.

In recording vital statistics it is customary to subdivide childhood into periods of 0–1, 1–4, 5–9, and 10–14 years. Such grouping, however, is founded more on statistical convenience than on biological considerations. For health workers, there is a need for a more specific classification.

The question of grouping is of particular importance with respect to nutritional needs. Following a proposal by the PAG *ad hoc* group on the subject of feeding the preschool child, the PAG secretariat has presented a classification that seems reasonable (Venkatachalam, 1975). This divides the period 0–5 years into the following four subgroups:

	Age range (months)
Young infants	0–6
Older infants	6–12
Younger preschool children	12–36
Older preschool children	36–60

This categorization corresponds fairly well to the progressive transition in the type of food, which for the large poverty groups of the Third World may, simplified, be given as shown in Table I.

Bo Vahlquist • Deceased. Department of Pediatrics, University Hospital, Uppsala, Sweden.

Table I. Types of Food According to Age Groupings

Age group	Type of food
Young infants	Human milk (alone or together with semisolids at 4–6 months)
Older infants	Human milk, gradual addition of more supplementary foods (semisolids and solids)
Younger preschool children	(Human milk) Weaning foods according to local conditions
Older preschool children	Progressive transition to adult foods

2. Anthropometric Data (Swedish Section of the CIE Longitudinal Growth Study)

During the last 25 years a large number of investigations on the physical development of young children have been carried out in various parts of the world. Of special interest are those which have applied a longitudinal technique, examining individual children over long periods of time. Only in this way is it possible to analyze individual growth patterns, to correlate maturational events, and to "link patterns of development and health in childhood with measures of health and sickness at later ages" (Falkner, 1973).

The most well known among this type of study is the now classical Longitudinal Growth Study coordinated by the International Children's Centre (CIE), Paris. This was begun in the mid-1950s in eight cities—Brussels, London, Paris, Stockholm, Zürich, Dakar, Kampala, and Louisville (Falkner, 1960). Today, in the first five of these cities, the study covers subjects ranging all the way from birth to adulthood. Its strength lies in the fact that a well-matched and highly experienced group of research workers has planned and coordinated the studies in the various sites and has thus guaranteed a basically uniform methodological approach.

In the following, brief reference will be made to the results obtained to date in the Swedish part of the study. Originally this comprised 212 children (122 boys and 90 girls), and so loyal has been the cooperation from all sides that 170 of them (84.4%), who had been investigated from birth, were still being followed up at 16 years of age (Karlberg *et al.*, 1976).

Only children with a birth weight of more than 2000 g were included (all except one were above 2400 g). With respect to social class at birth, the sample was representative of contemporary Swedish urban children, with 17% in the upper social stratum, 37% in the middle, and 46% in the lower (although hardly ever really poor).

A number of anthropometric variables were studied: weight, supine length, crown–rump length, standing height, sitting height, biacromial width, pelvis width, head circumference, anterior fontanel, femur and humerus bicondylar width, thorax width and depth, chest circumference, upper arm and calf circumference, biceps, triceps, subscapular, and suprailiac skinfold. In Figs. 1 and 2 the 10th, 50th, and 90th percentiles for height and weight are given, together with the corresponding velocity values. We recognize the well-known pattern with a physical growth which is relatively, and even absolutely, intense in the immediate postnatal period—the mean increment for supine length at 1–3 months of age being 37 cm and that for weight 10.6 kg. Even under favorable environmental conditions there are wide interindividual variations in height and weight development, as well as in other anthropometric measures. Other results from the Swedish study are briefly discussed in Sections 5 and 6.

3. Effect of Low Birth Weight on Subsequent Growth Pattern

The reasons for the discrepant results in earlier studies on this question are now becoming increasingly clear. First, a clear distinction was not made between a low birth weight due to a short length of gestation and that due to malnutrition *in utero*, and second, the postnatal feeding pattern was not clearly defined with respect to initial timing and nutritional adequacy.

If the growth velocity is compared between children with a birth weight which is low but appropriate for gestational age (AGA), and who have been fed in an optimal way from the immediate postpartum period, on the one hand, and full-term healthy children on the other, the following results may be

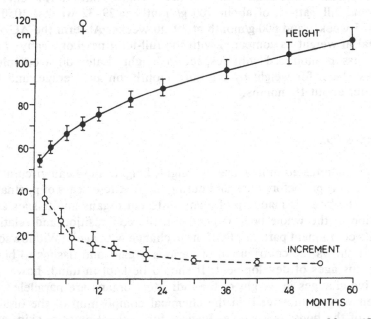

Fig. 1. Height (supine length)–distance values (cm) and increment (cm/year). Means (●) and 10th and 90th percentiles (⊺). Girls only (Karlberg *et al.*, 1976).

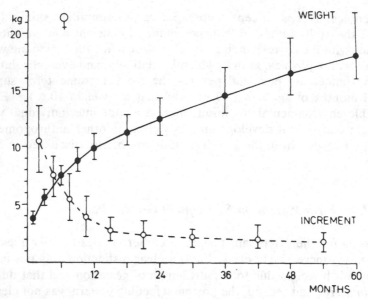

Fig. 2. Weight–distance values (kg) and increment (kg/year). Means (●) and 10th and 90th percentiles (⊺). Girls only (Karlberg *et al.*, 1976).

obtained (Brandt, 1976). In the period 31–40 weeks the extrauterine weight velocity of the AGA preterm infants before term shows a progressive increase from 380 to 890 g/month as compared with the normal " intrauterine curve" with its rise-and-fall pattern of about 700 g/month at 29–32 weeks, 1050 g/month at 33–35 weeks, and 600 g/month at 39–40 weeks. At term the AGA is still subnormal in weight as compared with the full-term newborn baby. This difference is less pronounced with respect to height. Later on a complete catch-up takes place, for weight by the third month, on an average, and for height only after about 18 months.

4. Body Composition

Growth, as manifested in increase in weight, height, and head circumference, is extremely rapid before birth and during the first few years of postnatal life (Fig. 3). "The body is made up of many different organs and tissues and the composition of the whole body depends on the composition and relative weight of all its component parts. Both of these change with age" (Widdowson, 1974). Table II shows the contribution of various organs and tissues to body weight at various ages of development. It should be kept in mind, however, that the relative changes in weight of the individual organs are paralleled by changes, often very pronounced, in the chemical composition of the organs and thus also of the body as a whole. Figures for skeletal muscle, skin, and liver are given as examples in Table III. This table also indicates that many of

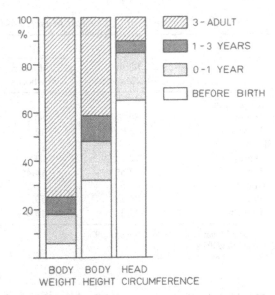

Fig. 3. Increase in body weight, body
height, and head circumference during
four periods of growth and development
(Vahlquist, 1975).

the changes in the proximate body composition during postnatal development
occur during the first year of life (see also Fomon, 1974).

The figures given in Tables II and III represent only gross averages, often
based on a rather limited number of subjects and derived by different analytical
methods. It is obvious that just as in the case of overall height and weight
development (Figs. 1 and 2), there are gross interindividual variations even in
seemingly healthy children. The main lesson to be learned from studies of
body composition is that an organism undergoing such fundamental changes,
anthropometrically as well as biochemically and indeed functionally, as that
of the preschool child and, more particularly, of the infant, must be extremely
vulnerable to the numerous stressing influences of the environment. It is worth
noting that even in well-nourished infants maturation parameters such as the

*Table II. Contribution of Various Organs and Tissues to Body Weight at Various
Ages[a,b]*

	Fetus 20–24 weeks	Full-term newborn	Adult
Skeletal muscle	25.0	25.0	40.0
Skin	13.0	4.0	6.0
Skeleton	22.0	18.0	14.0
Heart	0.6	0.5	0.4
Liver	4.0	5.0	2.0
Kidneys	0.7	1.0	0.5
Brain	13.0	12.0	2.0

[a] From Widdowson (1974).
[b] Values in percent body weight.

Table III. Composition of Skeletal Muscle, Skin, and Liver at Various Ages[a,b]

| | Fetus 20–22 weeks | Baby | | Adult |
		Full-term newborns	4–7 months	
Skeletal muscle				
Water	89	80	79	79.0
Protein	8.0	12.6	16.7	17.4
Skin				
Water	90	83	68	69
Protein	7.5[c]	14.4[c]	34.4[c]	33.1[c]
Liver				
Water	81	79	—	75
Protein	12.3	13.0	—	14.6

[a] Compiled from tables by Widdowson (1974).
[b] Values in g/100 g.
[c] Calculated as $g \cdot N \times 6.25$.

time of appearance of centers of ossification may be influenced by the composition of the diet (Mellander *et al.*, 1959).

5. Psychomotor and Mental Development

For normal development in this sphere, a child is dependent on continuous stimulus from and interaction with the environment. Abusive behavior, as well as inertia, on the part of the parents may seriously hamper a child's psychomotor and mental development. At the same time there are obviously large interindividual variations in this respect even among children reared under similar conditions. These and other aspects of the psychomotor and mental development of the child are discussed at length in many reviews (Illingworth, 1972; Klackenberg, 1971; Kagan *et al.*, 1966). Some of the milestones with respect to psychomotor development at an early age are given in Fig. 4.

6. Sexual Differences

At birth boys are slightly longer and slightly heavier than girls. In a Swedish study (Engström and Sterky, 1966) the values shown in Table IV were obtained. It is evident from the table that the differences observed cannot be explained by dissimilarities in mean gestational age.

The distance values continue to show a corresponding difference (the values for boys being higher), which, although modest in absolute terms, is statistically significant for weight up to the age of 2 years and for height up to 4 years (Karlberg *et al.*, 1976). The values of head circumference and skeletal dimension (measured as the bicondylar width of the femur and humerus) for

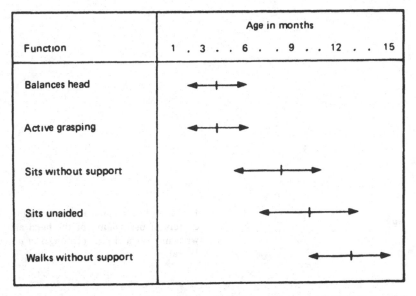

Function	Age in months
	1 . 3 . . 6 . . 9 . . 12 . . 15
Balances head	
Active grasping	
Sits without support	
Sits unaided	
Walks without support	

Fig. 4. Milestones in the motor development of children. Average ±2 SD [National Board of Health and Welfare Report (MEK-B), 1973].

boys are consistently higher throughout the preschool period and, indeed, during the entire period of growth (Karlberg *et al.*, 1976).

In the light of this background, it is interesting to note that the ossification of the various bones begins at an earlier age in girls than in boys (Fig. 5). There is little or no difference in dental development and, more particularly, the timing of dental eruption for the deciduous teeth, whereas for permanent dentition girls are definitely ahead of boys (Taranger *et al.*, 1976b).

In addition to these data related to the development of the skeletal system, a few words should be said about data concerning the muscular and fat compartments. Limb circumference values (arm and calf), which give at least a rough estimate of muscle mass, show no sexual difference in the preschool period. For skinfold measurements (biceps, triceps, and subscapular), girls are slightly ahead of boys throughout the preschool period (Karlberg *et al.*, 1976).

With respect to chemical composition it has been concluded (Owen *et al.*, 1966) that "fat comprises a greater percentage, and water a smaller percentage

Table IV. Birth Weight, Birth Length, and Gestational Age in Swedish Newborns[a]

Birth weight (g)	3596	±542	3408	±503
Birth length (cm)	51.4	±2.4	50.3	±2.3
Gestational age (days)	281.2	±13.5	281.2	±13.3

[a] From Engström and Sterky (1966).

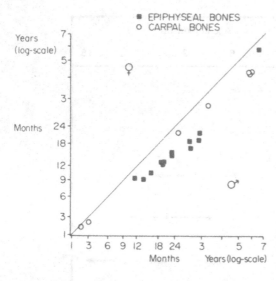

Fig. 5. Age at the appearance of various centers of ossification of the hand and wrist in boys and girls (Taranger *et al.*, 1976a).

of body weight of girl than of boy infants . . . . The demonstration of sex-related differences in body composition during infancy together with well-recognized sex-related differences in neonatal mortality and in the incidence of certain diseases, strongly suggests that fundamental physiological processes differ between infant boys and girls.''

The findings with respect to mental development are controversial. In the Swedish longitudinal study referred to earlier, evidence of a more rapid development in girls was found at 2 and 3 years of age. This was especially true in the language and personal–social spheres and was most pronounced in the lowest socioeconomic class (Klackenberg-Larsson and Stensson, 1968).

7. Ethnic Differences

Adult height and weight still differ considerably in different parts of the world. Beginning with the observations by Greulich (1957), Mellbin (1962), and others, more and more evidence has accumulated to the effect that these differences are largely, if not predominantly, due to differences in environment, particularly the prenatal and early postnatal milieu, rather than to differences in genetic inheritance (Habicht *et al.*, 1974; Eksmyr, 1970). This question has been discussed at some length in another context (Chapter 2). This recognition of the effect of environment must not lead to a total swing of the pendulum, however. There is good reason to believe that even in the most favorable environmental situation some differences in adult height and weight will remain, which must be regarded as ethnic in origin (the pygmies are *one* example—their dwarfism may be due to an anomaly in their cellular sensitivity to growth hormone or to the sulfation factor) (Royer, 1974).

The fact that an improvement of the environment at an early age, particularly with respect to nutrition and infectious diseases, has been proven to diminish, and often completely eradicate, differences in adult stature, raises the question: Could we use a single growth standard for all populations in the world? This would mean a true challenge for politicians on a national basis to close the gap between what there is and what might be achieved in the way of reaching truly "healthy" norms in growth and development. It seems fairly well documented that for children of preschool age such a universal reference for height and weight could be used. Indeed, growth charts with this aim have been produced in recent years (Morley, 1976). As for older children and adults certain reservations may have to be made. In a recent annotation Tanner (1976) expressed his views as follows: "If we want to follow only growth in height, then the same standards are fairly adequate for Europeans and Africans and their descendents in America and elsewhere. If we want to ask questions about body proportions . . . then different standards must be used. If we want to measure the height growth of Chinese or Japanese then it seems that European standards will not do."

With respect to psychomotor development it would seem that during the first year of life African children are more advanced in their psychomotor development than Caucasians (Geber and Dean, 1957 a,b; Faladé, 1955). This observation is of great interest and motivates further investigation. But, after 20 years, little more has been reported in this direction.

8. Forecasting of Growth and Development from Observations Made in Early Life

One of the advantages of longitudinal growth studies is that they allow some predictions to be made from one age group to another with respect to development. It was demonstrated long ago (Tanner *et al.*, 1956) that length at birth is poorly correlated to the same parameter in adult life (for boys $r =$ .25), whereas at 2 years of age the correlation is much higher (for boys $r =$ .79). This then must mean that in the first few years of life there is considerable "crossing over" between different growth channels, which subsequently diminishes. This is well illustrated in a diagram from the Swedish study (Fig. 6).

The possibility of prediction for psychomotor and mental development is no better. Klackenberg-Larsson and Stensson (1968) write: "as regards *normal* children, we cannot predict from the 3–12 month values anything about 3-year or 5-year values." Whether this is due solely to crossing over between channels or partly to lack of refinement of present-day test methods is not clear, however.

What we really should like to know, of course, is whether certain parameters (anthropometric, biochemical, mental, or other) could allow a forecast to be made in individual children at preschool age as to their subsequent development of illnesses which although very prevalent in the population are probably influenced by environmental factors, particularly nutritional (e.g., obesity, atherosclerosis, hypertension). In other words, we need to know

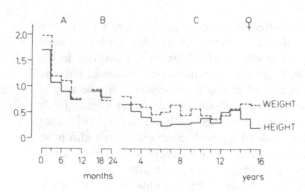

Fig. 6. Frequency of "channel crossing" in height and weight. Girls only (Taranger, 1976).

whether any correlations in individual subjects followed up over several decades might give some clue where other attempts to forecast disease by biochemical and other methods have failed. Up until now little has been achieved with respect to "children at risk" in this context. Even the claim that overweight in infancy may predispose to overweight later on in life has not been unchallenged (Mellbin and Vuille, 1973).

9. Physical Activity Pattern

The energy expenditure of an individual is dependent to a considerable extent on the physical activity. Much of what has been called overweight by excess food intake is in fact the result of low physical activity coupled with only a moderately high or even a normal or subnormal energy intake. What has now been said is in principle true also for young children (Spady *et al.*, 1976; Rose and Mayer, 1968).

The motivation for physical activity varies. An increasing number of individuals in industrialized countries today are engaged in sedentary work, and their degree of physical activity depends mainly on their mode of locomotion between home and work and on leisure occupations.

In the early years of life physical activity is a necessity for the child to learn, know, and control his own body, to become oriented in the environment, and to establish contacts with other individuals. Between the ages of $1\frac{1}{2}$ and 3 years movements in the form of walking, running, and climbing are in themselves a game. Children 3–4 years of age control their body movements sufficiently to incorporate these movements into other games. Large body movements characterize the play of preschool children (National Board of Health and Welfare Report (MEK-B), 1973)—"The preschool child who characteristically uses his large muscles during many hours of the day is continuing a self-imposed program of physical fitness" (American Academy of Pediatrics Committee Report, 1976).

Using a "radiotransmitter," the heart and respiratory rates of preschool children during spontaneous play have been monitored (Klimt, 1966). It was shown that the activity is intermittent, with "spikes" of high energy output followed by brief pauses. The young child consumes relatively more energy than the older child or adult. For a given activity the energy consumed is higher in a child than in an adult. The overall activity pattern and the total energy needed for physical activity differ considerably between individual children.

Even for preschool children the modern way of life, with easy access to vehicles for transport and more indoor entertainment (TV, etc.), has undoubtedly diminished the incentive for physical activities, but usually not to the same extent as in older age groups. There is a pressing need for a larger number of day care institutions and the like with ample space for moving around, both indoors and outdoors.

Sleep does not mean physical inactivity, but the activity occurs on a much lower level. From this point of view it is of interest to consider the number of sleeping hours at different stages of the preschool age period (Fig. 7). For a 1-year-old child the mean duration of sleep during a 24-hr period is 13.2 hr, and for a 5-year-old 11.6 hr (Klackenberg, 1971). During the period 6 months to 3 years, the decrease in the total number of sleeping hours averages no less than 2.5 hr per 24-hr period—mainly due to a reduction of sleep during the day. The individual variation with respect to sleeping hours is large, perhaps more so at an early age than later on.

Physical activity may, in principle, influence growth in two ways: by affecting the growth in stature and by causing changes in body composition. Of these, the latter correlation is more easily understood. Of the three main tissue components, skeleton, muscle, and fat, the fat component may be

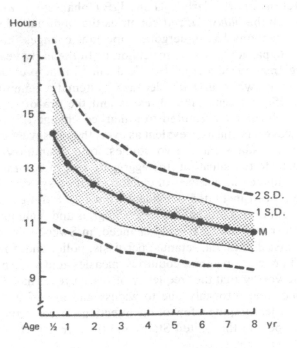

Fig. 7. Means and standard deviations for duration of sleep (total number of hours per 24 hr) (Klackenberg, 1971).

diminished as a result of increased activity, whereas the muscular tissue may increase. As for the skeleton it is less clear whether, for example, height growth is influenced by physical activity. There is some indication that extreme inertia may give suboptimal growth (stunting), whereas the opposite is not clearly proven.

In conclusion, there is reason to believe that the effects of varying degrees of physical activity on the growth and development of the child deserve more attention in the future.

10. "Normal" Disease Pattern

No age is free from disease. In the age group 0–5 years infectious diseases play a leading role. No other age group is so prone to acute infections. These also dominate in a relative sense—in relation to other diseases, amounting to 80% of all illnesses (Miller *et al.*, 1960).

In *preindustrial* societies the number, duration, and severity of infectious diseases are all very great. Together with malnutrition they account for the vast majority of sickness and fatalities in preschool age. In a study in four villages in rural Guatemala (Martorell *et al.*, 1975), the mean percentage of time of illness was, in the age group 6–12 months, for diarrhea, fever ("hot skin"), and respiratory illness, 13, 5, and 37%, respectively (rounded-off figures) and in the age group 48–60 months, 2, 2, and 27%, respectively. The massive systematic effect of infectious and parasitic diseases under these conditions is reflected in the high serum levels of different immunoglobulin fractions, particularly IgG and IgE (Johansson *et al.*, 1968).

In the *industrialized* countries the "normal" disease pattern with respect to infections has undergone important changes during the last century, and this is particularly true for children. In the early years improved nutrition and hygiene caused a considerable decline in the prevalence and severity of such disease, while in later decades systematic immunization programs against specific communicable diseases and the era of sulfonamides and antibiotics have further contributed to a marked change of pattern. The common cold, however, is still as prevalent as ever, though severe complications and fatalities are nowadays rare. Two studies from Scandinavian countries are used to illustrate the situation. the aggregate risk of contracting one of the "common communicable diseases of childhood" if no systematic immunization programs have been instituted is evident from Fig. 8, reflecting the position in Sweden for children born before 1950 (Sjögren and Vahlquist, 1965). This "natural" pattern has now markedly changed, in Sweden through routine immunization against diphtheria, tetanus, pertussis, polio, and, for selected groups, rubella, and in certain other countries measles and (beginning) parotitis also. It is noteworthy that the frequency of scarlet fever has diminished markedly at the same time, probably due to widespread use of sulfonamides and antibiotics.

Clearly quite a few of the infectious diseases that used to ravage Northern Europe and the United States and thus were part of the "normal" experience

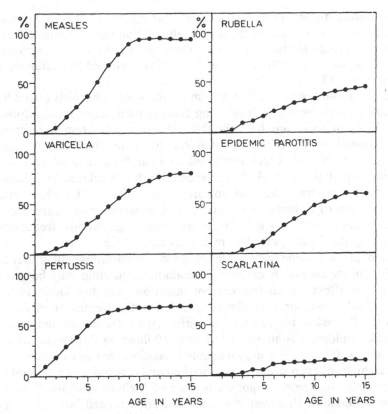

Fig. 8. Cumulative frequency of six common communicable diseases of childhood (with clinical manifestation) (Sjögren and Vahlquist, 1965).

of children in this part of the world have already been more or less exterminated (e.g., tuberculosis, diphtheria, whooping cough, polio, salmonellosis) or can be at any given time (measles, rubella, parotitis). However, the most common causes of disease in this age group, as in many following ones, the infections of the respiratory tract, remain essentially untouched, although their complications (sinusitis, otitis, pneumonia, etc.) and thus also to a large extent the fatalities caused by such diseases have been largely brought under control.

In a study in Finland (Bäckström-Järvinen *et al.*, 1966) of children 0–5 years old and followed up longitudinally the four dominating illnesses were febrile cold, tonsillitis, otitis, and gastroenteritis. The frequency of these four diseases combined was highest in the first year and then declined year by year. For the whole 0- to 5-year period the mean number of illnesses (all kinds) in boys was 10.7 (range 4–26) and in girls 9.5 (range 0–21).

And now to the question: To what extent may the infections of the "normal" child interfere with growth and development, more than, say, a small "dent" in the grid? In principle an infection may influence nutrient intake and nutrient utilization in several ways, including anorexia (especially in febrile conditions), vomiting and/or diarrhea, influence on nutrient absorption or metabolism, dietary food restrictions, and effects of drugs. Several studies have been carried out with the aim of elucidating the potential effects of frequently appearing infectious diseases. The results have been controversial as far as influence on growth and development is concerned (for references see Martorell *et al.*, 1975).

It would seem that a study conducted under the special hazards posed by the unfavorable environment of a developing country might have more informative results. In the investigation from the INCAP center in Guatemala referred to earlier (Martorell *et al.*, 1975), 716 children living in four villages were studied during a 2-year period and submitted to a careful statistical analysis. The conclusion was that diarrheal disease seemed to be correlated to reduced incremental growth, whereas this was not proven for "fever" (hot skin, temperature not measured) or respiratory illnesses. The accumulated differences in height and weight growth between the groups with a high and low frequency of diarrhea during the 2-year period were 3.5 cm and 1.5 kg.

It is reasonable to assume that the "premorbid" constitution will influence the effect of an acute illness. A chronically malnourished child may be more susceptible to the effects of an intercurrent infection. Another factor is the varying load of infection, for example as revealed in the frequency of manifestation. In the Finnish study referred to earlier (Bäckström-Järvinen *et al.*, 1966), individual children sustained up to nearly 30 illnesses during the whole 0- to 5-year period. It seems not unreasonable to assume that in exceptionally disease-prone children, even when well nourished and otherwise in a favorable position, there may be some temporary lagging of their genetically directed growth and development. However, there are not enough hard facts to verify this assumption.

A special word about possible effects of pharmaceutical drugs used widely in young children, e.g., antibiotics. Penicillin is metabolized and/or excreted rapidly from the body. Other drugs, particularly the tetracyclines, may remain for a long time in the skeleton and cause skeletal discoloration including the teeth in the young child. For this reason the use of tetracycline during pregnancy and in early childhood has been banned in most countries. In a longitudinal study in Czechoslovakia (Bakicova *et al.*, 1975) it was noted that children reared in institutions had received, on an average, 9.2 courses of antibiotics and children reared at home an average of 4.4 during the first 3 years of life. Ninety-four children were followed up for the period 0–6 years; they received antibiotics on an average of 9.3 times. In the girls there was a statistically significant correlation between number of antibiotic courses and height and weight development "with an upward trend." For boys no such correlation was found. It would seem that for the average "normal" child antibiotics will not have any marked influence on its growth and development. But again the possibility cannot be excluded that in an occasional child with a high susceptibility to infections and numerous periods of antibiotic therapy these may have some influence on growth and development. But in such a situation we are in the extreme range of "normality"!

We have discussed the infectious diseases at some length. Another group of diseases affecting the "normal" child to a large extent is that comprising the allergic conditions. In many affluent societies as many as 5–10% of all children are labeled allergic. The condition may vary from a slight eczema, an allergic rhinitis or a gastrointestinal upset, to a severe bronchial asthma. The allergic conditions often run a protracted course and, if severe, may definitely hamper the growth and development of the child. However, in this case, and particularly if treatment with steroids is involved, the question is no longer one of a "normal" child.

11. Home versus Institutional Environment

In countries with central planning and a high rate of employment for women in work outside the home, an extensive system of day care centers has long been established. This means institutional care of the majority of children from 4–6 months up to 3–4 years of age. In many Western countries in Europe and North America, an increasingly large number of children are looked after for part of the day outside their own home.

The effect of institutionalization with respect to the harmonious development of the children involved varies. For children from poverty groups with restricted means for creating a satisfactory home environment, day care centers with good resources may mean a positive experience in many ways, whereas the same may not be true for children from better situated families.

On the whole, the quality of the care provided by the day care institutions has progressively improved, parallel with an understanding that the staff must be sufficiently numerous and well trained to meet the needs of the children, physical, mental, and emotional. On this condition and provided the children are encouraged to take part in age-adapted activity and play, the risks of the collective group, e.g., in the form of an increased number of respiratory illnesses, should not be exaggerated. Nevertheless it is still a widely held opinion that the care in a family environment, when well composed and functioning satisfactorily, would seem the ideal solution in the most formative years of life.

Both in its own home and in institutional care a child may, if exposed to an unfavorable or even positively harmful environment, develop a "deprivation syndrome." It is widely accepted that this inherent depression may seriously hamper physical development (Bibliographie, 1976; MacCarthy, 1974). It is difficult to establish the mechanisms involved. In many such cases the transfer to a new, warm, and understanding environment will have a rapid favorable effect on the general well-being of the child, who will catch up in both height and weight. However, more often than not such a transfer may not only mean a change for the better in the emotional sense but will also make up for a previous more or less pronounced under- or malnutrition.

12. References

American Academy of Pediatrics Committee Report, 1976, Fitness in the preschool child, *Pediatrics* **58**:88.

Bäckström-Järvinen, L., Tiisala, R., Kantero, R.-L., and Hallman, N., 1966, Illnesses among normal Finnish children during the first five years of life, *Ann. Paediatr. Fenn.* **12**:13.

Bakicova, Z., Lipkova, V., Arochova, O., and Grunt, J., 1975, Medical treatment (in Czech), quoted from *Excerpta Medica Pediatrics* **32**:368.

Bibliographie, Psychosocial growth retardation (deprivation dwarfism), Recent references, 1976, *Courrier* **26**:565.

Brandt, I., 1976, Growth dynamics before and after term including catch-up growth in weight, length and head circumference of preterm infants, in: *Perinatal Medicine* (G. Rooth and L.-E. Bratteby, eds.), pp. 221–229, Almqvist & Wiksell, Stockholm.

Eksmyr, R., 1970, Anthropometry in privileged Ethiopian preschool children, *Acta Paediatr. Scand.* **59**:157.

Engström, L., and Sterky, G., 1966, Standardkurvor för vikt och längd hos nyfödda barn, *Läkartidn.* **63**:4922.

Faladé, S., 1955, *Le Développement Psychomoteur de Jeune Africain Originaire du Sénégal au Cours de Sa Première Année*, Imprimerie R. Foulon, Paris.

Falkner, F., 1960, The history and conception of the coordinated studies, in: *Modern Problems in Paediatrics*, Vol. 5 (F. Falkner, eds), p.1, Karger, Basel.

Falkner, F., 1973, Long term developmental studies: A critique, in: *Early Development*, Res. Publ. A.R.N.M.D. Vol. 51, Chap. 22, pp. 412–421.

Fomon, S. J., 1974, *Infant Nutrition*, 2nd ed., Saunders, Philadelphia.

Geber, M., and Dean, R. F. A., 1957a, The state of development of newborn African children, *Lancet* **15**:1216.

Geber, M., and Dean, R. F. A., 1957b, Gesell tests on African children, *Pediatrics* **20**:1055.

Greulich, W. W., 1957, A comparison of the physical growth and development of American-born and native Japanese children, *Am. J. Phys. Anthropol.* **15**:489.

Habicht, J. P., Martorell, R., Yarbrough, C., Malina, R. M., and Klein, R. E., 1974, Height and weight standards for preschool children. How relevant are ethnic differences in growth potential? *Lancet* **I**:611.

Illingworth, R. S., 1972, *The Development of the Infant and Young Child. Normal and Abnormal*, Churchill–Livingstone, Edinburgh.

Johansson, S. G. O., Mellbin, T., and Vahlquist, B., 1968, Immunoglobulin levels in Ethiopian preschool children with special reference to high concentrations of immunoglobulin E (IgND), *Lancet* **I**:1118.

Kagan, J., Wright, J. C., and Bayley, N., 1966, Psychological development of the child, in: *Human Development* (F. Falkner, ed.), pp. 326–407, Saunders, Philadelphia.

Karlberg, P., Taranger, J., Engström, I., Karlberg, J., Landström, T., Lichtenstein, H., Lindström, B., and Svennberg-Redegren, I., 1976, I. Physical growth from birth to 16 years and longitudinal outcome of the study during the same age period, *Acta Paediatr, Scand. Suppl.* **258**:7.

Klackenberg, G., 1971, A prospective longitudinal study of children. Data on psychic health and development up to 8 years of age, *Acta Paediatr. Scand. Suppl.* **224**:1.

Klackenberg-Larsson, I., and Stensson, J., 1968, The development of children in a Swedish urban community. A prospective longitudinal study. IV. Data on the mental development during the first five years, *Acta Paediatr. Scand. Suppl.* **187**:67.

Klimt, F., 1966, Telemotorische Herzschlagfrequenzregistrierungen bei Kleinkindern während einer körperlichen Tätigkeit, *Dtsch. Gesundheits.* **21**:599.

MacCarthy, D., 1974, Effects of emotional disturbance and deprivation (maternal rejection) on somatic growth, in: *Scientific Foundations of Paediatrics* (J. A. Davis and J. Dobbing, eds.), pp. 56–67, Heinemann, London.

Martorell, R., Habicht, J.-P., Yarbrough, C., Lechtig, A., Klein, R. E., and Western, K. A.,

1975, Acute morbidity and physical growth in rural Guatemalan children, *Am. J. Dis. Child.* **129:**1296.

Mellander, O., Vahlquist, B., and Mellbin, T., 1959, Breast feeding and artificial feeding. A clinical, serological, and biochemical study in 402 infants, with a survey of the literature. The Norrbotten study, *Acta Paediatr. Scand. Suppl.* **116:**57.

Mellbin, T., 1962, The children of Swedish nomad Lapps. A study of their health, growth and development, *Acta Paediatr. Scand. Suppl.* **131:**1.

Mellbin, T., and Vuille, J.-C., 1973, Physical development at 7 years of age in relation to velocity of weight gain in infancy with special reference to the incidence of overweight, *Br. J. Prev. Soc. Med.* **27:**225.

Miller, F. J. W., Court, S. D. M., Walton, W. S., and Knox, E. G., 1960, *Growing up in Newcastle upon Tyne*, The Nuffield Foundation–Oxford University Press, London.

Morley, D., 1976, The design and use of weight charts in surveillance of the individual, in: *Nutrition in Preventive Medicine* (G. H. Beaton and J. M. Bengoa, eds.), pp. 520–529, WHO Monogr. Ser., No. 62.

National Board of Health and Welfare Report (MEK-B), 1973, Diet and Physical Activity in Swedish Children, pp. 32–45, National Board of Health and Welfare Committee on Health Education, Stockholm.

Owen, G. M., Filer, L. J., Maresh, M., and Fomon, S. J., 1966, Part II: Sex-related differences in body composition in infancy, in: *Human Development* (F. Falkner, ed.), pp. 246–257, Saunders, Philadelphia.

Rose, H. E., and Mayer, J., 1968, Activity, calorie intake, fat storage and the energy balance of infants, *Pediatrics* **41:**18.

Royer, P., 1974, Growth and development of bony tissues, in: *Scientific Foundations of Paediatrics* (J. A. Davis and J. Dobbing, eds.), pp. 376–399, Heinemann, London.

Sjögren, I., and Vahlquist, B., 1965, Insjuknandeålder för de vanligaste smittosamma "barnsjukdomarna," *Läkartidn.* **62:**2757.

Spady, D. W., Payne, P. R., Picou, D., and Waterlow, J. C., 1976, Energy balance during recovery from malnutrition, *Am. J. Clin. Nutr.* **29:**1973.

Tanner, J. M., 1976, Population differences in body size, shape and growth rate: A 1976 view. Annotation, *Arch. Dis. Child.* **51:**1.

Tanner, J. M., Healy, M. J. R., Lockhart, R. D., Mackenzie, J. D., and Whitehouse, R. H., 1956, Aberdeen growth study. I. The prediction of adult body measurements from measurements taken each year from birth to 5 years, *Arch. Dis. Child.* **31:**372.

Taranger, J., 1976, Based on Karlberg *et al.*, (1976), Personal communication.

Taranger, J., Bruning, B., Claesson, I., Karlberg, P., Landström, T., and Lindström, B., 1976a, IV. Skeletal development from birth to 7 years, *Acta Paediatr. Scand. Suppl.* **258:**98.

Taranger, J., Lichtenstein, H., and Svennberg-Redegren, I., 1976b, III. Dental development from birth to 16 years, *Acta Paediatr. Scand. Suppl.* **258:**83.

Vahlquist, B., 1975, The drama of infant growth, in: *Action for Children—Towards an Optimum Child Care Package in Africa* (O. Nordberg, P. Phillips, and G. Sterky, eds.), pp. 71–78, The Dag Hammarskjöld Foundation, Uppsala.

Venkatachalam, P. S., 1975, Nutritional nuances of the term *pre-school child* and the need for precise definition, *PAG Bull.* **V(3):**27.

Widdowson, E. M., 1974, Changes in body proportions and composition during growth, in: *Scientific Foundations of Paediatrics* (J. A. Davis and J. Dobbing, eds.), pp. 153–163, Heinemann, London.

The Young Child: Failure to Thrive

Roy E. Brown

1. Introduction

The term "failure to thrive" (FTT) refers to the child who is "not up to expectations in general health," and more specifically for the practicing pediatrician it refers to the child who fails to gain as expected in weight and height (Luzzatti, 1964). FTT is a syndrome that represents a combination of growth and developmental failure, often associated with social and emotional disturbances (Barbero and Shaheen, 1967).

On occasion, FTT reflects the presence of an underlying disease, the symptoms of which may be either obvious or hidden. Each individual infant and child is affected by the interplay of a variety of influences in his environment including nutrition, hormones, emotions, and illnesses that may affect his growth and his developmental progress singularly, or more often, collectively (Fiser *et al.*, 1975). It is the task of the practitioner to determine whether the poor growth and/or development results from an underlying disease, represents a genetic variation, or is the result of environmental deprivation.

In order to discuss growth, there are a number of general terms that must be considered. These include the following

- Chronologic age—time from date of birth in weeks, months, and years.
- Bone (biologic) age—measure of physical maturity determined by X-ray of bones or teeth when reference is made to standard ossification centers (Hoerr *et al.*, 1962; Watson and Lowrey, 1962; Schour, 1960).
- Height age—height or length as compared to the mean or 50th percentile on the reference standard.
- Weight age—weight as compared to the mean or 50th percentile on the reference standard.

Roy E. Brown • Department of Community Medicine, Mount Sinai School of Medicine, New York, New York.

To appreciate fully the dynamics of growth changes, the practitioner should routinely employ a standard growth chart for height and weight, such as the one from Harvard University or the one from the University of Iowa. Table 1 illustrates standard height and weight from birth to 3 years of age. By the time he is 3, the average boy is approximately one-half of his adult height, after which time his growth rate starts to level off, proceeding at a relatively constant rate of 2.5 in. (6.4 cm) each year until puberty.

The reference curves can be used either to describe weight or height in terms of percentile for age whereby the "optimal or ideal" growth pattern falls somewhere between the 10th and 90th percentiles, or in terms of standard deviations from the mean, the "acceptable" growth being measurements within 2 SD from the arithmetic mean or average for that age. The reference standards in general use have been developed by prospectively following cohorts of Caucasian children and may not necessarily relate to other population groups such as blacks, Orientals, American Indians, or others. The growth patterns of children from any of these other ethnic groupings may deviate from the reference norms for Caucasian children and still be well within the normal range for that particular group (Luzzatti, 1964).

Every textbook of pediatrics has comprehensive tables that describe the patterns of behavior during the first 5 years of life (Nelson *et al.*, 1969). Observation and testing of infants and young children can be simply managed by the trained practitioner, indicating the child's performance in the social, motor, adaptive, and language areas of development to detect any major deviation from the accepted norms. FTT may first be evident by a lagging in development.

2. Causes of Growth Failure

There are many reasons for an infant's or child's failure to grow (Smith, 1967). Before a definitive diagnosis of FTT can be made, the practitioner must rule out a number of possibilities (Lacey and Parkin, 1974) that may mimic FTT, but whose origins are based on the presence of genetic defects or other pathology.

Table I. U.S. Mean Growth Milestones for Average U.S. Male Infant Born at Full Term[a]

Age	Weight		Height	
	Pounds	Kilograms	Inches	Centimeters
Birth	7.5	3.4	20	50.8
4–6 months	14	6.4	25	63.5
12 months	22	10	30	76
24 months	28	12.5	35	89
36 months	32	14.6	38	96.5

[a] Adapted from Stuart charts, Nelson *et al.* (1969).

2.1. Congenital and Genetic Anomalies

Congenital malformations, usually found in association with cardiac or renal anomalies, as well as constitutional dwarfism, when the child is small due to hereditary factors, are causes for failure to grow.

2.2. Enzymatic Defects

Defects of intermediary metabolism, such as diabetes, galactosemia, maple sugar urine disease, glycogen storage diseases, phenylketonuria, and various types of hepatic dysfunction (Stanbury *et al.*, 1960), can contribute to growth failure.

2.3. Endocrine Deficiencies

Although the prevalence rate of various endocrinopathies is relatively low, endocrine deficiency may be a factor in growth failure. Thyroid insufficiency can significantly impede growth as evidenced by height and bone age retardation. Pituitary deficiency, although difficult to evaluate in prepubertal children, must be ruled out along with other target gland deficiencies.

2.4. Chronic Diseases

Children with chronic illnesses such as tuberculosis, repeated and frequent upper and lower respiratory infections, related to or independent of hypogammaglobulinemia or cystic fibrosis of the pancreas, are usually retarded in growth. Those with malabsorption from a variety of causes will be small, particularly if the appropriate therapy is not employed early in life during the period of most rapid growth rate.

2.5. Miscellaneous Problems

Children with a primary central nervous system deficit, including those with epilepsy or mental retardation, or those with independent intrauterine growth retardation, may demonstrate reduced overall growth (Warkany *et al.*, 1961). A number of studies report a persistence of growth failure in children with low birth weight, especially in those with a low birth weight delivered at full term, the so-called small-for-date infant (Osofsky, 1975; Wingerd and Schoen, 1974; Beck and van den Berg, 1975; Shanklin, 1970).

3. Failure-to-Thrive Syndrome

Children who suffer from nutritional deprivation (Waterlow, 1972; Cheek *et al.*, 1970), or environmental deprivation (Richardson, 1974), or, as generally present, a combination of these two major factors constitute the FTT syndrome. Primary undernutrition may result from faulty feeding techniques and is commonly found in association with a disturbed infant–mother relationship which may contribute to a reduced caloric intake (Leonard *et al.*, 1966).

3.1. Nutritional Factors

Although severe caloric undernutrition is considered to be of low or rare incidence in developed countries many instances of FTT are identified by the alert practitioner with a high index of suspicion for this condition. Most often FTT is closely related to the designated pockets of poverty that exist in developed countries, for example, inner city ghettos, migrant worker camps (Chase *et al.*, 1971), isolated rural areas, and Indian reservations (Foman and Anderson, 1972; Brown, 1969). Malnutrition resulting mainly from proverty has been documented in the United States as a prevalent condition (Zee *et al.*, 1970), and conservative predictions indicate that with the continued rise in the cost of food, undernutrition due to poverty will increase in scope and severity (Mauer, 1975).

Prenatal growth retardation has been reported in infants born to poor families in the United States with birth weight 15% lower than infants of nonpoor, as well as smaller organ weights (Naeye *et al.*, 1969). An increasing prevalence of teenage pregnancies and out-of-wedlock pregnancies is correlated with low-birth-weight infants in the large urban centers of the United States (Lash and Sigal, 1976).

Poverty and the associated nutritional problems may be a major component of FTT. Inability to purchase sufficient food for the growing child, lack of nutrition and health education, as well as lack of emotional nurturing can all impede sound developmental progress.

The parent need not know the actual accepted recommendations for daily caloric intake of approximately 120 kcal/kg for young infants, falling slightly to 90–100 kcal/kg for children between 1 and 3 years of age (Ten-State Nutrition Survey, 1972; Committee on Nutrition, 1969; National Research Council, 1974). Nor is it essential for the average parent to be familiar with daily nutrient needs that should be distributed among proteins, carbohydrates, and fats for each age group (Straub *et al.*, 1969). However, it would be appropriate for the parent to be aware that the infant or young child is growing very rapidly and has certain nutritional needs for growth that must be met by eating a variety of foods including meat, milk, vegetables, fruits, and cereal grains and that micronutrients including vitamins and minerals are essential for health, normal growth, and protection.

Despite the common association between poverty and poor feeding (Owen *et al.*, 1974), even families without financial limitations often provide poor diets for their children, either through ignorance or lack of concern for the children's nutritional requirements. A large proportion of the infant feeding habits are as much culturally determined as they are modified by financial considerations (Fomon and Anderson, 1972; Fomon, 1974).

Just as the care of the young child, including nutritional considerations, is left to a sibling when there are working parents, the financially well-off parents may have other demands on their time and interests, and may leave their child to be looked after by a hired person, who may not necessarily have the best interests of the child at heart, or who may not be aware of his nutritional needs.

3.2. Infections

Infectious diseases complicate the problems of malnutrition. Poorly nourished infants, including the prematurely born and low-birth-weight infants, are particularly susceptible to infectious diseases of all sorts, especially respiratory and gastrointestinal infections (Balderrama-Guzman and Tantengco, 1971; Scrimshaw *et al.*, 1968; Gordon *et al.*, 1968). Not only in the developing countries, but in pockets of poverty in developed countries adequate diets are not the rule. The most vulnerable group at the highest risk of infections and complications from infections is comprised of the undernourished infants and young preschool children. Even such ordinarily mild contagious illnesses such as measles and whooping cough may be complicated secondarily by bronchitis and pneumonia.

Infections are nearly always associated with anorexia (Frood *et al.*, 1971), which in conjunction with the parent's modification of the diet will contribute to suboptimal nourishment. In the case of gastroenteritis, dietary restrictions, often prolonged, are the rule rather than the exception, and the use of folk medicine is found throughout the world, including home remedies in many parts of developed countries (Jelliffe, 1968). Parasitic diseases, particularly in rural areas and in migrants from rural areas where sanitation facilities and water supplies are substandard, further undermine children's nutritional status.

3.3. Environmental Factors

An association exists between growth failure and the social and emotional environment of the infant and young child. The dynamic and synergistic relationship between physical, mental, and emotional development should suggest to the clinician that FTT must be considered when somatic and/or psychological growth is delayed (Hannaway, 1970; Leonard *et al.*, 1966).

The severe emotional deprivation and devastating effects of institutional life on young infants have been well documented (Gesell and Amatruda, 1941; Spitz, 1946; Bakwin, 1949; Bowlby, 1951; Provence and Lipton, 1962). For example, a study by Widdowson (1951) portrays the effects of psychological stress due to harsh and unsympathetic handling that resulted in curtailment of heights and weights of institutionalized school-age orphans.

The practitioner must be aware that psychosocial factors within the family may have a direct bearing on FTT. Excessive infant crying or colic has been closely correlated with parental attitudes and growth failure (Stewart *et al.*, 1954). The terms "maternal deprivation" coined by Patton and Gardner (1962) and "psychologic malnutrition" (Talbot, 1963) both describe major influences

on growth patterns of young infants. "Hospitalism" or "environmental retardation" in infants has been related directly to psychological problems of the
mothers (Coleman and Provence, 1957).

In early infancy, the baby's environment consists primarily of his mother,
so there may be cause to suspect the mother when FTT difficulties arise
(Elmer, 1960). There are many reasons for the mother to be neglectful, and it
is essential to observe, probe, and attempt to discover the cause(s) of the
inadequate nurturing. Inconsistent parental attitudes are strongly correlated
with FTT (Stewart *et al.*, 1954). Particularly disastrous are the effects on the
infant of parental depression (Fabian and Donohue, 1956) and of other forms
of neurosis that result in maternal rejection of the infant (Elmer, 1960).

In the absence of effective family planning, there is a strong likelihood
that the woman who has children at frequent intervals may have begun her
reproductive life at an early age. The more young children in a household, the
greater are the demands on the mother's energy and skills to provide adequate
care for them (Brown and Wray, 1975). Viel (1973) describes the situation of
maternal neglect in the harsh terms of "unconscious infanticide" when a
mother attends less and less to each successive child. This associates a rising
infant mortality due to FTT with increasing birth order.

The child who suffers from environment deprivation from whatever cause
will become unresponsive to external stimuli. He tends to have solitary occupations, may develop rocking behavior, is unlikely to explore his immediate
surroundings or open possible avenues for future growth, and will generally
reject human companionship. Such a child may be stunted in physical, mental,
and emotional development and may pose a difficult problem for long-term
rehabilitation. The child's general apathy is intimately related to an absence
of appetite and poor growth.

The mechanisms by which the emotional factors influence growth are not
completely understood. Speculations include the direct effect on appetite and
food intake, gastrointestinal disturbances as evidenced by changes in motility
and absorption, alteration in intermediary metabolism, and certain indirect
endocrinopathies (Leonard *et al.*, 1966).

4. How Does FTT Present?

The child may reach the practitioner at any age in the preschool period,
from early infancy onward. The presenting complaint on the part of the parents
rarely is that of growth failure (Leonard *et al.*, 1966). Many parents are
unaware of their child's retardation in growth or development, even when his
weight and/or height falls significantly below the average. Many children with
FTT have had no previous contact with the health professions. In most instances, the infants are brought with complaints of feeding difficulty, vomiting,
constipation, diarrhea, abdominal distension, or excessive crying.

Kreiger and Sargent (1967) describe the postural sign of "tonic immobility" in which the infant with FTT characteristically assumes a position of

flexion of the extremities and appears immobile and displaced from environmental stimuli. After the age of 4 months "tonic immobility" is not seen in normal infants; its presence should arouse a suspicion of deprivation in the sensory environment. In addition, there may be unusual alertness and watchfulness, minimal smiling, lack of cuddliness, and diminished vocalization of any sort (Provence and Lipton, 1962; Leonard *et al.*, 1966).

There might be poor body hygiene as evidenced by severe diaper rash, cradle cap, or other skin lesions (Barbero and Shaheen, 1967); the baby may demonstrate either various degrees of apathy, withdrawing behavior, a lack of response, or extreme irritability and delays in neuromotor development. There may be various food intake disorders ranging from anorexia to voracious appetite or even pica (Barbero, 1969). In fact, FTT may resemble either iron deficiency anemia or lead poisoning, with similar symptomatology including fatigue, irritability, vomiting, and other gastrointestinal problems.

If the child had been followed periodically with physical examinations including height and weight measurements, it becomes obvious that there is a tendency for the measurements to persist in the lower ranges, often far below the 3rd percentile. Given more than one point on the growth curve, such faltering or growth failure is apparent; in most cases weight is more affected than height.

In the older infant, delay in language development as well as in motor and coordination progress can be strong FTT indicators. Regression or the loss of already acquired motor skills often occurs. Such clues may come from concerned day care or nursery school teachers. The major point is that the parents may very well be unaware of these problems (Fischhoff *et al.*, 1971).

5. The Investigation

The first big decision that faces the practitioner with a child who is failing to thrive is whether to admit the child to the hospital for a complete workup, or, alternatively, to pursue the investigation of etiology on an ambulatory basis. There is the general feeling that if an infant less than 2–3 months is identified as suffering from growth failure, the child should be admitted to the pediatric service for investigation unless otherwise contraindicated. However, other pediatricians feel that each child with FTT will eventually require hospitalization and for this reason should be admitted as soon as the diagnosis is suspected. Elsewhere, the decision is made on an individual basis as jointly determined by the practitioner and the social service worker who should be involved prior to admission (Elmer, 1960; Barbero and Shaheen, 1967). The main purpose of the entire workup is to rule out any significant organic basis for retardation of growth and development and to specify all the contributory factors.

The most important aspect of the investigation (workup), whether it is to be started prior to hospitalization or after admission, is a thorough historical review. The complete history should have details of the prenatal period and the delivery, documentation of birth weight, and any problems faced immediately following delivery. A careful family history will include the possibilities of inherited diseases such as allergies and birth defects, as well as growth defects or infectious processes within the family, and gives the actual heights and weights of siblings, parents, and grandparents (Tanner *et al.*, 1970). A detailed nutritional history describes the nature and type of feeding, with specific information as to how, when, with whom, and where the child eats; a social history as to parental background in occupation, education, and habits, including possible drug or alcohol abuse, and details on the siblings and the home environment, provides important information for diagnosis. Also important is specific information as to developmental milestones, past records of height and weight, information about previous hospitalizations, operations, or injuries. The completeness of immunization records may serve as a useful index as to parental awareness and utilization of preventive health services (Kanawati and McLaren, 1973). It is of great help to have a report of the home environment based on the observations made on a home visit. Detailed information in the review of each body system is also needed.

The next step in the workup is a complete physical examination that includes anthropometry, neurological, and psychological evaluations, and a careful evaluation of the child's nutritional status. The physical examination may suggest the etiology of the growth failure or may provide clues as to psychological problems.

The next steps in the investigation are laboratory and radiological studies, keeping in mind that it would be preferable for the physician to be selective, rather than to employ a shotgun type of investigation to cover all of many remote diagnostic possibilities (Hannaway, 1970). Basic workup includes a complete blood count, a blood smear for anemia, a sickle cell screening test, and a blood lead level. There should be a urinalysis, a tuberculin skin test, and stools for ova, parasites, fat, and occult blood. Fasting blood sugar, blood urea nitrogen, serum electrolytes, tests for thyroid function, total serum protein with albumin/globulin ratio, and cultures of nose, throat, and urine are routine.

X-rays should include studies for bone age, usually of the wrists, a chest x-ray, and possibly a skull series. If indicated, more specific roentgenographic tests may include examination of the gastrointestinal tract and the kidneys to demonstrate both appearance and function.

More sophisticated tests may include immunoglobulin determinations, sweat test, and urinary as well as blood amino acids; electrocardiograms and electroencephalograms may be indicated. It should be urged, however, that the customary automatic multisystemic approach in children with FTT should be abandoned. Rather than have the house officer or attending physician write extensive and expensive orders, a careful and organized plan should be developed on admission so that the workup can proceed in sequence, starting with more common findings such as an inadequate diet. Each result needs to be evaluated before the next step is taken.

While the child is in the hospital, there is ample opportunity for nurses, social workers, and doctors to evaluate the child's socialization, behavior with parents when they visit, his sleeping and eating behavior, as well as behavior during contact with other children. These notes and observations should be carefully evaluated to summarize the nutritional habits and family–child interactions.

6. Treatment and Management

The management of FTT is both delicate and complicated. It is characteristic of the infant with FTT to have immediate and usually dramatic improvement of many symptoms upon being hospitalized (Barbero and Shaheen, 1967). Weight gain is often striking, and with improvement of general body hygiene the infant's irritability or apathy will change or disappear. The response of the infant to being held may be different once separated from parents, interest in eating will increase, and symptoms such as vomiting, diarrhea, or constipation may clear up.

It is important not to make drastic changes in the feedings and not to administer multiple medications that may obscure the clinical course. The initial management calls for a minimum of interference, simple maintenance and housekeeping procedures, and the assignment of a kindly nurse or aide as guardian or caretaker for provision of emotional support.

Once the various etiologic influences in the parent–child relationship are determined, it is essential to provide supportive guidance to the parents to assist them in understanding the pathogenesis of FTT and to make them aware of needed behavioral changes. It is important not to condemn or criticize the parents who may feel threatened by the hospital and medical personnel. Psychological or psychiatric support therapy for one or both parents may be considered and this should be entered into the infant's management requirements. From the time of admission, the parents must be incorporated into the diagnosis and management of the FTT infant almost as if they were members of the clinical team. The mother needs nurturing herself to promote her capacity and ability to nurture her baby (Leonard *et al.*, 1966).

The need for long-term management beyond the period of hospitalization is a common feature of FTT. The clinical team must maintain a supportive and secondary role if the parent is going to achieve success with the child. Close supervision, without taking over the parental role, is a difficult but essential role for the health team members. Supervision must continue even after the immediate symptoms clear up if relapses and future complications are to be prevented. The major determinant of growth and developmental catch-up may be the improvement of the infant's home environment (Graham, 1972). There is suggestive evidence to support the contention that up to a point children have a seemingly unlimited potential for catch-up growth if they are given adequate nutrition in a healthy and supportive environment (Senior, 1970).

Rarely does the child recovering from undernutrition return to a nurturing home environment. However, if such a "favored" home situation exists, the improvement in growth and development recovery is dramatic (Graham and Adrianzen, 1972). There is no exact age at which rehabilitation for FTT should not be attempted. In a recent unique follow-up study, severely malnourished Korean orphans, when adopted by middle class American families, managed to catch up and surpass the expected mean height and weight of Korean children (Winick *et al.*, 1975). For certain children with FTT, the clinical and social service team must face the difficult decision as to the possible benefits and advantages of foster home placement.

7. Prevention

When one considers the group of children at high risk for FTT, certain steps for prevention are apparent. The profile of children at risk for being abused or neglected closely resembles the FTT risk profile including low-birth-weight infants, children born into large families, children living in poverty, those in broken or single-parent households, unwanted or abandoned children, as well as environmentally deprived children (Barbero *et al.*, 1963; Smith, 1975; Silver *et al.*, 1969; Fenaroff *et al.*, 1972). Poverty and other forms of environmental deprivation should increase the practitioner's index of suspicion of potential FTT, but other evidence of an abnormal mother–child relationship in any socioeconomic class should trigger the question of FTT.

With a growth chart as part of the usual patient record to indicate the height and weight at each routine visit, over a period of time the pattern of growth will be apparent. Evaluation is made simple by relating the individual child's growth to reference children of the same age and sex. There should be routine office or clinic visits in which somatic growth patterns are charted and developmental milestones are evaluated by history and examination. Faltering patterns of growth or development must be considered as early evidence of possible FTT. In the close following of prematurely born, low-birth-weight infants, and particularly small-for-date infants, there is an increased concern for the pattern of growth and possible catching up, somatically and psychologically.

Even in the presence of significant psychosocial and environmental disturbances, the major problem is related to undernutrition due to poor intake. In the prevention of FTT, the health worker must pay attention to the nutritional status and the nutritional needs of the child. A simplified categorization of food types into body-building foods, energy-providing foods, and other foods needed for body protection may well serve even poorly educated parents in their understanding of nutritional needs. A suitable reference book should be recommended to provide basic and simple information.

There is little likelihood of the child developing FTT when under the close supervision of a concerned clinician. More commonly the child will be referred in consultation or be seen for what apparently is an unrelated condition. Evidence of parental neglect, recurrent infections, or an acute illness that may be insufficient to account for the degree of growth or developmental failure should alert the consulting doctor to the possibility of FTT. Similar problems in elder siblings should also be cause for concern about younger children in the family.

If there is any suggestion of environmental deprivation or of psychosocial disruption, with or without poverty, there should be an investigation for FTT and arrangements for social service contact with the family.

Sufficient information supports the close following of children with the suspicion of early FTT so as to avoid the syndrome becoming fully manifest. Families in which there is alcoholism, financial deprivation, parental incompatibility, promiscuity, serious familial illness, loss of one parent, job instability, and other conditions of strain, particularly the multiproblem family, raise the suspicion that something may be present to cause environmental failure and interference with thriving (Barbero and Shaheen, 1967). The information necessary for confirmation of these problems requires an in-depth and probing history.

Case finding in the preliminary stages of FTT is only possible by coordinated action of the clinician and available health and social service resources within the community, such as day care and Head Start centers, and the school health system (Karp *et al.*, 1976; Adebonojo and Strahs, 1973). With a high index of suspicion, early identification of high risk groups may prevent the full-blown clinical picture (Kempe, 1973). The practitioner will enjoy more success in prevention only by close coordination with other social welfare activity in the community.

8. Summary

FTT is a syndrome that occurs more commonly than is currently being diagnosed. The classical FTT syndrome is found in infants and preschool children with growth and/or developmental retardation unexplained by organic disease. Similar to the syndrome of child abuse or neglect, FTT is not limited to poverty families, but can be identified in any socioeconomic group where the child's growth and development are impaired by environmental deprivation. The missing ingredient often is a nurturing home environment. If the child with FTT is hospitalized for a complete workup, immediate and dramatic improvement may soon be evident. The practitioner should assist the parents in seeking help for their closely related problems.

With knowledge as to what constitutes the child at high risk for FTT, the clinician should enlist the assistance of community resources, particularly social agencies. Parents may require basic instruction as to what nutrients are required since the most evident cause of FTT is nutritional deprivation. However, the emotional climate of the home must be carefully investigated so that support and aid can be provided along all possible fronts. Perhaps the most direct approach would be for the practitioner to keep in mind those factors that contribute to placing a child in the high risk category and to take preventive action to avoid the full syndrome. The signs may be quite subtle and a high index of suspicion will assist in prevention, diagnosis, and management.

9. References

Adebónojo, F. O., and Strahs, S., 1973, The state of nutrition of urban black children in the U.S.A.: The role of day care services in the prevention of nutritional anemia, *Clin. Pediatr.* **12**:563.

Bakwin, H., 1949, Emotional deprivation in infants, *J. Pediatr.* **35**:512.

Balderrama-Guzman, V., and Tantengco, V. O., 1971, Effect of nutrition and illness on the growth and development of Filipino children (0–4 years) in a rural setting. *J. Phil. Med. Assoc.* **47**:353.

Barbero, G. J., 1969, Gastrointestinal disturbances, in: *Textbook of Pediatrics* (W. E. Nelson, V. C. Vaughan, and R. J. McKay, eds.), p. 774, Saunders, Philadelphia.

Barbero, G. J., and Shaheen, E., 1967, Environmental failure to thrive: A clinical view, *J. Pediatr.* **71**:639.

Barbero, G. J., Morris, M. C., and Redford, M. T., 1963, Malidentification of mother, baby, father relationships expressed in infant failure to thrive, in: *The Neglected-Battered Child Syndrome*, Child Welfare League of America, New York.

Beck, G. J., and van den Berg, B. J., 1975, The relationship of the rate of intrauterine growth of low-birth-weight infants to later growth, *J. Pediatr.* **86**:504.

Bowlby, J., 1951, WHO Monograph Series, p. 67, No. 2, Geneva.

Brown, R. E., 1969, Poverty and health in the United States: Some significant considerations, *Clin. Pediatr.* **8**:495.

Brown, R. E., and Wray, J. D., 1975, The starving roots of population growth, in: *Ants, Indians, and Little Dinosaurs* (A. Ternes, ed.), Scribner, New York.

Chase, H. P., Kumar, V., Dodds, J. M., Sauberlich, E. H., Hunter, R. M., Burton, R. S., and Spalding, V., 1971, Nutritional status of preschool Mexican-American migrant farm children, *Am. J. Dis. Child.* **122**:316.

Cheek, D. B., Graystone, J. E., and Read, M. S., 1970, Cellular growth, nutrition and development, *Pediatrics* **45**:315.

Coleman, R. W., and Provence, S., 1957, Environmental retardation (hospitalism) in infants living in families, *Pediatrics* **19**:285.

Committee on Nutrition, 1969, Iron balance and requirements in infancy, *Pediatrics* **43**:134.

Elmer, E., 1960, Failure to thrive: Role of the mother, *Pediatrics* **25**:717.

Fabian, A. A., and Donohue, J. F., 1956, Maternal depression: A challenging child guidance problem, *Am. J. Orthopsychiatry* **26**:400.

Fanaroff, A. A., Kennell, J. H., and Klaus, M. H., 1972, Follow-up of low birth weight infants— The predictive value of maternal visiting patterns, *Am. J. Dis. Child.* **49**:287.

Fischhoff, J., Whitten, C. F., and Pettit, M. G., 1971, A psychiatric study of mothers of infants with growth failure secondary to maternal deprivation, *J. Pediatr.* **79**:209.

Fiser, R. H., Meredith, P. D., and Elders, M. J., 1975, The child who fails to grow. *Am. Fam. Physician* **11**:108.

Fomon, S. J., 1974, *Infant Nutrition*, 2nd ed., Saunders, Philadelphia.

Fomon, S. J., and Anderson, T. A. (eds.), 1972, *Practices of Low-Income Families in Feeding Infants and Small Children: With Particular Attention to Cultural Subgroups*, DHEW Publication (HSM) 72-5605.

Frood, J. D. L., Whitehead, R. G., and Coward, W. A., 1971, Relationship between pattern of infection and development of hypo-albuminemia and hypo-beta-lipoproteinemia in rural Ugandan children, *Lancet* **2**:1047.

Gesell, A., and Amatruda, C. S., 1941, *Developmental Diagnosis*, Harper (Hoeber), New York.

Gordon, J. E., Ascoli, W., Mata, L. J., and Guzman, M. A., 1968, Nutrition and infection field study in Guatemalan villages, 1959–1964: VI. Acute diarrheal disease and nutritional disorders in general disease incidence, *Arch. Environ. Health* 16:424.

Graham, G. G., 1972, Environmental factors affecting the growth of children, *Am. J. Clin. Nutr.* 25:1184.

Graham, G. G., and Adrianzen, T. B., 1972, Late "catch-up" growth after severe infantile malnutrition, *Johns Hopkins Med. J.* 131:204.

Hannaway, P. J., 1970, Failure to thrive: A study of 100 infants and children, *Clin. Pediatr.* 9:96.

Hoerr, N. L., Pyle, S. I., and Francis, C. C., 1962, *Radiographic Atlas of Skeletal Development of the Foot and Ankle,* Charles C Thomas, Springfield, Ill.

Jelliffe, D. B., 1968, *Infant Nutrition in the Sub-Tropics and Tropics,* 2nd ed. WHO Monograph Series, No. 29, Geneva.

Kanawati, A. A., and McLaren, D. S., 1973, Failure to thrive in Lebanon, *Acta Paediatr. Scand.* 62:571.

Karp, R. J., Nuchpakdee, M., Fairorth, J., and Gorman, J. M., 1976, The school health service as a means of entry into the inner-city family for the identification of malnourished children, *Am. J. Clin. Nutr.* 29:216.

Kempe, C. H ., 1973, A practical approach to the protection of the abused child and rehabilitation of the abusing parent, *Pediatrics* 51:804.

Krieger, I., and Sargent, D. A., 1967, A postural sign in the sensory deprivation syndrome in infants, *J. Pediatr.* 70:333.

Lacey, K. A., and Parkin, J. M., 1974, Causes of short stature: A community study of children in Newcastle upon Tyne, *Lancet* 1:42.

Lash, T. W., and Sigal, H., 1976, *State of the Child: New York City,* Foundation for Child Development, New York.

Leonard, M. F., Rhymes, J. P., and Solnit, A. J., 1966, Failure to thrive in infants, *Am. J. Dis. Child.* 111:600.

Luzzatti, L., 1964, Failure to thrive: A diagnostic approach, *Postgrad. Med.* 35:270.

Mauer, A. M., 1975, Malnutrition—Still a common problem for children in the United States, *Clin. Pediatr.* 14:23.

Naeye, R. L., Diener, M. M., Dellinger, W. S., and Blanc, W. A., 1969, Urban poverty: Effects on prenatal nutrition, *Science* 166:1026.

National Research Council, 1974, *Recommended Dietary Allowance Publication 2216,* 8th ed., Food and Nutrition Board of the National Research Council, National Academy of Science, Washington, D. C.

Nelson, W. E., Vaughan, V. C., and McKay, R. J., 1969, *Textbook of Pediatrics,* Saunders, Philadelphia.

Osofsky, H. J., 1975, Relationships between prenatal medical and nutritional measures, pregnancy outcome, and early infant development in an urban poverty setting: I. The role of nutritional intake, *Am. J. Obstet. Gynecol.* 123:682.

Owen, G. M., Kram, K. M., Garry, P. J., Lowe, J. E., and Lubin, A. H., 1974, A study of nutritional status of preschool children in the United States, 1968–1970, *Pediatrics* 53:597.

Patton, R. G., and Gardner, L. I., 1962, Influence of family environment on growth: The syndrome of maternal deprivation, *Pediatrics* 30:957.

Provence, S., and Lipton, R. C., 1962, *Infants in Institutions,* International Universities Press, New York.

Richardson, S. A., 1974, The background histories of schoolchildren severely malnourished in infancy, *Adv. Pediatr.* 21:167.

Schour, I., 1960, *Noyes' Oral Histology and Embryology,* 8th ed., Lea & Febiger, Philadelphia.

Scrimshaw, N. S., Guzman, M. A., Flores, M., and Gordon, J. E., 1968, Nutrition and infection field study in Guatemalan villages, 1959–1964: V. Disease incidence among preschool children under natural village conditions with improved diet and with medical and public health services, *Arch. Environ. Health* 16:223.

Senior, B., 1970, Practical approaches to growth retardation, *Hosp. Prac.* 5:67.

Shanklin, D., 1970, The influence of placental lesions on the newborn infant, *Pediatr. Clin. North Am.* **17**:25.

Silver, L. B., Bublin, D. D., and Lourie, R. S., 1969, Child abuse syndrome: The gray areas in establishing a diagnosis, *Pediatrics* **44**:594.

Smith, D. W., 1967, Compendium on shortness of stature, *J. Pediatr.* **70**:463.

Smith, S. M., 1975, *The Battered Child Syndrome*, Butterworths, London.

Spitz, R., 1946, Hospitalism: An inquiry into the genesis of psychiatric conditions in early childhood, *Psychoanal. Stud. Child.* **2**:113.

Stanbury, J. B., Wyngaarden, J. B., and Frederickson, D. S. (eds.), 1960, *The Metabolic Basis of Inherited Disease*, McGraw-Hill, New York.

Stewart, A. H., Weiland, I. H., Leider, A. R., Mangham, C. A., Holmes, T. H., and Ripley, H. S., 1954, Excessive infant crying (colic) in relation to parent behavior, *Am. J. Psychiatry* **110**:687.

Straub, C. P., Sister Jeanette Marie, and Sister Miriam Teresa, 1969, Nutritional intake of infants. I: Calories, carbohydrate, fat and protein, *J. Am. Diet. Assoc.* **54**:53.

Talbot, N. B., 1963, Has psychologic malnutrition taken the place of rickets and scurvey in contemporary pediatric practice? *Pediatrics* **31**:909.

Tanner, J. M., Goldstein, H., and Whitehouse, R. H., 1970, Standards for children's height at ages 2–9 years allowing for height of parents, *Arch. Dis. Child.* **45**:755.

Ten-State Nutrition Survey 1968–1970, Volumes I–V, 1972, U. S. Department of Health, Education and Welfare, HSMHA, Center for Disease Control, Atlanta, Georgia, DHEW Publication Nos. (HSM) 72-8130-34.

Viel, B., 1973, *La Explosión Demográfica*, Editorial Pax-Mexico, Liberia Carlos Cesarman, S.A. Mexico.

Warkany, J., Monroe, B. B., and Sutherland, B. S., 1961, Intrauterine growth retardation, *Am. J. Dis. Child.* **102**:127.

Waterlow, J. C., 1972, Classification and definition of protein–calorie malnutrition, *Br. Med. J.* **3**:566.

Watson, E. H., and Lowrey, G. H., 1962, *Growth and Development of Children*, 4th ed., Yearbook, Chicago.

Widdowson, E. M., 1951, Mental contentment and physical growth, *Lancet* **1**:1316.

Wingerd, J., and Schoen, E. J., 1974, Factors influencing length at birth and height at five years, *Pediatrics* **53**:737.

Winick, M., Meyer, K. K., and Harris, R. C., 1975, Malnutrition and environmental enrichment by early adoption, *Science* **190**:1173.

Zee, P., Walters, T., and Mitchell, C., 1970, Nutrition and poverty in preschool children, *J. Am. Med. Assoc.* **213**:739.

The Young Child: Protein–Energy Malnutrition

J. Michael Gurney

1. Definitions

Protein–energy malnutrition (PEM) has recently been defined (WHO and FAO, 1973) as "a range of pathological conditions arising from coincident lack, in varying proportions, of protein and Calories, occurring most frequently in infants and young children, and commonly associated with infections." The term has the great advantage of bringing forward the main cause while avoiding an emphasis on clinical signs (Jelliffe, 1959, 1969a). It is recognized that the etiology of the entire spectrum of infant and child malnutrition is not fully reflected in the term PEM and that not only protein and energy deficiencies but also other nutrient deficiencies and infections are involved (Joint WHO/ FAO Expert Committee on Nutrition, 1971). The time is apposite for changing from the antecedent term, protein–calorie malnutrition (PCM), as the joule is replacing the calorie as the measure of energy in nutrition; energy is the property whereas calories or joules are merely units to measure it.

PEM differs from most diseases in that it is neither a response to an invading organism nor an endogenous breakdown of an internal mechanism or structure. It is a series of responses to an environmental deprivation. The first is one of adaptation; only later, if the deprivation persists or becomes more severe, does the physiological and behavioral adaptation become a maladaptation and eventually a breakdown associated with overt clinical signs and symptoms.

Despite the Utopian statement in the WHO Constitution that "health is a state of complete physical, mental and social well-being and not merely the absence of disease or infirmity," it is in fact a compromise with the environment, with nature, and with nurture—perfection in health cannot be achieved

J. Michael Gurney • Caribbean Food and Nutrition Institute, Kingston, Jamaica, West Indies.

and eventual death is a fact of life. Nowhere is this compromise more apparent than when we are considering responses to undernutrition. "The ability to survive under varying conditions depends on adaptation; beyond the range of adaptation, however, survival and avoidance of disease can be achieved only by control of environmental factors" (Wadsworth, 1975). Where does an adaptation to dietary limitations become a maladaptation? This problem is central to much of the discussion that follows. The concept of "normal" and of physiological, behavioral, and societal adaptation and the cost of such adaptation are among the subjects discussed by the PAHO Advisory Committee on Medical Research (1971).

1.1. Mild and Moderate PEM

1.1.1. Activity

Dietary energy deficiencies diminish energy output as activity. This has been widely noted in individuals in day care centers, schools, concentration camps, their homes, and elsewhere; a low activity level is hard to define and measure, and even harder to prove as a direct result of dietary restrictions. Thus, activity levels, despite their importance, are not used as objective clinical indicators of mild-moderate malnutrition. However, the law of conservation of energy supports the clinical finding that children whose diets are inadequate in energy are less active than those who are well fed.

1.1.2. Growth

A second adaptation to reduced dietary energy and protein intakes is, in children, a reduction of rate of increase in size. When the deprivation is slight the growth failure may be minimal and difficult or impossible to measure. If it is small and balanced, the infant or child may be somewhat stunted but appear to the average observer well proportioned and normal. However, as the deficit becomes more severe and more unbalanced so does the growth failure. Wasting of skeletal muscle, the body's main pool of reserve protein (Waterlow, 1963), and of fat, the main energy reserve (Jelliffe, 1966), occurs; body weight actually decreases as these tissues are used up in the maintenance of homeostasis and obviously abnormal physical and biochemical signs begin to appear. Finally the full-blown syndromes of severe PEM—marasmus, kwashiorkor, and intermediate severe stages—make themselves apparent.

The mechanism and control of human physical growth is a complex and still not fully understood process. Simple measurements of growth suitable for clinical or public health practice do exist (Jelliffe, 1966; Gurney *et al.*, 1972). Weight is the most generally useful, and the mid-upper arm circumference (Jelliffe and Jelliffe, 1969), triceps fatfold, and height are also much used. These and other measurements are discussed in Chapters 17–19 in this volume. Difficulties over interpretation of measurements of size, shape, and growth remain unresolved and involve many complexities (LeGros Clark and Medawar, 1945).

1.1.3. The Internal Biochemical Environment

The constancy of the internal environment is maintained in mild malnutrition but where the adaptive process comes under more severe strain internal biochemical changes appear. When the deficiency is mainly of total energy (including energy derived from protein) and is therefore balanced, biochemical changes rarely become apparent. Whitehead (1971) and Whitehead and Alleyne (1971) have suggested the biochemical mechanisms that are involved and have summarized these as shown in Fig. 1. If the deficiency is primarily of protein, the internal environment may become distorted quite early. For example, the serum albumin level has been reported to start to fall before clinical signs, including growth failure, become apparent (Whitehead *et al.*, 1971).

A moderate hypochromic anemia is commonly found in association with mild-moderate PEM (DeMaeyer, 1976). The hemoglobin level is often around 10 g/100 ml of blood. Protein deficiency, usually associated with a reduced availability of dietary iron, is thought to contribute toward this PEM-associated anemia. However, views on the epidemiology of nutritional anemias are changing with the rapid increase in knowledge of the subject (WHO, 1975).

The ratio of nonessential to essential amino acids in the serum may rise in early protein deficiency before physical signs other than growth retardation have occurred (Whitehead and Dean, 1964). This sign is not constant and can be modified in the presence of an associated energy deficiency (Whitehead and Alleyne, 1971). It is of little value in the initial assessment of individual children but can help to identify on a community scale the relative importance of total energy lack and of protein deficiencies in diets (Gurney *et al.*, 1973).

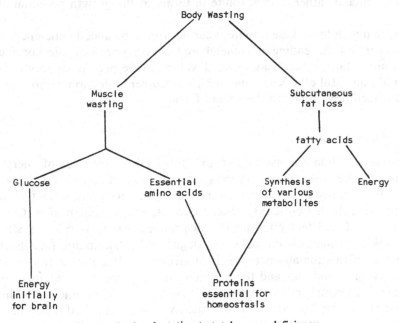

Fig. 1. Basic adaptation to total energy deficiency.

The search for biochemical indices of early PEM has continued for some decades without real success; anthropometric measurements still provide the best indications. However, biochemical tests are useful in monitoring progress in established and severe cases. We can look forward to the development of simple field biochemical tests of some value in detecting children on the kwashiorkor route to malnutrition. The most hopeful are perhaps measurement of blood levels of transferrin, a plasma protein (McFarlane *et al.*, 1969), or of prealbumin (Ingenbleek *et al.*, 1972). However, it seems unlikely that any biochemical measurement will prove more sensitive than anthropometry for measuring early PEM.

1.2. Severe Forms of PEM

Severe protein–energy malnutrition is made up of a spectrum of conditions. The two extremes—starvation marasmus and the sugar-baby variant of kwashiorkor—comprise separate clinical syndromes (Waterlow, 1948). Between these two fall the intermediate categories including marasmic kwashiorkor.

Severe PEM of any form is always a sequel to periods of mild then moderate undernourishment. On the marasmic route to malnutrition there is a deficiency in total dietary energy—whether derived from carbohydrates, fats, or protein. The kwashiorkor route implies a diet deficient in protein with a relative sufficiency of total energy. Growth utilizes both protein and energy; therefore, when dietary protein is laking and growth is not taking place less energy is needed; hence, the energy sufficiency in kwashiorkor may be relative to the protein intake rather than absolute in terms of the growth potential of the child.

In fact, many children take a route, veering from a balanced deficiency to primarily a protein lack, ending up somewhere between pure kwashiorkor and pure marasmus. Many others may never develop severe protein–energy malnutrition at all, and still others may die of an intercurrent infection without the malnutrition itself developing to the severe form.

1.2.1. Nutritional Marasmus

The marasmic child has used up all or almost all his reserves of energy and protein. His skeletal muscles are thus severely wasted and his fat depots are empty. He is an inactive, wizened, shrunken, and very underweight child with thin limbs, shriveled buttocks, sunken cheeks, and prominent ribs. Consequently his head and feet often appear disproportionately large. The skin hangs in wrinkles around his buttocks and thighs and a thumb and forefinger pinch of skin confirms the absence of subcutaneous fat. His abdomen may or may not be swollen and taut and the hair may or may not have retained its normal thickness, texture, and color. The child is most likely ravenous, but he may well need coaxing to eat. He is commonly irritable or fretful but may appear apathetic. If edema is present (unless it is obviously not of nutritional origin), the child does not have nutritional marasmus.

The marasmic child may have diarrhea and may present at hospital or clinic with gastroenteritis. If a doctor is more interested in acute and quickly treatable (or quick-ending) disease, this diagnostic preference may be reflected in the hospital or clinic statistical returns as numerous admissions, and perhaps deaths, from gastroenteritis and few if any from protein–energy malnutrition (Aykroyd, 1968). Nutritional marasmus occurs at all ages but the highest prevalence is between 6 and 18 months or even earlier.

1.2.2. Kwashiorkor

The link between malnutrition and the syndrome now called kwashiorkor was long unrecognized. It was Dr. Cicely Williams who in 1933 showed its relationship to diet (Williams 1933) and introduced to the wider world its name in the Ga language (from Ghana) which means "the disease of the deposed baby" (Williams, 1963; Williams and Jelliffe, 1972).

The child with kwashiorkor always has edema; this is most often detected on the dorsum of the feet and around the ankles, but it can also be found on other dependent parts including the forearms and trunk, and in severe cases it can be seen in the cheeks and around the eyes. Edema is an essential sign in diagnosis of the condition. (However, in the author's experience, slight edema over the shins, i.e., pretibeal edema, is sometimes found in apparently healthy babies and thus should not be considered in the diagnosis of kwashior-kor.) Death is a frequent outcome of kwashiorkor and may occur within a few weeks of the onset of these clinical signs. In children that die from kwashiorkor "there are almost invariably signs of terminal infection—bronchopneumonia and enteritis—but these must be regarded as contributory causes only. It seems probable that the cause of death is biochemical failure at the cellular level." This statement of Waterlow *et al.* (1960) remains appropriate.

The child with kwashiorkor is stunted, and his skeletal muscle is wasted. However, he still retains subcutaneous fat. He may even appear plump and may, to an inexperienced observer, appear healthy—this is not so for a child with severe marasmus, whose wasted appearance and distressed manner makes his condition obvious. The extensive edema often seen in a child with severe kwashiorkor likewise makes obvious the seriousness of his condition.

The pathogenesis of the edema is not entirely clear. It used to be thought to be due to the low level of serum proteins that is a feature of the disease, but this simple hypothesis is now considered to be insufficient (Srikantia, 1969; Whitehead and Alleyne, 1971). The lack of dietary protein causes certain essential amino acids to be unavailable for maintenance of homeostasis. This may cause the fatty enlarged liver that is often found in the disease.

Derangement of the hormonal and enzymatic mechanisms of the body often occurs; plasma concentrations of cortisol, antidiuretic hormone, insulin, and growth hormone are all affected in kwashiorkor. How much of these changes is caused by disturbed homeostasis and how much is an adaptation to preserve the body's internal environment is not yet clearly understood.

Other changes are commonly, but not invariably found in kwashiorkor. Both hair and skin may become depigmented. The skin, particularly over the legs, may dry and darken and may crack to give the "flaky paint" effect; these dry flakes may peel, leaving hypopigmented skin beneath. The area around the perineum may become excoriated. Diarrhea is commonly found.

The child with kwashiorkor is often miserable and apathetic, responding neither to affection nor simple stimuli whether pleasurable or painful. The first smile is a good prognostic sign. He is usually anorexic and much harder to feed than the typical marasmic child who is ravenous. The signs of kwashiorkor can occur at all ages but by far the highest prevalence is in the second, third, and fourth year of life. Pereira and Begum (1974) have reviewed the manifestations and management of kwashiorkor.

1.2.3. Intermediate Severe Syndromes

Marasmus and kwashiorkor are extremes at either end of a sliding scale. Many infants and young children with severe PEM show a mixture of the features of both conditions. Should edema be present with concomitant wasting of subcutaneous fat and a lower weight than is found in kwashiorkor, the designation marasmus with edema (WHO and FAO, 1973) or marasmic kwashiorkor is often applied.

Some children are seen who are extremely short in stature for their age but well proportioned (they are stunted but not wasted) and apparently none the worse otherwise. Such children, if the cause is an apparently successful adaptation to a balanced dietary deficiency, are referred to as nutritional dwarfs (Jelliffe, 1959; Downs, 1964).

1.3. Interrelationships of PEM with Infections

A preschool-age child in the developing world may suffer from an infectious illness for as much as a fifth or a quarter of his life (Mata *et al.*, 1967; Scrimshaw *et al.*, 1968). What is the effect of PEM on susceptibility to acquire infections, and resistance to infections already acquired? And how do infections affect PEM?

The FAO/WHO Expert Committee on Nutrition, meeting in 1970, concluded unequivocally that "there is extensive evidence that even a mild degree of PCM (PEM) in the preschool child increases susceptibility to diarrheal and respiratory and to other serious infections of childhood" (Joint FAO/WHO Expert Committee on Nutrition, 1971). However, a more recent review concludes that "it may be stated that a poor diet does not necessarily reduce the efficiency of the body's protective mechanisms against invasion by microorganisms and so lead to increased susceptibility to infections" (Davidson *et al.*, 1975). Unless the word susceptibility is used with different meanings, these two statements conflict.

There is clearer understanding over the role of nutrition in enabling the body to overcome an infection once established (i.e., resistance to infection already acquired). "There can be no doubt that the previous state of nutrition may be of great importance in determining the results of an infection" (Davidson *et al.*, 1975). Even here the position is not very simple. A 1968 monograph on nutrition and infection states that moderate to severe nutritional deficiencies interact synergically with infections making both conditions more severe. This synergism functions when the infecting organisms do not enter the cells of the host and are not dependent on the host's metabolic processes (i.e., does not occur with some viruses and protozoa) (Scrimshaw *et al.*, 1968). The 1970 Expert Committee supports this conclusion with regard to *severe* malnutrition but emphasizes that our knowledge of the mechanism of such synergism is incomplete and that "the relative importance of these mechanisms in mild or moderate PCM [PEM] needs clarification" (Joint FAO/WHO Expert Committee on Nutrition, 1971). Recent findings have been discussed by Behar (1975).

Resistance to infection is dependent on appropriate antibody formation, leukocyte activity, epithelial integrity, suitable intestinal flora, and endocrine function. There is accumulating evidence that all these mechanisms may become impaired in severe PEM, particularly kwashiorkor, and that some may be affected even in the milder forms of PEM.

Two main types of lymphocytes, B cells and T cells, appear to be actively involved in resistance to infections. B cells are lymphocytes derived from bone marrow. They can synthesize and secrete antibodies, including the immunoglobulins, and are thus associated with humeral immunity. A single such lymphocyte can secrete into its surroundings about 3000 antibody molecules per second (Ramalingaswami, 1975). Clearly the lymphatic system has considerable nutritional needs to maintain this output. T cells arise in the bone marrow but require the influence of the thymus gland to mature; they do not produce antibodies directly but do participate in their production and in the production of macrophages to resist infection, and are associated with cell-mediated immunity. There is evidence that protein and energy deficiencies inhibit both types of immunity, thus rendering the child with PEM more susceptible to serious infection (Awdeh *et al.*, 1972; Ramalingaswami, 1975).

Severely malnourished children, and even children who are not severely malnourished but who come from an environment in which malnutrition flourishes, often have a markedly thin intestinal mucosa, similar to that found in intestinal malabsorption. Intestinal flora that are normally found in the lower area are often found in large numbers in the upper gastrointestinal tract in these children (Behar, 1975). Changes in the structure and function of endocrine glands are commonly found in severe PEM (Whitehead and Alleyne, 1971). All these alterations are likely to lower the resistance to infection of such children.

Repeated infections tend, in a nutritionally precarious situation, to worsen the nutritional status of a young child (Puffer and Serrano, 1973). The mechanisms by which this happens are many, and include reduced food intake caused both by loss of appetite and by withholding of nutritious foods; increased nutrient requirements caused by the diseases themselves—pyrexia, for example, increases requirements for energy; reduced absorption of nutrients which often occurs when diarrhea is present; and increased metabolic loss of nitrogen in the urine which occurs in many infections.

The working capacity of individuals can be reduced by infectious disease, resulting in poverty and reduced food intake of the family. For example, malaria and guineaworm strike hardest during the rainy season which coincides with the crucial and short planting period. At this time the farmer needs to work at maximal intensity, expending 14.2 MJ (3400 kcal) or more per day (Weiner, 1964). His work output falls because of his illness and he may be unable to plant enough crops to sustain his family over the succeeding year. PEM may consequently manifest itself in his family.

Those countries in which PEM is commonly found are also those in which infectious organisms flourish. Even though the mechanism involved in the relationship is not fully understood, there is no doubt that PEM and infectious diseases must be considered together, in terms of practical measures to reduce their prevalence and impact.

1.4. Classification of PEM

The principal signs found in marasmus and kwashiorkor are illustrated in Fig. 2, which is taken from Jelliffe (1969b). Table I, which is adapted from Latham (1965), McLaren (1976), DeMaeyer (1976), Whitehead and Alleyne (1971), and Davidson *et al.* (1975), summarizes many of the principal findings in kwashiorkor and marasmus. It should be recalled that many children show intermediate signs. Advances in the field of chemical pathology are rapid so Table I will need modification as new findings occur and are confirmed.

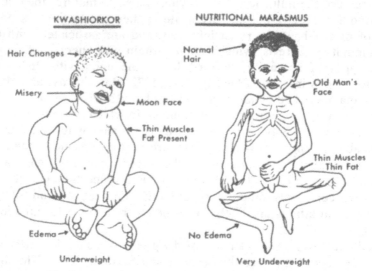

Fig. 2.   Principal signs of marasmus and kwashiorkor.

Table I. Some Typical Features Differentiating Marasmus and
Kwashiorkor

Clinical signs	Marasmus	Kwashiorkor
Growth failure	Severe	Present but less than in marasmus
Wasting of muscles	Always	Always
Subcutaneous fat	Absent	Present
Edema	Never	Always
Face	Emaciated	May be blubbery
Dermatosis	Rare	Common
Hair changes	Common	Very common
Liver enlargement	Sometimes	Very common
Appetite	Often ravenous	Usually poor
Mental state	Usually fretful	Apathetic
Physical activity	Low	Low
Diarrhea	Frequent	Usual
Internal and biochemical findings		
Serum albumin	Normal	Low
Extracellular water	Some increase	More increase
Body potassium	Low	Very low
Anemia	Often	Always
Liver	Normal or atrophic	Fatty infiltration
Serum branched-chain amino acids	Normal	Lowered
Thyroid	Hypofunction	May be normal
Serum growth hormone	Low or normal	High
Malabsorption	Some	More

Mild and moderate degrees of PEM are usually classified using anthropometry. It is not possible, in normal circumstances, to compare each child with his own optimal measurements. The natural variation of human growth makes it necessary to select references. The most commonly used reference values have long been the "Harvard standards" (Stuart and Stevenson, 1959). The phrase "reference value" is preferable to the word "standard" in this context (Kevany, 1974); an alternative term to reference weight for age is "average theoretical weight" (WHO and FAO, 1973).

Gomez *et al.* (1955, 1956) developed a classification that divided underweight into three degrees. Originally the Gomez classification was devised as "a guide for clinical use" and to "simplify nomenclature of a large group of diverse and confusing ailments, the common basis of which is deficient nutrition," using Mexican "average theoretic weights" as reference weights for age (Gomez *et al.*, 1955). The Mexican references used vary between 60 and 97% of the Harvard standards. The Gomez classification is as follows:

First degree: 76–90% of reference (originally 75–85%)
Second degree: 61–75% of reference (originally 60–75%)
Third degree: 60% of reference or below (originally less than 60%)

Later the degrees of malnutrition were adopted unchanged for community surveys using the Harvard standards as reference.

Jelliffe (1966) in a WHO monograph proposed classifying weight for age by 10% intervals around the Harvard reference weights. He avoided drawing diagnostic conclusions that may not be warranted in every case by suggesting that for public health purposes children could be classified into the following: first level underweight, 81–90% of standard; second level underweight, 71–80%; third level underweight, 61–70%; fourth level underweight, 60% and below.

An advantage of classification of weight for age by 10% intervals is that 80% of reference corresponds roughly to 2 SD below the mean (reference value) or the 3rd percentile of the reference population. The variance of other measurements is not necessarily the same.

The Classification of Infantile Malnutrition (1970) known as the Wellcome classification uses weight, age, and the presence or absence of edema to identify and separate in a simple way four categories of severe PEM. This is set out in Table II; it provides a framework for the more systematic analysis of other data (Waterlow, 1976).

The three classification systems just mentioned are widely used in community assessment and in individual diagnosis and monitoring. They are useful and quite simple but they suffer from two disadvantages. The first is that an accurate assessment of age is required. Jelliffe and Jelliffe (1971a) have reviewed this problem and suggested alternative measurements.

The second disadvantage of a classification using weight as the only body measurement is that it does not separate nutritional dwarfs from marasmic children and gives no indication of duration or acuteness of the condition found. A chronically underweight nutritional dwarf is unlikely to be particularly ill, whereas a child of the same age who has suddenly become very underweight is likely to be seriously unwell. The weight in each case may be the same.

Other classification systems have been proposed and used to a greater or lesser extent. Some of these have been reviewed by Jelliffe and Gurney (1974). Problems facing the differential diagnosis of PEM were discussed by Seoane and Latham (1971) and Gurney *et al.* (1972) in relation to type, extent, duration, and nutritional cause of PEM.

Waterlow (1976) has suggested that children could be classified by com-

Table II. The Wellcome Classification of Severe PEM

Weight for age (% of Harvard reference)	Edema	
	Yes	No
60–79	Kwashiorkor	Undernourished
<60	Marasmic kwashiorkor	Marasmus

bining indices of weight for achieved height (indicating any wasting) and height for age (indicating stunting). He discusses appropriate cutoff points and the most suitable tabulation for presenting such data derived from community surveys. The entire question of growth monitoring is discussed in detail in Chapters 16–18 in this volume.

2. Epidemiology

2.1. Nutritional Background

2.1.1. Growth and Requirements

The daily increase in body weight of a healthy baby is about 5–6 g/kg in the first 6 months of life. This declines to about 2–3 g in the second 6 months, and 0.5–0.6 g in the second year (Passmore *et al.*, 1974). Each gram of body weight gained costs about 21 kJ (5 kcal) (Payne and Waterlow, 1971).

The maintenance dietary energy requirements of a baby increase as he grows. However, with increasing age his rate of growth decreases, so the daily energy cost of growth declines. This is offset by the energy cost of the enormous increase in activity levels that occurs as children get older (Payne and Waterlow, 1971). In the presence of adequate dietary energy, activity does not use up protein significantly.

Hence, a 3-month-old infant needs almost three times more energy per unit weight than does his mother or his father. The protein requirements also are about three times as great per unit weight in an infant than either parent (Joint FAO/WHO Ad Hoc Expert Committee, 1973). This is not to say that absolute energy and protein requirements of infants are greater than those of their parents, or, indeed, their elder siblings; they are considerably less. However, the practical conclusion, in relation to PEM, is that infants and growing children need more food in relation to their size than do older people. The amount that can be taken in at one meal is presumably related largely to the size of the stomach, which is very small in infants. Consequently, infants and young children need to eat frequently and their food needs to be concentrated in energy with a content of the other nutrients in proportion.

Understanding of nutrient requirements is increasing rapidly and, particularly with regard to protein, is subject to controversy. A Joint FAO/WHO Ad Hoc Expert Committee (1973) discusses the issues in great detail and suggests recommended energy and protein intakes for various age and sex groups. Certain countries have drawn up their own tables of recommended intakes. All such recommendations will be subject to spasmodic changes over the years. These changes can disturb applied nutrition workers who may find they have to alter their dietary recommendations or that their evaluation of dietary survey findings become inappropriate. However, these tribulations should be born with fortitude as they are the price paid for increased knowledge.

2.1.2. First Feeding

The question of the adequacy of breast feeding has been reviewed by Jelliffe *et al.* (1975), Jelliffe and Jelliffe (1971b, 1978), and Harfouche (1970).

Unsupplemented human milk is all that is needed for feeding the young baby for the first 6 months of life in well-nourished communities; in less well-nourished communities breast milk alone is sufficient for optimal growth of the infant for 4–6 months if the mother is underfed. Hence, the severe forms of PEM are very rare in babies under 6 months of age who are fed only breast milk. Growth rates in such babies are usually good. A child with a low birth weight (often due to maternal undernourishment) may never fully "catch up" but will show a fast growth rate. Thus mild or moderate forms of PEM also are rare in totally breast-fed infants under 4 months of age in any community.

Cow's milk formulas can also support adequate growth. Recently, however, the inherent disadvantages involved in withholding breast milk have become more fully understood (Jelliffe and Jelliffe, 1971b, 1978; Jelliffe *et al.*, 1975; Petros-Barvazian, 1975), and consequently informed opinion worldwide now recommends breast-feeding (Twenty-Seventh World Health Assembly, 1974; Department of Health and Social Security, 1974). PEM can be caused by giving cow's milk formula in three major ways: (a) overdilution, with consequent inadequate energy and protein intakes; (b) contamination, leading to infectious organisms entering the body; and (c) displacement of breast milk which apart from its nutritional appropriateness protects against infectious disease organisms.

2.1.3. PEM and Staple Foods

The first major supplement to an infant's diet in most parts of the world is a gruel made from the local staple. The nature of the gruel may determine the type and extent of malnutrition found in many communities. The first point to note about a gruel is its concentration of energy. If it is too dilute, the limited capacity of the infant to drink great quantities will make it impossible for him to obtain enough energy, and marasmus can be expected. Force-feeding a child may protect him to an extent from such a deficiency caused by his inability, naturally, to take in enough food (Robson, 1972). However, frequent feeding is probably a more effective means of ensuring an adequate intake. Conversely, feeding a diluted gruel through a bottle causes an inadequate energy intake.

The staple foods contain little fat, a concentrated source of energy; most of their energy derives from starch. Starch on boiling takes up water so that the resulting cooked product cannot contain more than about 4.2 kJ (1 kcal) per gram. To receive half this daily energy requirements from such a cooked product, a 6-month-old infant would need to inbibe about 400 g each day. The only ways in which the energy concentration of a cooked starchy staple can be raised are either to drive off water or to add fat or sugar.

The second point about a gruel is its protein content. This is best expressed in relation to energy: as the proportion of total energy that is derived from protein (dietary protein–energy percent). A conversion for protein utilization can be made to arrive at a net figure.

Over the past decade knowledge (and the application of such knowledge) about protein requirements, and requirements for the amino acids from which protein are made, has improved (Joint FAO/WHO Ad Hoc Expert Committee, 1973). The cereals all contain enough, or almost enough, protein to ensure that an infant on a cereal gruel will not suffer from pure protein deficiency. The roots vary in their protein content expressed as protein–energy percent; yam (*Dioscorea* spp.) and the Irish potato (*Solanum tuberosum*) have a reasonable protein–energy percent. Cassava (*Manihot esculenta*), on the other hand, is very low in protein. The starchy fruits plantain and banana (*Musa* spp.) are also low in protein–energy (Gurney 1973, 1975; Payne, 1975).

In some societies protein is washed out of the staple during preparation; this can happen, for example, in parts of Yorubaland in West Africa, in the production of *eko*. Here dry maize (*Zea mays*) grains are soaked for at least 3 days, washed, ground, and sieved. This process removes most of the protein, leaving almost pure starch. This preparation is boiled in water and fed to babies as a diluted pap called *eko* (Dema, 1965).

These considerations lead us to certain conclusions:

1. A diet based on banana/plantain, cassava, overwashed maize, or other low protein products will tend to push an infant along the kwashiorkor route of development (Gurney *et al.*, 1973).

2. Most dilute gruels made from cereals, yam, or Irish potato will lead a child along the borderline route verging toward marasmus unless the energy content of the gruel is increased by use of fat or oil (Rutishauser and Frood, 1973). The staple food may not be the only factor influencing the commonest type of severe PEM in a community; the age of the child, the other types of food given before and during the weaning period, and the frequency and type of infections to which he is subject, all have their effect (Joint FAO/WHO Expert Committee on Nutrition, 1971).

3. A small child fed on a low fat starchy gruel needs a very large quantity of food to satisfy his dietary energy requirements. Hence, he needs frequent feedings and at each feeding he needs to eat a large amount (Nicol, 1971).

Clearly, an infant fed on a starchy staple alone is at risk of developing PEM. Even small additions of fat or oil will usefully increase the energy density of his diet. Similarly, small additions of high protein foods, such as legumes, which contain proteins that are complementary to those in cereals (Jelliffe, 1967), or foods of animal origin, enhance the protein content.

2.1.4. Consequences of Nutritional Background

Babies who are fed nothing but breast milk on demand usually grow normally for 4–6 months (Jelliffe *et al.*, 1975; Thomson and Black, 1975).

Should they continue to receive breast milk alone after this period, they will be receiving good quality food in insufficient quantity for their increased needs (Joint FAO/WHO Ad Hoc Expert Committee, 1973). Severe faltering of growth is likely after 6 months, and the infants will thus develop mild to moderate PCM and may become seriously malnourished later, probably toward the end of their second year of life. Such a late supplementation is common in rural areas of the developing world.

Babies fed cow's milk formulas in conditions of poverty are likely to have an inadequate supply of milk because of their mothers' attempt to "stretch" by overdilution of what is available. Hence, mild to moderate malnutrition rapidly turning to marasmus is a likely early consequence. Should the feed be prepared under unhygienic circumstances, episodes of infectious diarrhea are almost inevitable (made more likely and severe by the absence of the protection that would have been provided by breast milk). In this case marasmus, occurring in all probability before 6 months, will be a likely consequence. This picture of inadequate diet and weanling diarrhea is common in towns and cities of the developing world.

Should the child be breast-fed at first but then be weaned on to a gruel or porridge made from one of the staples that are particularly low in protein, kwashiorkor may well follow a long period of mild then moderate malnutrition. This has been described as a common picture in cassava-eating parts of the world or where bananas provide the porridge.

If the local gruel is cereal-based, its reasonably good protein content may well prevent the development of kwashiorkor but not of marasmus. PEM is particularly likely if the gruel is prepared in an energy-dilute form. In these circumstances, an infant previously adapting to his deficiency may be precipitated into biochemical breakdown by an infection and may develop marasmic kwashiorkor with edema (Srikantia, 1969).

An infant who is breast-fed for the first year and a half of his life, who starts other foods at about 6 months based on the multimix principle, and who, by the time he is 1 year old, is on much of the family diet, has a good chance of growing well. Care must be taken that the food is concentrated enough in its dietary energy and nutrients and fed often enough to ensure that the child receives adequate amounts. Despite good nutritional status, this child is still at risk of developing infections, which must therefore be prevented or reduced by adequate personal and food hygiene. Finally, good, appropriate health services can protect the child by immunization, education, monitoring of health and nutritional status, and the early treatment of infections.

2.2. Social and Economic Background of PEM

2.2.1. Poverty and Underdevelopment

PEM is a disease of poverty. It is not, however, caused solely by lack of money, although shortage of cash is a major immediate and basic cause. Aykroyd (1970) has discussed the history of PEM, which has clearly been with most of mankind for many centuries. However, it is possible that polygamy, prolonged lactation (with postpartum amenorrhea), and taboos against sexual intercourse, among other factors, have protected infants against severe forms of PEM to some extent by preventing birth intervals that are too short.

Rapid urbanization is one of the worldwide social changes taking place in this century. In the swelling and festering cities and towns of the world, we find a breakdown or replacement of traditional social and economic systems and the development of what Lewis (1962) calls "the culture of poverty." Undernutrition of small children is an almost inevitable consequence of the economic, social and psychological characteristics described by Lewis (1962):

> The economic traits which are most characteristic of the culture of poverty include the constant struggle for survival, unemployment and under-employment, low wages, a miscellany of unskilled occupations, child labour, the absence of savings, a chronic shortage of cash, the absence of food reserves in the home, the pattern of frequent buying of small quantities of food many times a day as the need arises. . . . Some of the social and psychological characteristics include living in crowded quarters, a lack of privacy, gregariousness, a high incidence of alcoholism, frequent resort to violence in the settlement of quarrels, frequent use of physical violence in the training of children, wife beating, early initiation into sex, free unions for consensual marriages, a relatively high incidence of the abandonment of mothers and children, a trend towards mother-centred families and a much greater knowledge of maternal relatives.

At about the time that Lewis was writing about the culture of poverty, others were concerning themselves with the political economics of underdevelopment. Guevara introduced the concept of the "Third World" as contrasted with the developed market economies and those that are centrally planned. He emphasized that "underdevelopment or distorted development carries with it a dangerous specialization of raw materials, containing a threat of hunger" (Guevara, 1961). He was speaking in the context of an island nation with an economy dominated by one crop: sugar, the sweet malefactor. However, many Third World countries are largely dependent on one export crop or industry and this gives them little control over their own development. The maldistribution of wealth between nations explains to a large extent why PEM is almost entirely concentrated in the nations of the Third World.

Information presented to the 1974 World Food Conference indicates that out of the 94 developing countries for which data were available, only 38 (40%) had a total dietary energy supply sufficient to meet the physiological requirements of their populations (including a 10% factor added to allow for wastage at the household level). The figures were estimated for the 1969–1971 period (Food and Agriculture Organization, 1974). Despite the considerable inaccuracies of such estimates of total food supplies, it can be concluded that many countries do not have enough food available to feed all the population. This statement is given more force if we accept the inevitability of some maldistribution of food resources between social classes and geographical areas within countries and between individuals within households.

Table III, which is derived from the Food and Agriculture Organization (1974), summarizes the evidence, based on estimates, that available food sup-

Table III. Average Daily Dietary Energy Supplies per Capita, World and Main Regions in 1970

	Population 1970 (millions)	kcal/capita	Percentage of requirements
Developed market economies	725	3050	120
Eastern Europe and USSR	349	3250	126
Developing market economies	1751	2180	95
Asian centrally planned economies	810	2360	100
World	3635	2500	105

plies are inequitably distributed between countries. It follows that PEM is an inevitable consequence of such maldistribution.

2.2.2. Cultural Influences

At birth, a human infant is totally dependent on others for his survival. He can be considered as an "exterogestate fetus" (Jelliffe *et al.*, 1975) linked nutritionally to his mother through her nipples and her loving care rather than the placenta and her uterus. Even though the toddler can reach out for food, he is still completely dependent on his family for his sustenance. However, at this stage his dentition and his alimentary tract can manage on the normal family diet, suitably prepared. The child in the transitional stage between early infancy and toddlerhood is extremely vulnerable nutritionally.

In contrast to the diet of the young of other species, the first foods introduced to young humans are to a large extent determined by their mothers' cultural background. The methods by which these foods are prepared for the child are also largely culturally determined. Should the cultural pattern not conform with nutritional needs, malnutrition can be expected.

In many traditional cultures, breast feeding continues from birth until well after the child's first or perhaps second or third birthday, but the first supplements are often given too late, too dilute, and too infrequently to fulfill the infants nutritional needs, even when the consumption of breast milk is taken into account. This late introduction of supplements is an important cause of PEM arising after 6 months of age and becoming severe in the child's second year in many rural areas.

In changing cultures, such as are found in many towns and cities of the Third World, dilute and perhaps contaminated, insufficiently nutritious foods are often given early; breast feeding may cease within a few weeks or months. PEM, arising within the first 6 months of life, is a likely consequence.

There are thus two major different behaviorally induced causes of PEM. The first (traditional), rural cause is associated with a too late introduction of supplementary foods, and the second, associated with the culture of poverty and with urban life, is linked with early introduction of dilute foods, replacing breast milk. Other, more local, culturally induced feeding patterns can influence the prevalence of PEM in different communities (Jelliffe, 1968).

It is not only political or religious despots who manipulate the minds of men and women. The purveyors of manufactured products also take advantage of transitional cultures, often for harm and sometimes for good, to create new appetites.

Developments in food science have on the whole greatly benefited those sections of mankind that can afford to make use of them. However, the promotion of seeming alternatives to breast milk has contributed to the decline of breast-feeding and to many cases of marasmus and deaths resulting from inadequate dietary energy and protein intake in conditions of poverty and poor hygiene. The use of expensive weaning foods and "tonics" instead of cheaper substitutes has similarly contributed to the prevalence of PEM. Among many studies and advocations those of Jelliffe (1971), McKigney (1968), Muller (1974), and Ledogar (1975) stand out as useful and factual.

Table IV, which is adapted from Jelliffe (1968), lists some of the major factors that combine to cause PEM.

Table IV. Miscellaneous Factors in the Etiology of PEM[a]

Geographicoclimatic	Unproductive soil
	Climate (high temperatures, extremes of rainfall)
Educational	Too few schools (illiteracy)
Social	Illegitimacy; family instability
	Absence of family planning (children too closely spaced)
	Poor communications (food distribution)
	Alcoholism
Economic	National poverty (low gross national product)
	Family poverty
	Unemployment
	Inadequate balance of world trade
Agronomic	Old-fashioned methods of agriculture
	Inadequate nutrient production (crops and livestock)
	Concentration of cash crop and poor food storage, preservation, and marketing
Medical	High prevalence of conditioning infections (measles, diarrhea, tuberculosis, whooping cough, malaria, intestinal parasites)
	Inadequate health facilities (too few, incorrect orientation)
	Inadequate staff (too few, ill trained in nutrition and maternal and child health)
Sanitational	Unclean, inadequate water supply
	Defective disposal of excreta and rubbish
Cultural	Faulty feeding habits of young children
	Recent urbanization (changing habits)
	Limited culinary facilities
	Inequable intrafamilial food distribution
	Overwork by women (limited time for food preparation for children)
	Sudden weaning (psychological trauma)

[a] Adapted from Jelliffe (1968, p. 184).

2.3. Prevalence of PEM

2.3.1. Historical Prevalence

Human communities have been subjected to periodic famines at least since the development of agriculture initiated by Neolithic man. Paleolithic man, who came earlier and was primarily a hunter, must also have suffered periods of severe hunger (Aykroyd, 1975). Infants and children in Europe and America have been larger for their age with succeeding generations over the past century and have been maturing earlier (Tanner, 1969); this is an indication of rising nutritional standards.

The slow increase in populations throughout the world until recent years, despite high birthrates, is clear indication that mortalities were very high. The contribution of PEM to such mortalities can only be quantified, and then approximately, for a few countries.

In England and Wales, one of the countries for which such data are available, the infant mortality rate fluctuated between 125 and 160 per 1000 births between 1841 and 1901 and then declined steadily to its present low level. During the nineteenth century, much of this mortality was in the second half of infancy. Second year mortality was also high, and has similarly declined. Infant and child mortality was worse in the cities than in rural areas. In 1911 the postneonatal infant mortality of the children of miners was four times that of the children of professional workers (Morris and Heady, 1955), indicating a social class effect. Aykroyd (1970), who has reviewed the matter, concludes that at the turn of the century in England and Wales, "something very like the complex of faulty diet and infection now called protein–calorie malnutrition was of key importance in the child health picture" and that although the part played by nutrition in the consequent improvement cannot be quantitatively assessed, it was unquestionably a large one.

The author has reviewed the situation in Guyana, a country for which data are available due to an efficient health service (J. M. Gurney, unpublished work). In the second half of the nineteenth century the infant and toddler mortality remained considerably higher than that in England and Wales. Marasmus and infections were major causes (with malaria taking an additional toll). A decline that started in the second decade of the century continued until the present era. It coincided with, and was probably to a large extent caused by, improved environmental sanitation and water supplies, infant welfare programs, and improved milk hygiene. Infant feeding practices in Georgetown, the capital of Guyana, were often extremely bad. They remained far from ideal up to the 1970s (Pan American Health Organization, 1976).

Neither Britain nor Guyana is fully representative of the wider world. Britain was relatively newly industrialized and Guyana had a plantation economy and was undergoing rapid social change. In both these societies, PEM was highly prevalent. Little or no information is known about the prevalence of PEM in culturally stable communities worldwide in the last century.

2.3.2. Current Prevalence

Bengoa has assessed the current evidence of the prevalence of PEM in the world today (Bengoa, 1972, 1974). He emphasizes that any estimates can be only approximations. The Joint FAO/WHO Expert Committee on Nutrition (1971) similarly found it "difficult to obtain even a rough estimate of the total number of malnourished children in the world" and made suggestions for improving the collection of prevalence data.

Using data provided by reasonably large-scale surveys carried out between 1963 and 1972, Bengoa (1974) estimates that "in Latin America the proportion of children with either severe or moderate malnutrition is of the order of 19%, in Africa of 26% and in Asia of 31%." It appears that in the world as a whole about 10 million children suffer from severe malnutrition and about 89 million from moderate malnutrition. These figures are extrapolated from weight for age data.

PEM is reflected in mortality statistics. The death rate of children aged between 12 and 59 months (1–4 years mortality rate) is considered to give an indication of the prevalence of PEM and infectious disease in a community (Wills and Waterlow, 1958). In areas in which malnutrition affects younger children, the second year mortality appears to give a better picture (Gordon *et al.*, 1967). The infant mortality rate provides an index of the whole complex of diseases (including nutrition) that can afflict the infant, and, except in those insanitary areas in which supplementary feeding is given soon after birth, gives a less direct indication of PEM than the second year or 1- to 4-year mortality rates. Unfortunately, 1- to 4-year mortality rates are not widely collected. In the least developed countries they may be between 10 and 50 times as great as in affluent countries. Recent figures for infant mortality in various regions of the World are given in Table V. They are derived from data collection systems of varying accuracy but nonetheless indicate that infant mortality in

Table V. World Infant Mortality Rates[a]

	Infant mortality rates	
Area	Per 1000 live births	As a proportion of that for North America
World	99	6.6
Asia	105	7.0
Africa	147	9.8
Europe	20	1.3
USSR	28	1.9
North America	15	1.0
Central and South America	84	5.6
Oceania	41	2.7

[a] From Kane (1978).

poor countries tends to be up to about ten times in excess of that in the affluent countries.

A very careful study of 34,197 children in seven countries in Latin America and the Caribbean has helped to clarify the role of nutrition among factors leading to the deaths of children under 5 years old (Puffer and Serrano, 1973). Nutritional deficiency of the child was found to be either the underlying cause of death or an associated cause in 52% of postneonatal deaths (35% of all deaths and 60% of infant deaths). This was almost all from PEM. Immaturity, often reflecting most probably a poor nutritional state of the mother, underlay or was associated with 63% of neonatal deaths (22% of all deaths). Diarrheal disease underlay or was associated with 41% of all deaths. Nutritional disease was an associated cause of death in 61% of all those deaths whose underlying cause was diarrheal disease.

Clearly in Latin America and the Caribbean PEM is a major, and often hidden, cause of death of infants and young children. There is reason to think that this conclusion applies in most other developing parts of the world.

It is probably not practicable to quantify usefully the effects of moderate rather than severe PEM on mortality, morbidity, or function except under experimental conditions. This is because of the adaptive nature of the condition and the complex and varied ecological contexts in which it occurs, further complicated by the synergistic interrelationships of PEM with a range of infectious diseases.

To cause death moderate PEM will either progress to the severe form or lead, through increased susceptibility, to a fatal infection. A moderately malnourished child is at greater risk of dying and of becoming ill than is a wellnourished child. His inactivity (combined with his likely cultural background) is likely to limit his exploratory behavior and thus his ability to learn. The moderately malnourished, inactive young child is likely to be understimulated and this may inhibit his mental and behavioral development (Read, 1973). Further, to quote Tizard (1974), "both chronic subnutrition and severe clinical malnutrition in childhood are statistically associated with subsequent growth retardation and with intellectual and scholastic backwardness."

It should be emphasized that a 7-year-old school entrant who, due to moderate PEM, appears superficially like a well-grown 4-year-old is not really comparable to either healthy 7-year-olds or healthy 4-year-olds. He has his own biological and behavioral characteristics and a distorted intersensorial organization (Beaton and Bengoa, 1976).

Stunting of growth, starting with moderate PEM in infancy or early childhood, may continue to adulthood. A mother who was malnourished as a child may be stunted. Low maternal heights are associated with low birth weight and high perinatal mortality; and low-birth-weight infants are at special risk of PEM. Mortality in elderly people from fractures of the long bones appears to be more common in short than in tall populations in the British Isles. Both these examples, cited by Eddy (1973), indicate that moderate PEM in a young child might have consequences in old age or even in the next generation.

3. Prevention and Treatment

Prevention of disease can be classified into three categories: primary prevention, including health promotion and specific protection; secondary prevention, including early diagnosis, prompt treatment, and limitation of disability; and tertiary prevention (rehabilitation) (Leavell and Clark, 1965). It is useful to consider PEM under these three headings and in relation to the epidemiology of the condition discussed in the preceding sections. Treatment of the failure of adaptation that constitutes severe PEM is included in secondary and tertiary prevention.

Any effective program for prevention of PEM, whether for an individual, a community, or a country, must incorporate the three types of prevention together. Consider a malnourished child who is brought to a health institution: The mother needs to be taught how to feed all her children so as to achieve cure and prevent relapses, and prevent similar disease occurring in siblings (primary prevention); the patient requires treatment of the condition by diet and control of infection and of any complications such as those resulting from dehydration (secondary prevention); the treated child needs to be brought back to a state at which cure can be consolidated at home without relapse (tertiary prevention). Both the improvement of the earning power of at-risk families (or communities or countries) and of learning power through education and training come under primary prevention. The distribution of supplementary foods is another case in which the three types of prevention may be involved: (a) primary prevention when the food goes to children in all at-risk families; (b) secondary prevention if food goes to children with anthropometric indication of early growth failure; and (c) tertiary prevention if food is given after clinical recovery from severe malnutrition. The use of industrially produced supplementary food mixtures for the protection of vulnerable groups is discussed by Chavez *et al.* (1975).

3.1. Primary Prevention

3.1.1. The Food Chain

Improvement of the nutritional, social, and economic background of PEM involves consideration of production, importation, marketing, distribution, and pricing of appropriate foods; all these factors influence availability to the consumer. It is useful in any environment or country to construct a diagrammatic "food chain" from production through to utilization. The concept of the food chain (or food path) is discussed simply and in detail as it relates to the family by King *et al.* (1972). Figure 3 gives an example of a food chain that links world conditions to nutritional status of individuals. Consideration of the diagram leads to identification of possible areas in which intervention will improve nutritional status and reduce the prevalence of PEM.

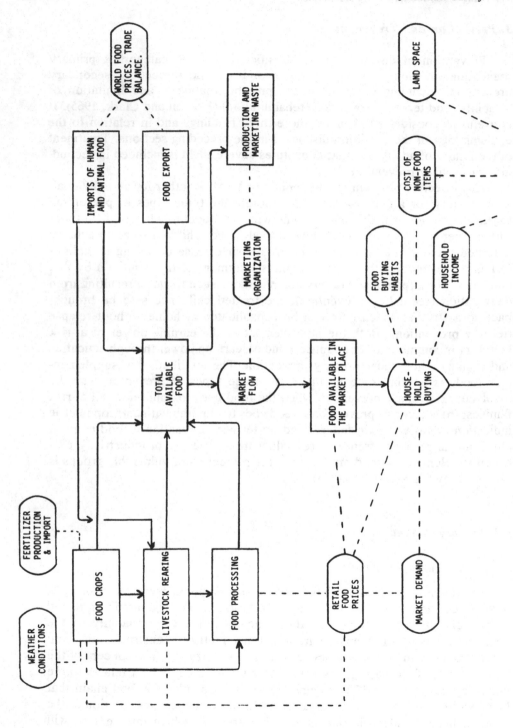

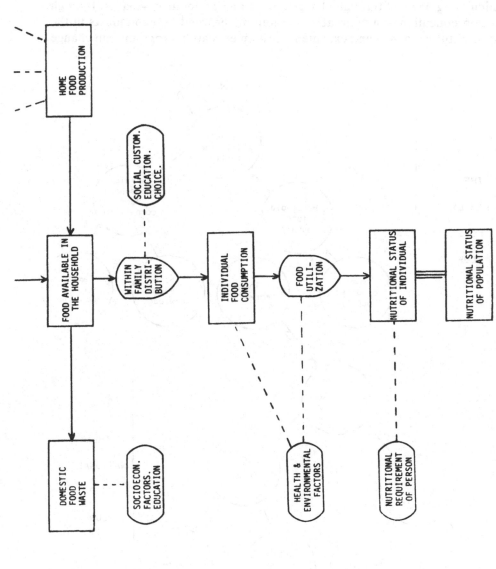

Fig. 3. An example of a diagrammatic food chain (compiled by P. Jutsum). Solid arrows indicate the major food flows. Broken lines indicate some of the major interactions between external controls and food activities and flows.

3.1.2. Modification of Demand

Figure 4 gives a simpler approach that is designed to relate national food flows (the lower section) to flows within the families that constitute the nation (upper section). The links between demand (in the economist's sense of the term) and supply are emphasized. The figure demonstrates the relevance of manipulating demand through changing taste and preference—that is, through nutrition education—an approach too often disregarded by economists in the past. Carefully planned and executed nutrition education programs can change

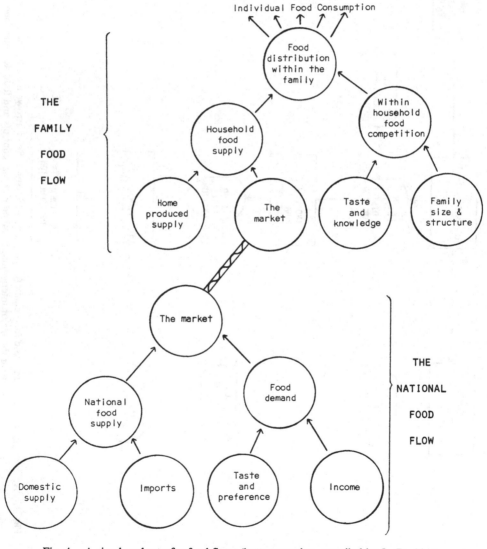

Fig. 4. A simple schema for food flows (bottom portion compiled by L. Rankine).

tastes and preferences. Such programs can be linked usefully with (a) price control, with the aim of influencing demand, and (b) manipulation of availability (e.g., by banning the importation of undesirable alternatives).

The content of nutrition education needs to be carefully thought out and related to needs and priorities. Reddy and Gurney (1977) have tried to identify such priorities in their nutrition "messages for mothers." Cameron and Hofvander (1976) have discussed the most appropriate methods of young child feeding around the world, particularly in relation to breast-feeding and the important multimix principle. Ritchie (1967) presents a useful text on learning better nutrition in which these and other principles are raised.

Modification of food habits with a view to improving nutrition needs to be undertaken with care. Often those who seek to do the modifying are themselves unaware of their own food prejudices. An obsessional belief in milk or hot dinners or even protein (McLaren, 1974) may inhibit rational nutrition education.

3.1.3. Food and Nutrition Policies and Programs

National food and nutrition policies fall mainly under the category of primary prevention, although a comprehensive policy also encompasses elements of secondary and tertiary prevention. The integration, likewise, of nutrition plans and programs into an overall national development plan is an important primary preventive measure. The planning and organization of a national food and nutrition policy has been well dealt with by Bengoa and Rueda-Williamson (1976), a Joint FAO/WHO Expert Committee on Nutrition (1976), and Cook and Yang (1974). An example of a national policy directed largely toward the elimination of PEM is provided by the government of Jamaica (1974).

Comprehensive, appropriately planned, and well-executed programs within the health services are essential in the prevention of PEM. An example of such a program is the Strategy and Plan of Action to Combat Gastroenteritis and Malnutrition in Children under Two Years of Age (1975), which was formulated in response to a request from the Ministers of Health of the English-speaking Caribbean. In many countries the most appropriate vehicle for such a program is a primary health care system (as the term is used by WHO). The basic principles of primary health care involve outreach to all communities, the use of appropriately trained auxiliary workers, a good backup, referral and supervisory systems, and community involvement in the program. These principles are well set out and amplified by Newell (1975).

Jelliffe (1968, 1969b) and Jelliffe and Jelliffe (1973) deal in detail with primary prevention of PEM in infants and young children. Guidelines to Young Child Feeding in the Contemporary Caribbean (1970) gives an example of a document that has led to specific government programs against PEM.

More recent views on what have been termed "Community Action–Family Nutrition Programs" have been summarized by UNICEF/IUNS (1977).

3.2. Secondary Prevention

The importance of regular monitoring of a young child's progress usually through regular attendance at a maternal and child health clinic cannot be overemphasized. The use of the weight chart as a monitoring device is discussed in Chapter 16 in this volume. It has recently been reviewed by Morley (1976). Through such regular monitoring early deviations from normal growth can be detected and dealt with promptly.

Bengoa (1967) has developed the concept of nutrition rehabilitation centers for the effective, inexpensive secondary and tertiary prevention of PEM, and has recently reviewed them in the light of experience and in relation to other forms of treatment (Bengoa, 1976). In well-run centers 62–80% of the children attending with malnutrition are recuperated, usually over a period of 4 months. This compares well with hospital treatment and has the advantage over most hospitals in that (a) the mother (or older female sibling) is educated on what comprises a proper diet for a young child; (b) the cost is much less (between one-tenth and one-fifteenth), and (c) hospital beds are released for other purposes. *A Practical Guide to Combating Malnutrition in the Preschool Child* (1970) provides useful information on setting up and managing nutrition services which involve such rehabilitation centers.

Many hospitals in the Third World are unsuitable places for treating PEM (Cook, 1971b). The tender loving care so necessary for the recuperation of severely malnourished children is often not present; food is often given too infrequently and the likelihood of cross-infection between patients is high. Such hospitals rarely give adequate education to prevent relapse among discharged patients or the same disease in younger siblings. They are often too far from the home for children to be brought early. Finally, hospitals are expensive.

Food is an essential part of secondary and tertiary prevention of PEM. Gueri (1977) writes that in Trinidad "food supplementation is, on the average, a satisfactory and comparatively cheap, yet somewhat slow, method of improving the nutritional status of children with primary malnutrition whether moderate or severe. The process can be speeded up by means of close supervision, but an increase of 30% in the rate of recovery increases the cost by over 300%." Supervision and nutrition education in Gueri's study did seem to have a beneficial effect on the quality of the diet of the children 6 months to a year after the educational process had ceased. A study in India, however, showed a poor response to food supplementation and nutrition education (Kumari *et al.*, 1975). Clearly primary prevention, early detection of growth failure, and prompt secondary prevention are more likely to be effective than late secondary prevention, and are much cheaper. Cook (1971a, 1972) has discussed the costs of malnutrition to governments and to parents.

In most countries with a high prevalence of severe PEM, skilled supervision is needed to treat complications such as dehydration or hypothermia and concurrent infections and to ensure that patients are protected against infectious diseases (e.g., by immunization). A useful manual on hospital treatment of PEM providing considerable detail is that of Picou *et al.* (1975). Oral rehydration is described in the WHO booklet "Treatment and Prevention of Dehydration in Diarrheal Diseases—A Guide for Use at the Primary Level" (1976).

3.3. Tertiary Prevention

The high energy costs of growth have been discussed earlier in this chapter. This is the basis for the currently popular use of very high energy density feeding of children suffering from all forms of severe PEM and during rehabilitation. A rapid catch-up to the expected weight for height of the child is thought to be beneficial.

Intakes in excess of 840 kJ of dietary energy per kilogram of body weight per day (200 kcal) are recommended along with 3–4 g of reference protein (DeMaeyer, 1976; Picou *et al.*, 1975). Such intakes appear to produce a fast and balanced growth of lean tissues as well as fat (Brooke and Wheeler, 1976). Mixtures of this kind obtain at least half their energy from oil or fat and have an energy density of around 500 kJ/100 ml (120 kcal).

Ashworth (1974) demonstrated that most children being treated for severe PEM develop voracious appetites (readily consuming the amounts mentioned in the preceding paragraph). After reaching their expected weight for their achieved height, their appetite falls so they consume an average of 515 kJ (123 kcal)/kg body weight each day of food of high energy density. This dietary pattern is accompanied by a rapid increase in weight, i.e., catch-up growth, followed by a dramatic slowing of weight gain when the expected weight for height is reached.

Often children treated for PEM revert to an inadequate diet on discharge from hospital or center. Therefore they fail to continue (at a slower rate than possible in the first weeks of treatment) to catch up in their expected weight or height for their age. They may well become nutritional dwarfs, and only achieve a measure of catch-up by continuing to grow in height longer than they would have were they properly nourished (with an associated delayed puberty). However, there is evidence that, should the diet become adequate, complete catch-up toward expected weight and height is possible even after severe malnutrition (Agricultural Research Council/Medical Research Council, 1974).

There is a theoretical risk that once a severely malnourished child has reached his expected weight for height further high energy feeding might result in obesity. However, this would seem unlikely in most circumstances, first because when children reach their expected weight for height, they voluntarily limit their intake when offered such very high energy feeds as are now recommended for treatment of severe forms of PEM; and second because the home environment of such children is usually so deprived as to make it most unlikely that they will be offered excessive dietary energy.

Children who have recuperated from severe malnutrition often relapse and their younger siblings may develop the same condition. Adequate follow-up and referral to health services is therefore necessary. Home visiting is usually needed. It may be some years, if ever, before the child exhibits catch-up growth in height as well as weight to within the optimal healthy range and thus achieve his genetic potential.

4. References

Agricultural Research Council/Medical Research Council, 1974, *Food and Nutrition Research*, Report of the ARC/MRC Committee, p. 121, H. M. Stationery Office, London.

Ashworth, A., 1974, *Ad lib* feeding during recovery from malnutrition, *Br. J. Nutr.* 31:109.

Awdeh, Z. L., Bengoa, J., Demaeyer, E. M., Dixon, H., Edsall, G., Faulk, W. P., Goodman, H. C., Hopwood, B. E. C., Jose, D. G., Keller, W. D. E., Rodriguez, J. K., Mata, L. J., McGregor, I. A., Miescher, P. A., Rowe, D. S., Taylor, C. E., and Torrigiani, G., 1972, Survey of nutritional-immunological interreactions, *Bull. WHO* 46:537.

Aykroyd, W. R., 1968, Mortality in infancy and early childhood and its relation to malnutrition, *Turk. J. Pediatr.* 10:50.

Aykroyd, W. R., 1970, *Conquest of Deficiency Diseases*. Achievements and Prospects, WHO, Geneva (FFHC Basic Study No. 24).

Aykroyd, W. R., 1975, *The Conquest of Famine*, Reader's Digest Press, New York.

Barnes, R. H., 1976, Dual role of environmental deprivation and malnutrition in retarding intellectual development. *Am. J. Clin. Nutr.* 29:912.

Beaton, G. H., and Bengoa, J. M., 1976, Nutrition and health in perspective: An introduction, in: *Nutrition in Preventive Medicine* (G. H. Beaton and J. M. Bengoa, eds.), pp. 13–20, WHO, Geneva (Monograph Series No. 62).

Behar, M., 1975, The role of feeding and nutrition in the pathogeny and prevention of diarrheic processes, *PAHO Bull.* 9:1.

Bengoa, J. M., 1967, Nutrition rehabilitation centres, *J. Trop. Pediatr.* 13:169.

Bengoa, J. M., 1972, Nutritional significance of mortality statistics, in: *Proceedings Third Western Hemisphere Nutrition Congress*, pp. 270–281, Futura, New York.

Bengoa, J. M., 1974, The problem of malnutrition, *WHO Chron.* 28:3.

Bengoa, J. M., 1976, Nutritional rehabilitation, in: *Nutrition in Preventive Medicine* (G. H. Beaton and J. M. Bengoa, eds.), pp. 321–334, WHO, Geneva (Monograph Series No. 62).

Bengoa, J. M., and Rueda-Williamson, R., 1976, Planning and organization of a national food and nutrition policy, in: *Nutrition in Preventive Medicine* (G. H. Beaton and J. M. Bengoa, eds.), pp. 419–444, WHO, Geneva (Monograph Series No. 62).

Brooke, O. G., and Wheeler, E. F., 1976, High energy feeding in protein–energy malnutrition, *Arch. Dis. Child.* 51:968.

Burgess, A., and Dean, R. F. A., 1962, Malnutrition and Food Habits—Report of an International and Interprofessional Conference, Tavistock, London.

Cameron, M., and Hofvander, Y., 1976, *Manual on Feeding Infants and Young Children*, 2nd ed., PAG, New York.

Chavez, A., Bourges, H., and Basta, S. (eds.), 1975, Protection of vulnerable groups through protein-rich mixtures, in: *Proceedings Ninth International Congress of Nutrition*, Vol. 4, *Approaches to Practical Solutions*, pp. 159–206, Karger, Basel.

Classification of infantile malnutrition (Editorial), 1970, *Lancet* 2:302.

Cook, R., 1971a, The cost of malnutrition in Jamaica, *Ecol. Food Nutr.* 1:61.

Cook, R., 1971b, Is hospital the place for the treatment of malnourished children? *J. Trop. Pediatr. Environ. Child Health* 17:15.

Cook, R., 1972, The primary costs of malnutrition and its impact on a society, in: *Proceedings Western Hemisphere Nutrition Congress III*, pp. 324–327, Futura, New York.

Cook, R., and Yang, Y. H., 1974, National food and nutrition policy in the commonwealth Caribbean, *Bull. PAHO* 8:133.

Davidson, S., Passmore, R., Brock, J. F., and Truswell, A. S., 1975, *Human Nutrition and Dietetics,* 6th ed., Churchill–Livingstone, Edinburgh.

Dema, I. S., 1965, *Nutrition in Relation to Agricultural Production,* Food and Agric. Org., Rome.

DeMaeyer, E. M., 1976, Protein–energy malnutrition, in: *Nutrition in Preventive Medicine* (G. H. Beaton and J. M. Bengoa, eds.), pp. 23–54, WHO, Geneva (Monograph Series No. 62).

Department of Health and Social Security, 1974, *Present Day Practice in Infant Feeding,* Report on Health and Social Subjects No. 9, H. M. Stationery Office, London.

Downs, E. F., 1964, Nutritional dwarfing: A syndrome of early protein–calorie malnutrition, *Am. J. Clin. Nutr.* 15:275.

Eddy, T. P., 1973, Past and present malnutrition and its effect on health today, *Roy. Soc. Health J.* 93:314.

Food and Agriculture Organization, 1974, Population, Food Supply and Agricultural Development, Food Agric. Org., Rome (W5/F0651, mineograph).

Food and nutrition policy for Jamaica with programmes for incorporation in the national development plan 1975/76–1977/78, Nutrition Advisory Council, Kingston, 1974.

Gomez, F., Galvan, R. R., Cravioto, J., and Frenk, S., 1955, Malnutrition in infancy and childhood, with special reference to kwashiorkor, *Adv. Pediatr.* 7:131.

Gomez, F., Galvan, R. R., Frenk, S., Cravioto Munos, J., Chavez, R., and Vasquez, J., 1956, Mortality in second and third degree malnutrition, *J. Trop. Pediatr.* 2:77.

Gordon, J. E., Wyon, J. B., and Ascoli, W., 1967, The second year death-rate in less developed countries, *Am. J. Med. Sci.* 254:357.

Gueri, M., 1977, The Management of Protein–Energy Malnutrition Outside Hospital—Report of a Project Carried Out in Trinidad, 1975–1976, Caribbean Food and Nutrition Institute, Kingston [CFNI-J-15-77].

Guevara, E., 1961, Translated quotation in Sinclair, A., 1971, *Guevara,* Collins, London.

Guidelines to Young Child Feeding in the Contemporary Caribbean, 1970, PAHO, Washington, D.C. (Sci. Publ. No. 217).

Gurney, J. M., 1973, Nutrition facts on staples, *Cajanus* (Newsletter of the Caribbean Food and Nutrition Institute) 6:213.

Gurney, J. M., 1975, Nutritional considerations concerning the staple foods of the English-speaking Caribbean, *Ecol. Food Nutr.* 4:171.

Gurney, J. M., Jelliffe, D. B., and Neill, J., 1972, Anthropometry in the differential diagnosis of protein–calorie malnutrition, *J. Trop. Pediatr. Environ. Child Health* 18:1.

Gurney, J. M., Ogbeide, M. I., Reddy, S., and McFarlane, H., 1973, Amino acid ratios in protein–energy malnutrition, *Trop. Geogr. Med.* 25:387.

Harfouche, J. K., 1970, The importance of breast-feeding, *J. Trop. Pediatr.* 16:135.

Ingenbleek, Y., De Nayer, P. H., and De Visscher, M., 1972, Measurement of prealbumin as index of protein–calorie malnutrition, *Lancet* 2:106.

Jelliffe, D. B., 1959, Protein–calorie malnutrition in tropical preschool children (a review of recent knowledge), *J. Pediatr.* 54:227.

Jelliffe, D. B., 1966, The Assessment of the Nutritional Status of the Community (with Special Reference to Field Surveys in Developing Regions of the World), WHO, Geneva (Monograph Series No. 53).

Jelliffe, D. B., 1967, Approaches to village-level infant feeding: (i) Multimixes as weaning foods, *J. Trop. Pediatr.* 13:46.

Jelliffe, D. B., 1968, *Infant Nutrition in the Tropics and Subtropics,* 2nd ed., WHO, Geneva (Monograph Series No. 29).

Jelliffe, D. B., 1969a, Letter to the Editor, *Am. J. Clin. Nutr.* 22:1159.

Jelliffe, D. B., 1969b, *Child Nutrition in Developing Countries—A Handbook for Fieldworkers,* revised ed., U.S. Govt. Printing Office, Washington, D.C.

Jelliffe, D. B., 1971, Commerciogenic malnutrition, *Food Technol.* 25:55.

Jelliffe, D. B., and Jelliffe, E. F. P., 1971a, Age-independent anthropometry, *Am. J. Clin. Nutr.* 24:1377.

Jelliffe, D. B., and Jelliffe, E. F. P. (eds.), 1971b, The uniqueness of human milk, *Am. J. Clin. Nutr.* 24:268.

Jelliffe, D. B., and Jelliffe, E. F. P. (eds.), 1973, *Nutrition Programmes for Preschool Children,* Inst. Public Health, Croatia, Zagreb.

Jelliffe, D. B., and Jelliffe, E. F. P., 1978, *Human Milk in the Modern World,* Oxford University Press, London.

Jelliffe, D. B., and Welbourn, H. F., 1963, Clinical signs of mild-moderate protein–calorie malnutrition of early childhood, in: *Mild-Moderate Forms of Protein–Calorie Malnutrition* (G. Blix, ed.), pp. 12–19, Swedish Nutrition Foundation, Uppsala.

Jelliffe, D. B., Gurney, M., and Jelliffe, E. F. P., 1975, Unsupplemented human milk and the nutrition of the exterogestate fetus, in: *Proceedings Ninth International Congress, Mexico, 1972,* Vol. 2 (A. Chavez, H. Bourges, and S. Basta, eds.), pp. 77–85, Karger, Basel.

Jelliffe, E. F. P., and Gurney, J. M., 1974, Definition of the problem, in: *Nutrition and Malnutrition* (A. F. Roche and F. Falkner, eds.), Plenum Press, New York.

Jelliffe, E. F. P., and Jelliffe, D. B. (eds.), 1969, The arm circumference as a public health index of protein–calorie malnutrition of early childhood, *J. Trop. Pediatr.* 15:177 (Monograph No. 8).

Joint FAO/WHO Ad Hoc Expert Committee, 1973, Energy and Protein Requirements, WHO, Geneva (Technical Report Series No. 522).

Joint FAO/WHO Expert Committee on Nutrition, 1971, Eighth Report—Food Fortification, Protein–Calorie Malnutrition, WHO, Geneva (Technical Report Series No. 477).

Joint FAO/WHO Expert Committee on Nutrition, 1976, Food and Nutrition Strategies in National Development, 9th rep., WHO, Geneva (Technical Report Series No. 584).

Kane, T. T., 1978, *1978 World Population Data Sheet,* Population Reference Bureau, Inc., Washington, D.C.

Kevany, J., 1974, Height and weight standards for preschool children, *Lancet* 1:993.

King, M. H., King, F. M. A., Morley, D. C., Burgess, H. J. L., and Burgess, A. P., 1972, *Nutrition for Developing Countries,* Chap. 9, Oxford University Press, London.

Kumari, V. K. K., Damodaran, M., and Rao, P., 1975, Evaluation of domiciliary management of protein–calorie malnutrition, *Trop. Geogr. Med.* 27:99.

Latham, M., 1965, Human Nutrition in Tropical Africa: A Textbook for Health Workers with Special Reference to Community Health Problems in East Africa, Food Agric. Org., Rome.

Leavell, H. R., and Clark, E. G., 1965, *Preventive Medicine for the Doctor in His Community— An Epidemiologic Approach,* 3rd ed., pp. 19–28, McGraw-Hill, New York.

Ledogar, R. J., 1975, *Hungry for Profits,* IDOC, New York.

LeGros Clark, W. E., and Medawar, P. B. (eds.), 1945, *Essays on Growth and Form,* presented to D'Arcy Wentworth Thompson, Clarendon Press, Oxford.

Lewis, O., 1962, *The Children of Sanchez,* Secker and Warburg, London.

Mata, L. J., Urrutia, J. J., and Gordon, J. E., 1967, Diarrhoeal disease in a cohort of Guatemalan village children observed from birth to age of 2 years, *Trop. Geogr. Med.* 19:247.

McFarlane, H., Ogbeide, M. I., Reddy, S., Adcock, K. J., Adeshina, H., Gurney, J. M., Cooke, A., Taylor, G. O., and Mordie, J. A., 1969, Biochemical assessment of protein–calorie malnutrition, *Lancet* 1:392.

McKigney, J., 1968, Economic aspects of infant feeding practices in the West Indies, *J. Trop. Pediatr.* 14:55.

McLaren, D. S., 1974, The great protein fiasco, *Lancet* 2:93.

McLaren, D. S., 1976, *Nutrition and Its Disorders,* 2nd ed., Churchill-Livingstone, Edinburgh.

Morley, D., 1976, The design and use of weight charts in surveillance of the individual, in: *Nutrition in Preventive Medicine* (G. H. Beaton and J. M. Bengoa, eds.), pp. 520–529, WHO, Geneva (Monograph Series No. 62).

Morris, J. N., and Heady, J. A., 1955, Mortality in relation to the father's occupation, *Lancet* 1:554.

Muller, M., 1974, *The Baby Killer,* War on Want, London.

Newell, K. W., 1975, *Health by the People,* WHO, Geneva.

Nicol, B. M., 1971, Protein and calorie concentration, *Nutr. Rev.* 29:83.

PAHO Advisory Committee on Medical Research, 1971, Metabolic Adaptation and Nutrition: Proceedings of the Special Session Held during the Ninth Meeting, PAHO, Washington, D.C. (Sci. Publ. No. 222).

Pan American Health Organization, 1976, The National Food and Nutrition Survey of Guyana, PAHO, Washington, D.C. (Sci. Publ. No. 323).

Passmore, R., Nicol, B. M., Narayana Rao, M., and Bocobo, D. L., 1974, *Handbook on Human Nutritional Requirements,* WHO, Geneva (Monograph Series No. 61).

Payne, P. R., 1975, Safe protein–calorie ratios in diets. The relative importance of protein and energy intake as causal factors in malnutrition, *Am. J. Clin. Nutr.* **28**:281.

Payne, P. R., and Waterlow, J. C., 1971, Relative energy requirements for maintenance, growth and physical activity, *Lancet* **2**:210.

Periera, S. M., and Begum, A., 1974, The manifestations and management of protein–calorie malnutrition (kwashiorkor), in: *World Review of Nutrition and Dietetics,* Vol. 19 (G. H. Bourne, ed.), pp. 1–50, Karger, Basel.

Petros-Barvazian, A., 1975, Maternal and child health and breast-feeding, *Mod. Probl. Pediatr.* **15**:155.

Picou, D., Alleyne, G. A. O., Kerr, D. S., Miller, C., Jackson, A., Hill, A., Bogues, J., and Patrick, J., 1975, Malnutrition and Gastroenteritis in Children: A Manual for Hospital Treatment and Management, Caribbean Food and Nutrition Institute, Kingston.

A Practical Guide to Combating Malnutrition in the Preschool Child—Nutritional Rehabilitation through Maternal Education, 1970, Appleton, New York.

Puffer, R. R., and Serrano, C. V., 1973, Patterns of Mortality in Childhood—Report of the Inter-American Investigation of Mortality in Childhood, PAHO, Washington, D.C. (Sci. Publ. No. 262).

Ramalingaswami, V., 1975, Nutrition, cell biology and human development, *WHO Chron.* **29**:306.

Read, M. S., 1973, Malnutrition, hunger and development, *J. Am. Diet. Assoc.* **63**:379.

Reddy, S. K., and Gurney, J. M., 1977, Family Food and Nutrition—A Manual of Priorities for the Eastern Mediterranean Region—Messages for Mothers, WHO, Alexandria (EM/Nut/76).

Ritchie, J. A. S., 1967, Learning Better Nutrition, Food Agric. Org., Rome, (Nutrition Studies No. 20).

Robson, J. R. K., 1972, *Malnutrition, Its Causation and Control (with Special Reference to Protein–Calorie Malnutrition),* Gordon & Breach, New York.

Rutishauser, I. H. E., and Frood, J. D. L., 1973, The effect of a traditional low-fat diet on energy and protein intake, serum albumin concentration and body-weight in Ugandan preschool children, *Br. J. Nutr.* **29**:261.

Scrimshaw, N. S., Taylor, C. E., and Gordon, J. E., 1968, *Interaction of Nutrition and Infection,* WHO, Geneva (Monograph Series No. 57).

Seoane, N., and Latham, M. C., 1971, Nutritional anthropometry in the identification of malnutrition in childhood, *J. Trop. Pediatr. Environ. Child Health* **17**:98.

Srikantia, S. G., 1969, Protein–calorie malnutrition in Indian children, *Indian J. Med. Res.* **57**:36.

Strategy and Plan of Action to Combat Gastroenteritis and Malnutrition in Children under Two Years of Age, 1975, *Environ. Child Health* **21**:23.

Stuart, H. C., and Stevenson, S. S., 1959, Physical growth and development, in: *Textbook of Pediatrics,* 7th ed. (W. Nelson, ed.), pp. 12–61, Saunders, Philadelphia.

Tanner, J. M., 1969, Growth of children in industrialized countries, with special reference to the secular trend, in: *Nutrition in Preschool and School Age Children* (G. Blix, ed.), pp. 9–28, Almqvist & Wiksell, Uppsala.

Thomson, A. M., and Black, A. E., 1975, Nutritional aspects of human lactation, *Bull. WHO* **52**:163.

Tizard, J., 1974, Early malnutrition, growth and mental development in man, *Br. Med. Bull.* **30**:169.

Treatment and Prevention of Dehydration in Diarrhoeal Diseases—A Guide for Use at the Primary Level, 1976, WHO, Geneva.

Twenty-Seventh World Health Assembly, 1974, Infant nutrition and breast-feeding, Resolution WHA 27.43.

UNICEF/IUNS, 1977, *Community Action—Family Nutrition Programs,* UNICEF, New Delhi.

Wadsworth, G. R., 1975, Nutrition in public health, in: *The Theory and Practice of Public Health,* 4th ed. (W. Hobson, ed.), Oxford University Press, London.

Waterlow, J. C., 1948, *Fatty Liver Disease in Infants in British West Indies,* Medical Research Council (Br.), Special Report Series No. 263, H. M. Stationery Office, London.

Waterlow, J. C., 1963, The assessment of marginal protein malnutrition, *Proc. Nutr. Soc.* **22**:66.

Waterlow, J. C., 1976, Classification and definition of protein-energy malnutrition, in: *Nutrition in Preventive Medicine* (G. H. Beaton and J. M. Bengoa, eds.), pp. 530–555, WHO, Geneva (Monograph Series No. 62).

Waterlow, J. C., Cravioto, J., and Stephen, J. M. L., 1960, Protein malnutrition in man, in: *Advances in Protein Chemistry,* Vol. 15, p. 131, Academic Press, New York.

Weiner, J. S., 1964, Human ecology, in: *Human Biology—An Introduction to Human Evolution, Variation and Growth* (G. A. Harrison, J. S. Weiner, J. M. Tanner, and N. A. Barnicot, eds.), p. 439, Oxford University Press, London.

Whitehead, R. G., 1971, Nutrition and gastroenterology, in: *Proceedings: XIII International Congress of Pediatrics, Vienna, 29 August–4 September 1971,* Vol. II, Pt. 1 (Weiner), p. 231, Medizinischen Akademie, Vienna.

Whitehead, R. G., and Alleyne, G. A. O., 1971, Pathophysiological factors of importance in protein–calorie malnutrition, *Br. Med. Bull.* **28**:72.

Whitehead, R. G., and Dean, R. F. A., 1964, Serum amino acids in kwashiorkor, 1–Relationship to clinical condition, *Am. J. Clin. Nutr.* **14**:313.

Whitehead, R. G., Frood, J. D. L., and Poskitt, E. M. E., 1971, Value of serum albumin measurements in nutritional surveys: A reappraisal, *Lancet* **2**:287.

WHO, 1975, Control of Nutritional Anaemia with Special Reference to Iron Deficiency, Report of an IAFA/USAID/WHO Joint Meeting, WHO, Geneva (Technical Report Series No. 580).

WHO and FAO, 1973, Food and Nutrition Terminology: Definitions of Selected Terms and Expressions in Current Use (compiled with the collaboration of the International Union of Nutritional Sciences), WHO, Geneva (NUTR/73.2).

WHO Scientific Group, 1973, Cell-Mediated Immunity and Resistance to Infection, WHO, Geneva (Technical Report Series No. 519).

Williams, C. D., 1933, Nutritional disease of children associated with a maize diet, *Arch. Dis. Child.* **8**:423.

Williams, C. D., 1963, The story of kwashiorkor, *Courrier* **13**:361.

Williams, C. D., and Jelliffe, D. B., 1972, *Mother and Child Health—Delivering the Services,* p. 49, Oxford University Press, London.

Wills, V. G., and Waterlow, J. C., 1958, The death rate in the age-group 1–4 years as an index of malnutrition, *J. Trop. Pediatr.* **3**:167.

The Young Child: Obesity

June K. Lloyd

Obesity is now accepted as being the most common nutritional disorder of infancy and childhood in the industrialized countries of the Western world. This chapter reviews its prevalence, natural history, etiology, effects, treatment, and prevention.

1. Prevalence

Studies of the prevalence and incidence of obesity in children are hampered, as they are in adults, by differences in definition and methods of measurement and lack of agreed standards for "normality." These aspects are discussed in detail in Chapters 4 and 5 in this volume, and therefore are not considered further here, but it must be emphasized that what may be considered "normal" in a society is not necessarily desirable in health terms. As Tanner and Whitehouse (1975) have commented in presenting revised standards for skinfold measurements in British children, these standards represent "what is, and not what ought to be." In the same way, prevalence data describe the situation at a given time, in a given population, studied by a specific method, and assessed against specific standards; the resulting figure may misrepresent the size of the actual problem and is unlikely to be strictly comparable with other findings.

1.1. Infancy

Early in the 1970s reports from England indicated that about 20–35% of babies were obese during their first year of life, obesity being defined as a weight which was 20% or more in excess of expected weight for length (Hutchinson-Smith, 1970; Shukla *et al.*, 1972).

More recently, however, a study from Sweden (Sveger *et al.*, 1975) using

June K. Lloyd • St. George's Hospital Medical School, University of London, London, England.

the same definition and standards reported a much lower prevalence of around 6%, and a report from London (Whitelaw, 1977) has also indicated a downward trend in prevalence at 1 year in the UK.

Weight gain from birth to 6 months of age has been described as the most sensitive index of fatness in 6-month-old babies (Crawford *et al.*, 1974) and thus a change in the birth weight doubling time, usually quoted as being achieved at 5–6 months, might be taken to indicate a change in the prevalence of obesity. A study of birth weight doubling time in the United States has recently shown that the mean age at which birth weight is doubled is 3.8 months (Neumann and Alpaugh, 1976). A more precise indication that babies are becoming fatter may be derived from a comparison of skinfold measurements; data collected in the United Kingdom show that the measurement corresponding to the 90th percentile for triceps skinfold at 6 months of age became over a 12-year period the measurement corresponding to the 50th percentile (Tanner and Whitehouse, 1975).

The relative paucity of data and the variation in the findings indicate a need for the continuous monitoring of fatness in the first year of life in order to determine the prevalence of obesity and the secular trends in various communities.

1.2. Childhood and Adolescence

Between infancy and school age there is little information regarding the prevalence of obesity. Estimates in schoolchildren, in the age range 6–14 years, in various countries vary between 2 and 15% with the prevalence increasing at adolescence. Obesity is more common among girls than boys and the difference in fatness between the sexes has been well documented in the Ten-State Nutrition Survey carried out in the United States between 1968 and 1970 (Garn and Clark, 1976) in which it was shown that, after the age of 3 years, females were fatter than males at all ages. This study also emphasizes the marked differences between boys and girls at adolescence; boys tend to lose their pubertal gain of fat, whereas girls continue to gain fat at an increasing rate from puberty through adolescence. This finding is in keeping with the high prevalence of obesity in 14-year-old English girls reported by Colley (1974) who found that 32% had triceps skinfolds greater than 25 mm, a value Garn and Clark (1976) would regard as indicating obesity. The corresponding prevalence in boys in Colley's study was 4%.

Differences in prevalence between different socioeconomic groups are well recognized in adults with fatness being more common in the so-called poorer social classes. Observations in children have given conflicting results; Whitelaw (1971), in a study of skinfold thickness in London schoolboys, found between 8 and 11% (depending on the method of definition) of boys from social classes IV and V to be obese compared with 5–7% of boys from classes I and II. He found obesity to become less frequent with increasing number of siblings. In the Ten-State study, however, children from higher socioeconomic groups were found to be fatter than those from lower income groups at virtually all ages through to puberty (Garn and Clark, 1976). Nevertheless, in later adolescence there was shown to be a reversal of this income-related fatness in the girls (though not in the boys), with the poorer girls becoming fatter and the girls in higher income families becoming thinner. Garn and Clark point out that the economic aspects of obesity in childhood are complex and changing and need to be observed in life-cycle context.

2. Natural History

In recent years the concept that fat babies grow into fat children who in turn become fat adults has been widely and somewhat uncritically accepted. As a result, there is a growing concern to prevent obesity in infancy together with a tendency to adopt a defeatist attitude toward the treatment of the older obese child. It is therefore important to examine the evidence upon which our present beliefs about the prognosis of obesity in childhood are based.

It has become customary to consider the natural history of obese infants separately from that of obese children and, although the division is somewhat artificial, the methods of collection of the data and the results vary so greatly that the distinction is probably valid at the present time.

2.1. Infancy

Infancy begins at birth and for the purposes of this chapter is taken to end at 1 year of age. Babies who are heavier than average at birth have long been known to remain heavier than average during childhood (Illingworth *et al.*, 1949) but this does not necessarily mean that they are fat. Studies of the birth weight of children found to be obese during childhood have given conflicting results; Wolff (1955) found that the birth weights of obese children did not differ significantly from those of normal weight children, whereas Mossberg (1948) and Börjeson (1962) found obese children to have higher than average birth weight, and more recently Shukla *et al.* (1972) and Sveger *et al.* (1975) have reported that obese babies had significantly higher birth weight than nonobese infants. As Whitelaw (1976) has shown, however, weight at birth may not be a particularly sensitive index of subcutaneous fatness; in a study of subcutaneous fat in infants he found no significant correlation between skinfold thickness at birth and at 1 year, nor any significant difference in skinfold thickness at 1 year between infants who were fat or thin at birth (Whitelaw, 1977).

The prognosis of fatness arising during the first year of life has been the subject of a number of studies. In England, Asher (1966) and Eid (1970) showed that babies who, at the age of 6 months, were overweight (and probably obese), were more likely to be overweight at the age of 5–7 years than were infants whose weight at 6 months was normal. However, despite a statistically significant difference between the prognosis for the groups of obese and nonobese infants, the majority (80%) of the children who had been overweight as babies were no longer overweight at school entry. This conclusion, which is not usually emphasized, is in keeping with the accepted figures for the prevalence of obesity in the two age groups (25–30% in babies and 3–6% during the school years). More recently, Poskitt and Cole (1977) in a follow-up of the infants originally reported by Shukla *et al.* (1972) found that although in infancy 14% of the children has been "obese" (more than 20% above expected weight) and a further 26% "overweight" (10–20% above expected weight), by 4–6 years most were of normal weight and fatness, with only 2.5% "obese" and 11% "overweight." Three of the five obese children were also obese as infants, but only one in nine obese infants was obese at 5 years. However, Sveger *et al.* (1975) in Sweden reported 50% of obese babies to be still obese at 1½–2½ years; in their study, though, the prevalence of obesity during the first year was about 6% compared with about 17% in the study by Shukla. It is possible, therefore, that there is a subgroup of obese infants whose prognosis differs from that of the larger number of fat babies reported in many studies. Court and Dunlop (1975) consider that some children whose obesity starts during the first 2 years of life do indeed fall into a distinct clinical entity. In these children the growth of fat follows a physiological pattern but at a level above the normal range of values; there is lack of weight reduction in response to long-term treatment; emotional problems are less marked than in other obese children; and plasma lipid concentrations are normal. These authors advocate that conventional treatment of weight reduction is inappropriate for this group of obese infants who will remain fat in spite of attempts at treatment.

That weight gain in early infancy is not necessarily a strong indicator of obesity at school entry is shown by the large and careful longitudinal study of Mellbin and Vuille (1973) in Sweden. They found only a weak correlation between velocity of weight gain in infancy and weight at 7 years and this was almost wholly confined to the boys.

A rather different perspective of the importance of obesity in the first year of life is given by retrospective studies of obese children. Several such studies have shown that nearly half of these children were already obese in infancy (Asher, 1966; Brook *et al.*, 1972; Court and Dunlop. 1975).

The degree to which obesity in infancy may influence the incidence of obesity in later childhood (as distinct from at school entry), adolescence, and even adult life has been the subject of much speculation and relatively little investigation. A retrospective study of a group of obese adolescent girls in the United States (Heald and Hollander, 1965) showed that these girls as a group had gained weight more rapidly during the first year of life, and had been heavier at the age of 1 year than their nonobese peers. A prospective study of children in Switzerland, however, showed little correlation between skinfold measurements at 1 year of age and those at puberty (Hernesniemi *et al.*, 1974). The conclusions of these two studies are not necessarily at variance because the Swiss workers studied an unselected population of children and their overall findings may not apply to obese children. In fact when the fatter babies in their study (skinfolds greater than the 75th percentile) were considered separately a positive correlation was found with tricep skinfolds for the girls at adolescence, and this is in keeping with the observations of Heald and Hollander. Recently a more ambitious investigation has studied the relationship between weight in infancy and obesity some 20–30 years later (Charney *et al.*, 1976). Although some criticisms may be made of the methodology, the conclusions that infant weight correlates strongly with adult weight independently of other factors such as education and social class level are probably valid. In terms of actual risk, 14% of heavy babies (greater than the 90th percentile) were found to be "obese" in adult life (more than 20% above standard weight) and a further 22% were "overweight" (10–20% above standard), this compared with figures of 5 and 11%, respectively, for babies of average weight (25th–75th percentile).

The important conclusions arising out of these various studies, and which are worth emphasizing, are that not all, and not even the majority, of fat babies will grow up to be fat children or fat adults. Nevertheless, the chance of a fat baby being fat in later childhood, adolescence, and adult life is significantly greater than that of a thin baby.

2.2. Childhood

Virtually all the information on the natural history and prognosis of obesity in childhood has come from studies on children referred to hospital clinics for treatment. Such samples are likely to be unrepresentative, a fact emphasized by the community study of Wilkinson *et al.* (1977) who found that only 4 out of 60 obese children age 10 years had been referred to hospital for treatment and only one-third of the group had sought advice from a doctor or had attempted dietary treatment.

In a recent review on current controversies on obesity Weil (1977) concludes that there is indeed a tendency for obese children to remain obese as adults but emphasizes that most adult obesity does not in fact arise in childhood.

Studies on children referred to hospital clinics have all given similar results; in the majority of these children obesity persisted into adult life, even though weight reduction was initially achieved with treatment (Haase and Hosenfeld, 1956; Lloyd *et al.*, 1961; Hammar *et al.*, 1971). In the study of Lloyd *et al.* (1961), 80% of their patients were obese when reexamined 9 years after their original attendance at a special clinic. There was a tendency for the degree of obesity to increase with increasing age so that the group of individuals over 20 years at follow-up was more overweight than the younger group aged 17–19 years.

3. Etiology

The increased storage of fat in obesity is due to an energy intake in the form of food in excess of the individual's requirements, including, in the case of children, requirements for growth. Many factors can disturb this energy balance and they differ in their importance, not only between individuals, but also in the same individual at different times. Although it is convenient to discuss the various etiological factors separately they seldom occur in isolation, and furthermore the etiology of obesity is complicated by the development of vicious circles, so that in an individual child it may be impossible to distinguish between cause and effect.

3.1. Genetic Factors

That there is an important genetic component in the etiology of childhood obesity is beyond dispute but the problems in separating the effects of heredity from those of the common family environment make it difficult to determine the size of the genetic influence. This difficulty applies to the interpretation of the classical study of Gurney (1936) who found that if both parents are obese, two-thirds of their children will be obese, whereas if one of the parents is obese, only half of their children will be obese, and to the more recent studies of Garn and Clark (1976) and Wilkinson *et al.* (1977). In the Ten-State Study (Garn and Clark, 1976) sibling correlations for fatness were 0.4 and parent–child correlations 0.3, and in the latter study the children of two obese parents were found at the age of 17 years to be three times as fat as the children of two lean parents. Garn and Clark, however, point out that lean wives tend to have leaner husbands, and fatter wives fatter husbands, and these husband–wife similarities may express common attitudes toward food, eating, and exercise, all of which factors may be more important in determining fatness (or leanness) in their children than heredity.

Accurate measurements of heritability (i.e., the proportion of the total variance of a characteristic in a population due to genetic causes) can only be derived from studies of twins. Studies based on weight show a close correlation between the weights of identical twins (Newman *et al.*, 1937; Bakwin, 1973). Brook *et al.* (1975) have examined factors in determining skinfold thickness in 222 pairs of like-sex twins of whom 78 were monozygotic and 144 dizygotic. The conclusions reached were that taking all ages together and for both sexes genetic factors appeared more important in determining trunk fat (estimated by subscapular skinfolds) than limb fat (estimated by triceps skinfolds). Genetic factors played a greater part in determining limb fat in girls than in boys. Above 10 years of age it was found that heritability was high for both trunk and limb fat in boys and girls, but for younger children environmental influences appeared to be of greater importance, and only in the trunk fat of younger boys was the degree of heritability high. The study contained only few twins who were actually obese and no special investigation of these was possible; the results of the study, therefore, strictly apply only to heritability of skinfold thickness over the normal population range, and could be different for individuals who are obese. Similar caution is needed in the interpretation of the other reported twin studies. The mechanisms whereby genetic factors influence fat deposition are not known.

3.2. Energy Intake

Energy is consumed as food. The composition of the food, and the periodicity with which it is eaten, as well as the absolute amount of energy derived from it, are all relevant to the etiology of obesity.

There is much speculation about the role of individual nutrients in causing obesity. Carbohydrates are generally considered to be "fattening," and the increased prevalence of obesity of children in the lower socioeconomic groups has been attributed to increased consumption of these relatively cheap foods (Whitelaw, 1971). Cook *et al.* (1973) found that children from the lower social classes and larger families had a higher proportion of their energy intake from carbohydrate and added sugar; however, they also found that the heavier children in their study had a lower sugar intake than the lighter children but considered that some of the restriction by the heavier children may have been intentional.

Periodicity of eating may be important in the regulation of energy balance, with a correlation between obesity and infrequent meals. Nibbling (the taking of small frequent meals) appears less likely to lead to obesity than gorging (the taking of large infrequent meals). Studies in adults have given discrepant findings (Garrow, 1974); in schoolchildren between 10 and 16 years of age one study has shown that those who ate three meals daily were fatter than those who ate five or seven meals a day (Fabry *et al.*, 1966). There is a clinical impression that many obese children do not eat breakfast and epidemiological studies are needed to investigate this association.

In normal individuals, at all ages and in both sexes, there is a large variation in energy intake (Widdowson, 1962) but the reasons for this wide range of nutritional requirements are not understood. The concept of nutritional individuality needs to be stressed and its neglect may result in the overfeeding of some children whose needs happen to be less than the "average standard requirement."

3.2.1. Infants

Babies who are breast-fed are less likely to become overweight than those who are bottle-fed (Fomon *et al.*, 1971; Shukla *et al.*, 1972; Neumann and Alpaugh, 1976). The reasons for this difference are not fully understood. One explanation is that artificially fed infants are more likely to have their nutritional individuality ignored by being urged to consume a predetermined volume of feed based on the standard "recommended intake" which may be inappropriate for the individual baby. Overconsumption of nutrients in cow's milk feeds can also arise because of inaccuracies in the reconstitution of the feed (Wilkinson *et al.*, 1973). Increased energy intake in infancy has also been attributed to the giving of nonmilk solids at an early age. There has been a tendency to start these foods, usually in the form of cereals, during the first 3 months and often before the age of 1 month (DHSS, 1974). Although Hutchinson-Smith (1970) found that early introduction of cereals was equally common among breast-fed and bottle-fed infants, Neumann and Alpaugh (1976) reported a much earlier introduction of solids in bottle-fed infants (mean 1.9 months) than in breast-fed infants (mean 3.9 months). Few actual attempts have been made to show whether there is a positive correlation between the early introduction of solids and the subsequent development of obesity; the study of Davies *et al.* (1977) failed to demonstrate any effect of the early introduction of solid foods on the growth of bottle-fed infants in the first 3 months of life.

There is some evidence in normal babies (Fomon *et al.*, 1971) and in malnourished children (Ashworth, 1974) that appetite can be adjusted to the nutritional status. Possibly the mechanism for this adjustment fails in some individuals; for instance, studies in heavy newborn infants have suggested that they have an appetite-regulating system which is relatively insensitive to internal cues of hunger and satiety (Nisbett and Gurwtiz, 1970). Hall (1975) postulated that in the newborn the appetite control mechanism is sensitive to the change in the composition of breast milk which occurs during the feed. This mechanism cannot operate in babies fed on cow's milk preparations whose composition does not alter during the feed.

A further factor may be related to differences in activity between breast- and bottle-fed infants. Breast-fed infants have been found to spend more time sucking at the breast (which is in itself hard work), more time awake, and more time being stimulated by their mother than bottle-fed infants (Bernal and Richards, 1970).

3.2.2. Older Children

For older children there is little evidence that obese children as a group eat more food than their nonobese peers. Population studies have shown that heavier children have a higher average daily energy intake than lighter children (Cahn, 1968; Cook *et al.*, 1973), but the heavy children were not necessarily fat. In groups of 14-year-old children Durnin *et al.* (1974) found that between 1964 and 1971 body fat had increased in the boys while over the same period there had been a decrease in daily energy intake of between 0.8 and 1.0 MJ (200–250 kcal). Among the girls, the fattest consumed consistently less energy than the thinnest. This study suggests that, at least in this age group, decreased energy output is of greater importance in determining fatness than excess energy intake.

3.3. Energy Output

Reduced energy output probably plays an important part in the etiology of obesity even though this is difficult to quantify. Bradfield *et al.* (1971) showed that adolescent girls in the United States spent on average less than 1 hr a day in moderate or strenuous physical activity, and of the remaining 15 waking hours, 9 were spent in light activity such as sitting while being transported to school, while at school, or when at home watching television. Durnin *et al.* (1974) showed that adolescent boys in Glasgow were fatter than a comparable group studied 7 years previously, despite a reduced energy consumption, and suggested that this was due to a reduced level of activity. It is possible that a similar pattern could be found in younger children.

Studies of energy output in children are hampered by methodological difficulties, and investigations in obese children have the further problem that reduced energy expenditure and/or physical activity does not necessarily imply etiological importance. In order to clarify the role of diminished energy expenditure in the causation of obesity children should be studied when no longer obese, or ideally even before becoming obese.

In conditions in which physical activity is grossly limited, for example meningomyelocele and myopathy, obesity is common and the contribution of low energy output to its causation is clear.

3.4. Metabolic Factors

In spite of extensive study of the pathways involved in the synthesis, storage, and lipolysis of adipose tissue triglyceride, it has not proved possible to establish a primary metabolic defect in the vast majority of obese individuals. The numerous metabolic abnormalities which have been described tend to be corrected by weight loss and are, therefore, likely to be results rather than causes of the obesity.

There are, however, a few disorders with an endocrine or metabolic basis in which obesity occurs and may even be the presenting feature in childhood. Juvenile hypothyroidism and Cushing's syndrome are both rare, but their diagnosis should be considered when an obese child is found to be below average height or below the height expected for his family; other clinical features will usually also be present. Children with growth hormone deficiency tend to be relatively heavy for their height and have increased skinfold thickness; loss of body fat occurs during treatment with growth hormone (Brook, 1973).

The obesity that frequently occurs in children receiving long-term therapy with corticosteroids is well recognized and measures should be taken to prevent it in all such patients.

Certain syndromes which may have a metabolic basis, although none has yet been demonstrated, present with obesity during childhood. In the Prader-Willi syndrome (Dunn, 1968) obesity is associated with mental retardation, short stature, and hypogonadism. In infancy hypotonia and feeding difficulties are the outstanding features and obesity does not usually develop until the second year of life. During the later years of childhood or adolescence diabetes mellitus may occur. Treatment of the condition is complicated by the fact that a low calorie diet may further slow down growth in height. Even rarer conditions which are associated with obesity are pseudohypoparathyroidism and the Laurence-Moon-Biedl syndrome.

3.5. Emotional Factors

The relative importance of emotional factors in the etiology of obesity is often difficult to assess in the individual child. In cases of extreme obesity, or where there is serious emotional disturbance, the situation is usually clear-cut, but in most children, once obesity has developed, the child's appearance often leads to teasing, which itself creates an emotional problem and thus sets up a vicious circle.

There are several psychological mechanisms whereby emotional disturbance can lead to a food intake in excess of physiological requirements. In early infancy the giving and taking of food is central to the mother–infant relationship. Normally, as the child grows older, the psychological significance of food becomes less, though even in the older child food continues to have important psychological meaning in addition to satisfying physiological needs. For obese children, eating may be a means of dealing with emotions which cannot be dealt with in other ways. The child may subconsciously wish to revert to babyhood, a stage of life when food was of prime importance in the mother–infant relationship.

Family eating patterns form an important part of emotional life and excessive eating may be part of the pattern. In the life of some families eating together has a particularly important place and any attempt to change the eating habits can meet with deep resistance.

Psychological disturbance in the parents (especially in the mother) is likely to be present when the obesity is gross. Some mothers feel themselves most adequate in their maternal role when their children are still babies, and may use food subconsciously as a means of maintaining the kind of relationship that existed when the child was an infant. Other mothers feel guilty because they believe that they do not love their child sufficiently and subconsciously give food in place of love. Whether the primary emotional disturbance is in the mother or child, it is bound to have secondary effects on the mother–child relationship and may also affect other members of the family.

4. Effects

4.1. Growth and Puberty

Overnutrition accelerates linear growth and thus the height of obese children tends to be above that expected for their family (Wolff, 1955); Brook (1972) has suggested that this effect is most marked when overnutrition occurs during the early years of childhood. Skeletal development also tends to be advanced and on average the bone age is about 1 year ahead of the chronological age (Mossberg, 1948). Likewise in both sexes the onset of puberty occurs about 1 year earlier (Wolff, 1955). The early onset of puberty is associated with early cessation of growth and the ultimate height of obese children is unlikely to be increased and may even be slightly below average (Lloyd et al., 1961).

4.2. Metabolic Effects

Many metabolic changes have been described in obese children; these include abnormalities of glucose tolerance and hyperinsulinemia, deficient secretion of growth hormone, increased secretion of cortisol with the appearance of an excess of metabolites in the urine, and abnormalities of serum lipids and lipoproteins. All these abnormalities can be reversed by weight reduction. Changes in glucose tolerance and serum insulin, and in serum lipids, are perhaps of special relevance as they may play a role in the later development of diabetes mellitus or atherosclerosis.

4.2.1. Glucose Tolerance and Serum Insulin

Between 15 and 30% of obese children show abnormalities in blood glucose response after an oral glucose load which can be classified as "chemical diabetes" (Chiumello *et al.*, 1969; Martin and Martin, 1973). Virtually all obese children show hyperinsulinemia after oral glucose; a positive correlation has been reported between the degree of insulin response on the one hand and the age of the children and duration of the obesity on the other (Parra *et al.*, 1971; Martin and Martin, 1973). The relationship between the hyperinsulinemia and the abnormality of blood glucose response is less clear; Paulsen *et al.* (1968) and Martin and Martin (1973) found hyperinsulinemia to be most marked in the children with the most abnormal blood glucose response, but Chiumello *et al.* (1969) found the opposite relationship.

Studies in adults have shown a strong positive correlation between the size of adipose cells and hyperinsulinemia (Stern *et al.*, 1972). Because weight loss is associated with a fall in serum insulin and a return of glucose tolerance to normal, and also with a decrease in the sie of adipose cells, it has been inferred that the large adipose cells are responsible for the insulin resistance. In obese children, however, Brook and Lloyd (1973) found only a weak correlation between the degree of hyperinsulinemia and the adipose cell size, and during treatment with a low calorie diet the hyperinsulinemia lessened rapidly, well before any significant reduction in cell size could have occurred. Reduced levels of serum insulin were only maintained while weight was being lost on the low calorie diet and, even after there had been considerable reduction in adipose cell size, hyperinsulinemia still recurred as soon as the intake of calories and carbohydrates was increased. These findings suggest that the enlargement of the adipose cells in obesity is not the cause of the hyperinsulinemia but rather that both abnormalities may be due to excess carbohydrate ingestion.

A relationship between obesity and clinical diabetes mellitus is well established in adults. Obese children, most of whom will become obese adults, are therefore predisposed to the later development of diabetes and the risk has been shown to be greatest for those most grossly overweight (Abraham *et al.*, 1971). Thus the prevention and treatment of obesity in childhood plays a part in the prevention of diabetes mellitus in the adult, although as Mann (1974) has pointed out such measures will have only a small effect on the overall incidence of diabetes in the community.

4.2.2. Serum Lipids

In obese children, as in obese adults, serum concentrations of nonesterified fatty acids (NEFA) in the fasting state are higher than in nonobese individuals (Theodoridis *et al.*, 1971; Fosbrooke *et al.*, 1971). Serum cholesterol and triglyceride concentrations are usually within the normal range (Spahn and Plenert, 1968; Fosbrooke *et al.*, 1971). A significant positive correlation has been found between serum cholesterol and obesity in a population study of 9- to 12-year-old schoolchildren in Holland (Uppal, 1974), and a similar correlation has been reported in obese adults (Montoye *et al.*, 1966). Correlations have also been found between cholesterol synthesis and adipose cellularity (Nestel *et al.*, 1973) and it is suggested that cholesterol is stored in fat tissue together with triglyceride (Nestel and Goldrick, 1976).

4.3. Cardiorespiratory Effects

The most dangerous effect of obesity is cardiorespiratory failure. This complication, which is rare in children, is also known as the Pickwickian or obesity-hypoventilation syndrome. The typical features are breathlessness, somnolence, cyanosis, tachycardia, raised central venous pressure, cardiac dilatation, hepatomegaly, and peripheral edema.

All children with gross obesity probably have a low chest wall compliance and reduced functional residual capacity, and a high oxygen cost of breathing. The lungs empty more completely than normal at the end of each breath, and as a result there is a high closing volume with some shunting of blood through poorly ventilated alveoli. This may cause intermittent but easily reversed cyanosis at rest. In some patients secondary alveolar hypoventilation supervenes with retention of carbon dioxide and a reduced ventilatory response to the increased partial pressure of carbon dioxide. Eventually secondary cardiac failure occurs, and deaths have been reported in childhood (Ward and Kelsey, 1962). This situation is a medical emergency demanding hospitalization, and its treatment consists of measures to relieve heart failure and reduce weight. Care must be exercised in the use of oxygen which may lead to severe underbreathing and oxygen narcosis. Respiratory function may not return to normal until a near normal weight has been achieved.

4.4. Emotional Effects

Obese children tend to be teased about their appearance at school and sometimes also at home. Many dislike organized games and other physical activities, not only because their performance is likely to be poor, but also because they may be embarrassed at changing in front of their peers. The accumulation of fat on the chest of boys may give the impression of breast development and this causes distress and anxiety that pubertal development may be abnormal. Many obese children deny that they are embarrassed by their appearance, and insist that they do not mind being teased or indeed that they are not teased. Beneath such denial, however, there usually hides much unhappiness which may show itself as aggression, depression, or withdrawal. Underachievement at school is probably common and leads to further emotional problems. School refusal may occur.

Lack of attractiveness, especially to the opposite sex, and difficulties in finding "fashionable" clothes aggravate the emotional problems for the adolescent.

4.5. Adipose Tissue Cellularity

When the amount of adipose tissue is increased, or in other words when an individual is obese, there must be an increase either in the total number of adipose cells, or in the size of the cells, or in both number and size. During the past decade much work has been undertaken in both experimental animals and humans to determine the effects of overnutrition on adipose tissue cellularity and the role, if any, of changes in number and size of adipose cells in the genesis of obesity. Major methodological problems have limited the acquisition of knowledge in the human, and especially in young children, and it is outside the scope of this chapter to discuss these here. The current position has been admirably reviewed by Hirsch and Batchelor (1976) who conclude that all human obesity is characterized by adipocyte hypertrophy, and that with increasing severity of the obesity hyperplasia becomes progressively manifest. Severity and hypercellularity are often found together in children and it has been suggested that these individuals are particularly difficult to treat although this supposition has not yet been proven. Hypercellularity can, however, also be demonstrated in adults with later onset of obesity although to what extent such individuals may have had their facility for adipocyte profileration determined in childhood is not known. As Hirsch and Batchelor (1976) conclude, "until the details of cellular development in man are more fully understood, the precise timing of 'critical' periods for cellular development must remain speculative." Unfortunately such speculation, although of great academic interest, does not at present advance the practical management of obesity and may even hinder it by introducing the nihilistic concept that nothing can be done for an obese person if, as an infant, he has already laid down too many adipose cells.

5. Treatment

It is generally accepted that obesity in children, as in adults, is a difficult and disappointing condition to treat, with 80% of patients relapsing in the long term (Lloyd *et al.*, 1961). Evaluation of treatment is, however, almost entirely confined to studies based on children attending hospital clinics and the outcome for children treated in other settings may be different. Because a "cure" is so uncommon it is pertinent to consider whether an attempt should be made to treat *all* obese children, but a good case can probably be made for an attempt at treatment in every instance in view of the role of obesity in the pathogenesis of much adult disease and disability including diabetes mellitus, osteoarthritis, and cardiovascular disorders, and because of the potentially serious emotional and cardiopulmonary consequences in childhood. It could be argued that for some children with associated defects, such as mental retardation, primary psychological disorders, meningomyelocele, or myopathy, treatment of the obesity imposes an additional and unnecessary burden on the child and his family. For such severely handicapped children, however, obesity undoubtedly makes management of the primary condition more difficult and thus it seems fully justifiable to institute treatment for it. Much more difficult is the decision of whether to abandon treatment in those children in whom it is unsuccessful, or is found to pose an intolerable burden.

In the management of the obese child, before explaining the details of treatment, reassurance should be given to the child and parents that no primary endocrine abnormality is present; such reassurance can usually be given without the need for special investigations if the child is above average height (taking into account of course the parents' height) and physical examination shows no abnormality which is not directly attributable to the obesity. An explanation of the probable causes of the obesity in the individual child should be given, and the concept of nutritional individuality explained with the corollary that not all obesity is due to gluttony. In view of the strong probability that other members of the family will also be obese (Garn and Clark, 1976) the opportunity should be taken to discuss management in a family context.

The principle of treatment is to ensure that intake of energy is less than the output. A reduction in food intake is the mainstay of treatment but an increase in energy output through increasing physical activity is also important, and the two approaches should be combined.

The practice of weight reduction is difficult; many methods have been described and successes claimed. No method can be regarded as entirely satisfactory for all patients, and there is need for a critical evaluation of current practice as well as of new methods. For any method it is necessary to know in what proportion of children weight loss is achieved, how much of the excess weight is lost, for how long satisfactory weight is maintained, what proportion of children relapse and at what interval after starting treatment, and how many children fail to attend follow-up appointments. Many reports purporting to show success of a particular treatment regime do not fulfill these criteria.

5.1. Diet

Diets must be adjusted to individual needs. The amount and type of food allowed will depend on the age, the degree of obesity, and individual requirements and preferences. Helpful information can be obtained by taking a careful dietary history. If this is done by a dietitian, a more accurate assessment will be obtained, but even a record obtained by a doctor will indicate whether the child has been eating excessively, or if his intake has been relatively small. When taking a dietary history inquiries must be made about drinks as well as about foods and about the number and frequency of meals and snacks.

For most obese children a diet providing 3.3–4.1 MJ/day (800–1000 kcal/day) will result in weight loss in most cases. Such a diet should contain approximately 60 g of protein and 40 g of fat; the main restriction will thus fall on carbohydrate which will be reduced to about 100 g. The food should be divided between not less than three meals, and children who do not eat breakfast should be encouraged to do so. Within the limits of the prescribed energy intake, diets should allow as much choice as possible, and rigid diet sheets which suggest foods the child does not like and the family cannot afford should be avoided. Since the ultimate aim is to achieve a pattern of eating which will enable the child, after the excess weight has been lost, to maintain a normal rate of weight gain, education of child and family in the principles involved should start as early as possible. The contribution of a dietitian, preferably with pediatric experience, is invaluable.

For preschool children, for whom there is less information about the management of obesity, a diet providing 2.5–3.3 MJ/day (600–800 kcal) is probably suitable. For the obese toddler between the ages of 1 and 2 years the aim should be to prevent further weight gain unless the obesity is gross. Because children at this age are growing rapidly in height, it is possible for an obese toddler whose weight is kept stable by moderate dietary restriction to "grow into his height" and "out of his fatness" in about 6–12 months.

For infants under 1 year of age it is less easy to give dogmatic advice, and because many will in any case lose their fatness spontaneously rigid diets need not be used. Advice is given about the general principles of infant feeding, and a diet suitable for a normal infant of that age is described with the aim of preventing further weight gain rather than achieving weight loss.

For children with another disability in whom obesity further limits physical activity, for example meningomyelocele or myopathy, very low calorie intakes may be needed to achieve weight loss. Children requiring intakes as low as 1.5 MJ/day (350 kcal) should probably be admitted to hospital, as such diets are difficult to provide at home.

An initial period of total starvation has been tried in the treatment for obese adults, but long-term results have been disappointing (Maage and Mogensen, 1970). Starvation continued until normal weight is achieved has also been attempted (Forbes, 1970; Rooth and Carlström, 1970; Munro *et al.*, 1970). Such treatment can only be carried out in hospital, may have undesirable side effects, and can even result in death (Spencer, 1968; Lloyd-Mostyn *et al.*, 1970); it cannot be recommended for children.

Ileal-bypass operations, which result in malabsorption of food, have been used in the treatment of gross obesity unresponsive to ordinary treatment. In adults encouraging results have been reported with the patients losing weight without the need for severe dietary restriction and often deriving much psychological and social benefit (Gazet *et al.*, 1974; Solow *et al.*, 1974). Side effects can, however, be serious especially in the immediate postoperative period, and deaths have been reported (McGill *et al.*, 1972). Experience with children is limited but in carefully selected cases considerable weight loss has been achieved and maintained during the first postoperative year, without impairment of growth in height (Randolph *et al.*, 1974).

The use of drugs to curb appetite and thereby reduce food intake is seldom advisable during childhood and adolescence. All available drugs have some effect on the central nervous system and may be habit-forming. In any case their therapeutic effect is limited and usually of short duration.

5.2. Physical Activity

Encouragement to become physically more active should be part of the management. Regular daily exercise taken as part of normal life, for example walking to school and walking up stairs, is more likely to be effective than short and irregular bursts of more strenuous activities. In the individual child details of physical activities obtained during the taking of the history may suggest ways in which energy output can be increased.

5.3. Psychiatric Treatment

Many obese children need psychiatric help, irrespective of whether they have a primary emotional disturbance or one resulting from the obesity. When the obesity is gross the associated emotional disturbance is usually also gross and generally requires urgent help. The relationship between mother and child is often severely disturbed and the mother may not be able to cooperate in efforts to reduce the child's weight. Irrespective of whether or not a psychiatrist is involved in the treatment, the pediatrician should be aware of the emotional problems, and should not regard the child who, despite dietary advice, fails to lose weight as "naughty" or "uncooperative."

Psychotherapy, although often successful in helping with the emotional problems and in improving the family relationships, is not necessarily effective in achieving weight reduction. The value of behavior modification in the treatment of obesity has not yet been critically assessed in the long term. Short-term studies in adults suggest that it deserves further evaluation (Stuart, 1967; Penick *et al.*, 1971; Levitz and Stunkard, 1974), and trials of behavior modification in children, as well as evaluation of other forms of treatment such as group therapy, including groups for the parents, are needed.

6. Prevention

Because treatment of obesity is difficult and the long-term results disappointing prevention should be the aim (Weil 1977). Preventive measures may be instituted in the prenatal period, in infancy, and in later childhood.

6.1. Prenatal Period

The avoidance of excessive weight gain during pregnancy is desirable from the obstetric point of view and may also play a role in preventing fetal overnutrition (Whitelaw, 1976). Antenatal care should include advice on infant feeding, and in particular the advantages of breast feeding should be emphasized; among these is the fact that obesity is less common in breast-fed babies than in those artificially fed. The mother's intention to breast-feed and the encouragement of her advisors are probably the most important factors in ensuring successful breast feeding. Some of the problems encountered in attempts to promote breast feeding and how they can be overcome have recently been summarized (Breast Feeding Symposium, 1976).

The disadvantages of overfeeding need to be explained to the parents and their relatives and the concept of nutritional individuality explained; what is an adequate intake for one baby may be excessive for another of the same age. Other points to be made are that a baby may cry because of thirst rather than hunger, and that it is usually unnecessary to give foods other than milk before the age of 4–5 months.

6.2. Infancy

Weight records are usually kept by all clinics and by many mothers; the original aim was the prevention of underfeeding, and in the developing world the weight chart is still an important tool in the prevention of malnutrition (Morley, 1977). Because in most of the so-called developed countries undernutrition has become uncommon, it has been suggested that there is no longer a need for routine weight records. However, the value of weight charts in the prevention of overnutrition deserves investigation. If an infant is found to be gaining weight at an excessive rate, his mother can be given appropriate advice, for example to make up more dilute feeds or give less solid food, especially cereals. In our anxiety to prevent infant obesity, however, we must not forget that undernutrition still poses a greater threat to health than overnutrition and that even apparently normal breast-fed babies may occasionally fail to thrive (Evans and Davies, 1977).

6.3. Childhood

In childhood, prevention of obesity consists mainly in establishing sensible eating habits and encouraging regular physical activity. When nutritional advice is given, cultural and social differences in eating patterns must be respected, and the expense and time involved in the provision and preparation of meals remembered. In the prevention of obesity broad guidelines can be suggested for all children and are not out of place for the rest of the family. Meals should be spaced throughout the day, and excess of carbohydrate, especially of refined products such as sweets, biscuits and the various proprietary sweetened drinks, should be avoided. For children in families with a high incidence of obesity, and for those who are beginning to become too fat, additional advice may include the avoidance of snacks between meals, and encouragement to satisfy appetite with vegetables and fruit rather than with foods of high energy density.

Prevention of obesity is particularly important for children with a disorder which interferes with physical activity, such as meningomyelocele and myopathy, and for children with severe emotional disorders and mental retardation.

6.4. Adolescence

A fairly rapid increase in body fat is a normal feature of puberty; however, as has been shown in the Ten-State Nutrition Survey (Garn and Clark, 1976), whereas boys in general tend to lose the fat put on in puberty, for girls this is unlikely to occur. Thus special attention needs to be paid to the prevention of obesity in girls during the years of adolescence. As at other ages, a family history of obesity is of particular relevance in concentrating the preventive effort on to these families; Garn and Clark (1976) have shown that by the age 17 years the children of two obese parents are three times as fat as the children of two lean parents. A more detailed discussion of the nutritional problems of adolescence and their prevention can be found in Chapter 6 in this volume.

7. References

Abraham, S., Collins, G., and Nordsieck, M., 1971, Relationship of childhood weight status to morbidity in adults, *Public Health Rep.* **86:**273.

Asher, P., 1966, Fat babies and fat children, *Arch. Dis. Child.* **41:**672.

Ashworth, A., 1974, Ad libitum feeding during recovery from malnutrition, *Br. J. Nutr.* **31:**109.

Bakwin, H., 1973, Body-weight regulation in twins, *Dev. Med. Child Neurol.* **15:**178.

Bernal, J., and Richards, M. P. H., 1970, The effects of bottle and breast feeding on infant development, *J. Psychosom. Res.* **14:**427.

Börjeson, M., 1962, Overweight children, *Acta Paediatr. Scand.* **51:**Suppl. 132.

Bradfield, R. B., Paulos, J., and Grossman, L., 1971, Energy expenditure and heart rate of obese high school girls, *Am. J. Clin. Nutr.* **24:**1482.

Breast Feeding Symposium, 1976, *J. Hum. Nutr.* **30:**223.

Brook, C. G. D., 1972, Evidence for a sensitive period in adipose cell replication in man, *Lancet* **2:**624.

Brook, C. G. D., 1973, Effect of human growth hormone treatment on adipose tissue in children, *Arch. Dis. Child.* **48:**725.

Brook, C. G. D., and Lloyd, J. K., 1973, Adipose cell size and glucose tolerance in obese children and effects of diet, *Arch. Dis. Child.* **48:**301.

Brook, C. G. D., Lloyd, J. K., and Wolff, O. H., 1972, Relation between age of onset of obesity and size and number of adipose cells, *Br. Med. J.* **2:**25.

Brook, C. G. D., Huntley, R. M. C., and Slack, J., 1975, Influence of heredity and environment in the determination of skinfold thickness in children, *Br. Med. J.* **2:**719.

Cahn, A., 1968, Growth and calorie intake of heavy and tall children, *J. Diet Assoc.* **53:**476.

Charney, E., Goodman, H. C., McBride, M., Lyon, B., and Pratt, R., 1976, Childhood antecedents of adult obesity: Do chubby infants become obese adults? *N. Engl. J. Med.* **295:**6.

Chiumello, G., Guercio, M. J. Del, Carnelutti, M., and Bidone, G., 1969, Relationship between obesity, chemical diabetes and beta pancreatic function in children, *Diabetes* 18:238.

Colley, J. R. T., 1974, Obesity in school children, *Br. J. Prev. Soc. Med.* 28:221.

Cook, J., Altman, D. G., Moore, D. M. C., Topp, S. G., Holland, W. W., and Elliott, A., 1973, A survey of the nutritional status of school children, *Br. J. Prev. Soc. Med.* 27:91.

Court, J. M., and Dunlop, M., 1975, Obesity from infancy: A clinical entity, in: *Recent Advances in Obesity Research* (A. Howard, ed.), p. 34, Newman Publ., London.

Crawford, P. B., Keller, C. A., Hampton, M. C., Pacheco, F. P., and Huenemann, R. L., 1974, An obesity index for six-month old children, *Am. J. Clin. Nutr.* 27:706.

Davies, D. P., Gray, O. P., Elwood, P. C., Hopkinson, C., and Smith, S., 1977, Effects of solid food on growth of bottle-fed infants in first three months of life. *Br. Med. J.* 2:7.

Department of Health and Social Security, 1974, *Present Day Practice in Infant Feeding*, Report on Health and Social Subjects No. 9, H. M. Stationery Office, London.

Dunn, H. G., 1968, The Prader–Labhart–Willi syndrome: A review of the literature and report of nine cases, *Acta Paediatr. Scand.* Suppl. 186:1–38.

Durnin, J. G. V. A., Lonergan, M. E., Good, J., and Ewan, A., 1974, A cross-sectional nutritional and anthropometric study with an interval of 7 years, on 611 young adolescent school children, *Br. J. Nutr.* 32:169.

Eid, E. E., 1970, Follow-up study of physical growth of children who had excessive weight gain in first 6 months of life, *Br. Med. J.* 2:74.

Evans, T. J., and Davies, D. P., 1977, Failure to thrive at the breast: An old problem revisited, *Arch. Dis. Child.* 52:974.

Fabry, P., Hejda, S., Cerny, K., Osancora, K., and Pechar, J., 1966, Effect of meal frequency in school children: Changes in weight:height proportion and skinfold thickness, *Am. J. Clin. Nutr.* 18:358.

Fomon, S. J., Thomas, L. N., Filer, L. J., Ziegler, E. E., and Leonard, M. T., 1971, Food consumption and growth of normal infants fed milk-based formulas, *Acta Paediatr. Scand. Suppl.* 233.

Forbes, G. B., 1970, Weight loss during fasting: Implications for the obese, *Am. J. Clin. Nutr.* 23:1212.

Fosbrooke, A. S., Brook, C. G. D., and Lloyd, J. K., 1971, Plasma lipids in obese children treated with 350 kcal diets, *Postgrad. Med. J. (June Supplement)*, 444.

Garn, S. M., and Clark, D. C., 1976, Trends in fatness and the origins of obesity, *Paediatrics* 57:443.

Garrow, J. S., 1974, in: *Energy Balance and Obesity in Man*, p. 150, North-Holland Publ., Amsterdam.

Gazet, J.-C., Pilkington, T. R. E., Kalucy, R. S., Crisp. A. H., and Day, S., 1974, Treatment of gross obesity of jejunal bypass, *Br. Med. J.* 4:311.

Gurney, R., 1936, Hereditary factor in obesity, *Arch. Intern. Med.* 57:557.

Haase, K.-E., and Hosenfeld, H., 1956, Zur Fettsucht im Kindesalter, *Z. Kinderheilk.*, 78:1.

Hall, B., 1975, Changing composition of human milk and early development of appetite control, *Lancet* 1:779.

Hammar, S. L., Campbell, V., and Wooley, J., 1971, Treating adolescent obesity: Long-range evaluation of previous therapy, *Clin. Pediatr.* 10:46.

Heald, F. P., and Hollander, R. J., 1965, The relationship between obesity in adolescence and early growth, *J. Pediatr.* 67:35.

Hernesniemi, I., Zachmann, M., and Prader, A., 1974, Skinfold thickness in infancy and adolescence, *Helv. Paediatr. Acta* 29:523.

Hirsch, J., and Batchelor, B., 1976, Adipose tissue cellularity in human obesity, in: *Clinics in Endocrinology and Metabolism*, Vol. 5, No. 2 (M. T. Albrink, ed.), p. 299, Saunders, Philadelphia.

Hutchinson-Smith, B., 1970, The relationship between the weight of an infant and lower respiratory infection, *Med. Off.* 123:257.

Illingworth, R. S., Harvey, C. C., and Gin, S. Y., 1949, The relation of birth weight to physical development in childhood, *Lancet* 2:598.

Levitz, L. S., and Stunkard, A. J., 1974, A therapeutic coalition for obesity: Behaviour modification and patient self-help, *Am. J. Psychiatry* **131**:423.

Lloyd, J. K., Wolff, O. H., and Whelen, W. S., 1961, Childhood obesity: A long-term study of the height and weight, *Br. Med. J.* **2**:145.

Lloyd-Mostyn, R. H., Lord, P. S., Glover, R., West, C., and Gilliland, I. C., 1970, Uric acid metabolism in starvation, *Ann. Rheum. Dis.* **29**:553.

Maage, H., and Mogensen, E. F., 1970, Effect of treatment on obesity: A follow-up of material treated with complete starvation, *Danish Med. Bull.* **17**:206.

Mann, G. V., 1974, The influence of obesity on health, *N. Engl. J. Med.* **291**:226.

Martin, M. M., and Martin, A. L. A., 1973, Obesity, hyperinsulinism, and diabetes mellitus in childhood, *J. Pediatr.* **82**:192.

McGill, D. B., Humphreys, S., Baggenstoss, A., and Dickson, E. R., 1972, Cirrhosis and death after jejunoileal shunt, *Gastroenterology* **63**:872.

Mellbin, T., and Vuille, J.-C., 1973, Physical development at 7 years of age in relationship to velocity of weight gain in infancy with special reference to the incidence of overweight, *Br. J. Soc. Prev. Med.* **27**:225.

Montoye, H. J., Epstein, F. H., and Kjelsberg, M. O., 1966, Relationship between serum cholesterol and body fatness, *Am. J. Clin. Nutr.* **18**:397.

Morley, D., 1977, Growth charts, "curative or preventive," *Arch. Dis. Child.* **52**:395.

Mossberg, H. O., 1948, Obesity in children: A clinical–prognostical investigation, *Acta Paediatr. Scand.* **35**:Suppl. 2, 1.

Munro, J. F., Maccuish, A. C., Goodall, J. A. D., Fraser, J., and Duncan, L. J. P., 1970, Further experience with prolonged therapeutic starvation in gross refractory obesity, *Br. Med. J.* **4**:712.

Nestel, P., and Goldrick, B., 1976, Obesity: Changes in lipid metabolism and the role of insulin, in: *Clinics in Endocrinology and Metabolism*, Vol. 5, No. 2 (M. J. Albrink, ed.), p. 313, Saunders, Philadelphia.

Nestel, P., Schreibman, P. H., and Ahrens, E. H., 1973, Cholesterol metabolism in human obesity, *J. Clin. Invest.* **48**:982.

Neumann, C. G., and Alpaugh, M., 1976, Birth-weight doubling time: A fresh look, *Pediatrics* **57**:469.

Newman, H. H., Freeman, F. N., and Holzinger, K. J., 1937, *Twins, a Study of Heredity and Environment*, University of Chicago Press, Chicago.

Nisbett, R. E., and Gurwitz, S., 1970, Weight, sex and eating behaviour of human newborns, *J. Comp. Physiol. Psychol.* **73**:245.

Parra, A., Schultz, R. B., Graystone, J. E., and Cheek, D. B., 1971, Correlative studies in obese children and adolescents concerning body composition and plasma insulin and growth hormone levels, *Pediatr. Res.* **5**:605.

Paulsen, E. P., Reichenderfer, L., and Ginsberg-Fellner, F., 1968, Plasma glucose, free fatty acids and immunoreactive insulin in 66 obese children, *Diabetes* **17**:261.

Penick, S. B., Filion, R., Fox, S., and Stunkard, A. J., 1971, Behaviour modifications in the treatment of obesity, *Psychosom. Med.* **33**:49.

Poskitt, E. M. E., and Cole, T. J., 1977, Do fat babies stay fat? *Br. Med. J.* **1**:7.

Randolph, J. G., Weintraub, W. H., and Rigg, A., 1974, Jejunal bypass for morbid obesity in adolescents, *J. Pediatr. Surg.* **9**:341.

Rooth, G., and Carlström, S., 1970, Therapeutic fasting, *Acta Med. Scand.* **187**:455.

Shukla, A., Forsyth, H. A., Anderson, C. M., and Marwah, S. M., 1972, Infantile overnutrition in the first year of life: A field study in Dudley, Worcestershire, *Br. Med. J.* **4**:507.

Solow, C., Silberfarb, P. M., and Swift, K., 1974, Psychosocial effects of intestinal bypass surgery for severe obesity, *N. Engl. J. Med.* **290**:300.

Spahn, W., and Plenert, W., 1968, Investigations on the treatment of obesity in childhood by total starvation. III Serum lipid patterns during drastic calorie restriction and total starvation, *Z. Kinderheilk.* **103**:13.

Spencer, I. O. B., 1968, Death during therapeutic starvation for obesity, *Lancet* **1**:1288.

Stern, J. S., Batchelor, B. R., Hollander, N., Cohn, C. K., and Hirsch, J., 1972, Adipose cell size and immunoreactive insulin levels in obese and normal weight adults, *Lancet* 2:948.

Stuart, R. B., 1967, Behavioural control of overeating, *Behav. Res. Ther.* 5:357.

Sveger, T., Lindberg, T., Weibull, B., and Olsson, U. L., 1975, Nutrition, overnutrition and obesity in the first year of life in Malmo, Sweden, *Acta Paediatr. Scand.* 64:635.

Tanner, J. M., and Whitehouse, R. H., 1975, Revised standards for triceps and subscapular skinfolds in British children, *Arch. Dis. Child.* 50:142.

Theodorodis, C. G., Albutt, E. C., and Chance, G. W., 1971, Blood lipids in children with the Prader–Willi syndrome: A comparison with simple obesity, *Aust. Paediatr. J.* 7:20.

Uppal, S. C., 1974, *Coronary Heart Disease, Risk Pattern in Dutch Youth,* New Rhine Publ., Leiden.

Ward, W. A., and Kelsey, W. M., 1962, The Pickwickian syndrome, *J. Pediatr.* 61:745.

Weil, W. B., 1977, Current controversies in obesity, *J. Pediatr.* 91:175.

Whitelaw, A. G. L., 1971, The association of social class and sibling number with skinfold thickness in London schoolboys, *Hum. Biol.* 43:414.

Whitelaw, A. G. L., 1976, Influence of maternal obesity on subcutaneous fat in the newborn, *Br. Med. J.* 1:985.

Whitelaw, A., 1977, Infant feeding and subcutaneous fat at birth and at one year, *Lancet* 2:1098.

Widdowson, E. M., 1962, Nutritional individuality, *Proc. Nutr. Soc.* 21:121.

Wilkinson, P. W., Noble, T. C., Gray, G., and Spence, O., 1973, Inaccuracies in measurements of dried milk powders, *Br. Med. J.* 2:15.

Wilkinson, P. W., Pearlson, J., Parkin, J. M., Philips, P. R., and Sykes, P., 1977, Obesity in childhood: A community study in Newcastle-upon-Tyne, *Lancet* 1:350.

Wolff, O. H., 1955, Obesity in childhood: A study of the birth weight, the height, and the onset of puberty, *Q. J. Med.* 24:109.

The Adolescent

Felix P. Heald

1. Introduction

Unlike the child who is growing at a constant rate or the adult who has reached physical and nutritional stability, the teenager is a rapidly changing biological organism. The different growth patterns and the individual biological makeup both exert a major influence on the nutritional needs of adolescents and further compound the general complexity of the growth and nutrition of adolescents. The basic problem confronting nutritionists who advise the teenager on nutritional matters is to develop a system whereby these individual variations can be taken into account.

Energy requirements are dependent on (a) physical activity, (b) body size and composition, (c) age, (d) climate and other ecological factors, and (e) sex. The teenager, however, must contend with an additional factor—that is, those nutrients that are necessary for growth. Since in-depth discussions of these five factors as they relate to energy and protein needs are fully discussed in standard text, this discussion focuses on the uniqueness of the adolescent and his particular nutrient requirements for growth.

Nutrient requirements are strongly influenced by the velocity of growth during adolescence. Until the ninth or tenth birthday, the schoolchild has a fairly steady increment in weight; a gain averaging 2.3–2.7 kg/year. Thereafter, there is a slow and steady increase in the weight increment curve. This represents the beginning of the adolescent growth spurt and is the only time in extrauterine life that the velocity of growth actually increases. Girls demonstrate this acceleration in growth earlier than boys. The most rapid spurt in linear growth in girls is between the 10th and 12th year, and in boys about 2 years later. This is frequently termed the year of maximum growth which for both height and weight is greatest in girls in the year prior to menarche. The

Felix P. Heald • Division of Adolescent Medicine, University of Maryland Hospital, Baltimore, Maryland.

weight increment curve for both boys and girls has a lower peak than height and lasts longer. In contrast to weight, the annual increments in height continually diminish from birth to maturity, except for the short period of the adolescent growth spurt.

During childhood, the height increment is approximately 5 cm/year. This diminishes slightly as age increases. The adolescent growth spurt begins in girls somewhere between the ages of 10 and 12 and in boys between ages 12 and 14. Linear growth ceases in girls at or somewhat later than the 17th year, but in boys may continue through and beyond the 20th year. The most rapid portion of the growth curve occurs in early to mid-pubescence, and growth thereafter gradually slows down. It is important to consider the fact that the linear spurt during adolescence contributes about 15% to final adult height in sharp contrast to the adolescent's contribution to the adult weight which is about 50%. Therefore, it is obvious that nutrition plays a significant role in the doubling of body mass during pubescence. Since nutritional requirements are quite closely related to the rapid increase in body mass, it is of little surprise that peak nutritional requirements appear to occur during the year of maximum growth.

There are three characteristics of puberty which influence the nutritional requirements of the adolescent: (a) the doubling of body mass during pubescence; (b) calorie and protein requirements are higher than at almost any other time of life; and (c) the adolescent appears to be more sensitive to caloric restriction than does the child or adult.

The National Research Council (1974) and the World Health Organization (1973) have considered the data available on teenagers and have proposed certain levels of nutrient intake. These recommendations represent the best judgment of nutritional experts for the nutrients required to meet the growth and health needs of most adolescents in the population. Frequently data on adolescents are not available from human experience and either animal or extrapolated data from other ages of the human must be relied upon. Therefore, recommendations are frequently the best estimates of intakes which can apparently support good health.

2. Secular Trends in Growth

The rate of pubertal maturation is a complex set of interrelationships between genetic endowment and its surrounding environment. If the environment is stable, the major variability in pubescence is genetic (Tisserand-Perrier, 1953). In the past century Europe, Japan, and North America have participated in a systematic lowering of the chronological age at which puberty occurs (Tanner, 1975). For example, the mean menarcheal age for Norwegian girls was 17 years in 1840, and by 1970 it was just over 13 years. Confirmation of this trend is evidenced by the fact that children and adults are getting taller and heavier. Fortunately, there are some signs that the secular trends have stopped in certain societies. In Oslo there has been no change in menarche from 1952 to 1970, and in London very little change from 1959 to 1966. In the United States Damon (1968) has published data which indicate a leveling off of adult height in well-off families. Thus, we may be witnessing maximum attainment of genetic growth when the environment is most favorable.

3. Menarche, Nutrition, and Growth

That nutrition exerts a major influence on growth, particularly in infancy, is well known from animal and human studies. The effects of long-term nutritional influences are less well documented. Overnutrition (obesity) is consistently characterized in adolescence by earlier maturation and menarche and greater skeletal maturation and growth (Bruch, 1939). Chronic undernuttrition during the entire growth period results in slowed skeletal growth and maturation, delayed menarche, and a prolonged growth period. The most pronounced retardation occurred just prior to menarche (Dreizen et al., 1967). However, these malnourished girls, followed into their 20th year, ended up just as tall as a group of well-nourished girls. Thus, it is clear that under- and overnutrition influence growth and maturation in the adolescent.

Frisch studied a large amount of data on growth and nutrition from South America and Asia (Frisch and Revelle, 1969). These authors concluded that the mean age of maximum increment in growth of height or weight of adolescent boys is a useful index of nutritional status of a region or country. Subsequently, Frisch developed a hypothesis that a critical body weight may trigger spurts in weight and height so characteristic of pubescence. Then the authors presented data to suggest that a critical weight (48 kg) (Frisch and Revelle, 1970) is associated with onset of menstruation. That is, regardless of chronological age, when the individual in general weighs 48 kg, menstruation begins. Recently a critique of the critical weight at menarche suggests that we exert some caution in accepting this hypothesis (Johnston et al., 1975). Though Frisch and Revelle's (1970) proposal is attractive, its validity must await the development and testing of further data.

4. Energy

There have been numerous attempts in the past to relate energy needs to height, weight, size, and body surface area, yet there is still no agreed upon expression of the energy needs of individual juveniles. This critical issue was discussed extensively in a recent review by Wait et al. (1969). Conclusions supported by data developed from a group of adolescents were that the relationship of total calories to height or calories per unit of height per age were preferred indices for determining caloric needs. Data of Widdowson and McCance (1936) lend general support to these observations. Therefore, it is possible to speculate that the increments in height during adolescence may best represent the anabolic effect of the juvenile growth period. During childhood and even more so in adolescence one must emphasize the large individual variation in caloric intake. This has been noted by a number of workers and is of special importance to those interested in nutrition in children and adolescents. Dukes, as cited by Widdowson and McCance (1936), was familiar with this phenomenon, for he says, "Individual differences require different amounts of food and this point is scarcely ever regarded. Some children are capable of consuming and do not seem to thrive without a large quantity of food while the same amount in others would prove a positive poison and cause ill health and disease." Though physical activity is a major determinant in total energy requiements, it may not be the whole explanation. For example, a manual worker of the same size and age may eat less than a sedentary worker. There is some correlation between size and caloric intake but this does not apply to every child. For example, Widdowson and McCance (1936) have observed that a small boy often eats more than a big one. Therefore, in considering the energy requirements of adolescents, the importance of individual variation from one adolescent to another in making recommendations for nutrition must be kept in mind. Further, previous growth and nutritional status must be considered in making current nutrient recommendations.

Sex difference between caloric intake is already apparent at age 6 when boys consume approximately 110 cal/day more than girls; at age 10, 200 more; at age 12, 300 more; at age 14, 400 more; at age 16, 630 more; and at age 18, 830 more.

5. Protein

There are many factors other than the physiological state of the individual and dietary intake which influence protein metabolism in the body. For example, there is the amino acid composition of the dietary protein, the adequacy of caloric intake, and the nutritional status of the teenager. The existence of an acute or chronic illness or disease state further alters the teenager's requirement for protein and amino acids. Different dietary proteins differ in their amino acid content, and the relative amounts of essential amino acids are a measure of the quality of the protein in food.

Although there have been a number of attempts to express dietary protein needs in terms of requirements for essential amino acids, inadequate information has made this task particularly difficult during adolescence. Data on amino acid requirements, although less complete than on total nitrogen requirements, have been used in making recommendations for dietary intake. The requirement for proteins is determined by the amount required for maintenance plus that needed for growth of new tissue, which during adolescence may represent a substantial portion of the total nitrogen needs. The current recommendations by WHO/FAO are listed in Table I and represent a safe level of protein intake. These values represent the total nitrogen requirements needed to meet the demands imposed by obligatory losses and growth plus a 30% increase on the basis of data from nitrogen balance studies and an addi-

Table I. Protein Requirements of Female and Male Adolescents[a]

Age (years)	Sex	Body weight (kg)	Obligatory N loss (mg/kg/day)	N need for growth (mg/kg/day)	Total	Safe level of intake[b] g protein/kg/day (milk or egg)
10	F	33.79	68	9	27	0.81
	M	33.93	72	6	78	0.82
11	F	37.74	64	8	72	0.76
	M	36.74	70	7	77	0.81
12	F	42.37	60	10	70	0.74
	M	40.23	66	8	74	0.78
13	F	47.04	57	7	64	0.68
	M	45.50	62	11	73	0.77
14	F	50.35	55	4	59	0.62
	M	51.66	59	9	68	0.72
15	F	52.30	54	2	56	0.59
	M	56.65	57	6	63	0.67
16	F	53.57	54	1	55	0.58
	M	60.33	57	4	61	0.64
17	F	54.20	53	1	54	0.57
	M	62.41	56	2	58	0.61

[a] From WHO (1973).
[b] Estimated total N requirement increased by 30% to coincide with balance data and further increased by 30% to cover individual variability.

tional 30% to cover individual variability. Recommendations made by FAO/ WHO will be generally adequate for all but perhaps 2.5% of individuals who might have physiological requirements above recommended levels. It is of some comfort to note that both the National Research Council and the FAO/ WHO recommendations for the protein needs of adolescents are quite similar.

6. Vitamins and Minerals

The vitamin and mineral requirements for adolescents are usually considered to be the smallest intake needed to prevent symptoms and deficiencies. Criteria that have been used to determine minimum requirements include (a) the intake of the nutrient in a population with no apparent deficiency symptom, (b) the observed levels of intake in populations in which deficiencies occur, and (c) the results of controlled experiments on human subjects to produce, prevent, or cure deficiencies. The amount of direct data on which the requirements for vitamins and minerals in teenagers are derived is very limited. Therefore, to a great extent the requirements are extrapolated from the data on infant and adult allowances. The reader is referred to the National Research Council and FAO/WHO publications for the specific allowances for each vitamin and mineral. Brief comments will be made about those vitamin and mineral requirements frequently discussed in the nutritional literature on adolescence.

6.1. Vitamin A

Vitamin A is frequently cited in nutritional surveys of teenagers as being considerably below the requirements for adolescents. However, there is little evidence of functional disability in teenagers with vitamin A deficiency. The problem, therefore, is whether the standards are set too high for this age group or whether we have not detected functional impairment of vitamin A deficiency.

6.2. Vitamin D

There are no data on vitamin D requirements in older children or adolescents. It is well known that vitamin D is needed to maintain the homeostasis of calcium and phosphate in the mineralization of bone. Part of the difficulty in recommended allowances is the variability of exposure to sunlight in meeting vitamin D needs. It should be noted that in situations in which teenagers are not exposed to the sun for long periods of time it may be necessary to increase exogenous sources of vitamin D.

6.3. Ascorbic Acid

There are some data on the adult requirement of vitamin C to prevent scurvy and the amount metabolized daily. However, there is a paucity of data for children and adolescents. Nevertheless, the recommendations for intake of vitamin C are somewhat higher than for other groups because of the unknown demands of growth.

6.4. Folacin

Because of its role in DNA synthesis, folacin is important during periods of increased cell replication and growth. The effectiveness of folacin depends in part on the adequacy of vitamin B_{12} stores in the body. Some data on the amount needed in adults to produce a hematologic response in individuals with a deficiency of the vitamin have been obtained. Values for adolescents can only be interpolated between the amount required by adults and children. There is little evidence in the literature of widespread folacin deficiency in the adolescent population.

6.5. Vitamin B_{12}

Vitamin B_{12} is required for the rapid growth of cells. Teenagers would appear to have increased needs for this vitamin. In addition, vitamin B_{12} has a role in fat, carbohydrate, and protein metabolism. Recommended amounts in the absence of any clear-cut data on adolescents are similar to those for adults.

6.6. Niacin, Riboflavin, and Thiamine

Niacin, riboflavin, and thiamine are all known to participate in energy metabolism. For this reason, the recommendations for intake of these vitamins have been based on caloric intake. Data on these vitamins are extremely limited, and there is still considerable debate about the amount required for normal adolescent growth.

6.7. Vitamin B_6

Vitamin B_6 is involved in a large number of enzyme systems associated with nitrogen metabolism. Its requirement in human beings is geared to protein intake. The recommendations for adolescents again are interpolated from data gathered from formula-fed infants or adults.

6.8. Minerals

The discussion of the mineral requirements of adolescents is limited to calcium, zinc, and iron. Obviously, the increment in skeletal mass associated with pubescence has a significant impact on the dietary requirements. It has been difficult to establish the mineral requirements for this age because current theoretical and technical knowledge has not allowed us to establish a data base with a high degree of precision.

An example is that our current information on calcium metabolism has to a large extent been based on calcium balance. Hegsted (1973) has suggested that calcium balance data merely reflect prior calcium intake and not calcium requirements. The recommendation of calcium intake from FAO/WHO is 700 mg/day for adolescents 11–15 years of age, and from 500 to 600 mg/day for ages 16–19.

Zinc is known to be essential for the growth process, and recent evidence has suggested that it is more important than we have previously realized. Zinc serves a number of important metabolic functions, including a role in insulin metabolism, nucleic acid and protein synthesis, gonadal development, wound healing, and several enzyme systems. Data from animal studies reveal that the earliest abnormality of experimental zinc deficiency is a decrease in growth rate. Thus, it may be very important to ensure an adequate zinc intake for rapidly growing children and adolescents.

Teenagers must consume enough iron to cover the losses in feces, urine, and skin, and provide for the growth of the red cell mass and other tissues during pubescence. The recommendations of the two committees are the same for adolescent girls, 18 mg/day, but the FAO/WHO set the standard for adolescent males at 12 mg/day, while the NRC's corresponding value for males is 10 mg/day. The extent of iron deficiency in adolescents is unknown in this country and may be higher than some of us have previously suspected.

7. Anemia

This is usually defined as below-normal hemoglobin concentration. It is argued that a more valid index of anemia is the evaluation of the total red cell mass. Hemoglobin concentration, however, represents the oxygen carrying capacity of the blood and thus it seems a more logical measurement. Hematocrit is an adequate screening test for anemia and in practice it usually is measured in conjunction with hemoglobin. From ages 6 to 12 there is a slight rise in hemoglobin concentration in boys and girls—to 14 g for boys and 13.7 g for girls by 13 years of age. Hemoglobin concentration in girls during the remainder of adolescence is slightly lower, whereas that of boys increases and achieves 15 g/100 ml by the 18th year compared to 12.0 g/100 ml in girls. Because of the changes in hemoglobin concentration as a function of age and sex during adolescence, it has been difficult to set a single level for anemia, particularly in males. Recently WHO has revised its definition of anemia (1968). For males above 14 years of age the value used is 13 g/100 ml and for nonpregnant adult females 12 g/100 ml. Ninety-five percent of normal individuals are believed to show hemoglobin levels higher than the values given, which are appropriate for all geographical areas. These values are for people residing at sea level, and must be modified for persons who reside at high altitudes. Although the concept of normal values is useful for population surveys, such screening does overlook the problem of the individual patient whose concentration may be within the population norm, but is subnormal when compared to their own normal values. This situation applies particularly to the early stages of iron, folate, and vitamin B_{12} deficiencies which are of nutritional significance in some teenage populations. In order to conduct appropriate nutritional surveillance throughout the world. it is necessary to develop centralized standards. The International Committee for Standardization in Hematology has fulfilled this function (WHO, 1968). For example, in the measurement of hemoglobin, cyanmethemoglobin solution is used as the standard, and WHO has established central reference laboratories to standardize the assays of iron, vitamin B_{12}, and folate.

7.1. Iron Deficiency Anemia

All indications are that iron deficiency has a very high incidence worldwide. In the United States its incidence has not changed over the past 30 years. Iron deficiency remains, therefore, a worldwide major nutritional deficiency in addition to the already documented energy and protein deficiencies. The sequential changes in the development of iron deficiency, particularly in adolescence and adults, are well known and have been widely described. With the development of clear-cut anemia and changes in the peripheral blood, the syndrome of iron deficiency is already in a fairly advanced stage. The practical problem has been the detection of cases of mild to moderate iron deficiency prior to the development of anemia. Recent studies have suggested that those individuals with mild iron deficiency may have subnormal mean corpuscular hemoglobin concentration (Natvig and Vellar, 1967). It has been demonstrated that when such subjects are fed medicinal iron their mean corpuscular hemoglobin concentration rises. Therefore, hemoglobin concentration and mean corpuscular hemoglobin concentration have been successfully applied to large population samples in extensive field testing for mild iron deficiency.

Iron deficiency in the developing countries is more than just a nutritional problem. The issue of general sanitation and its relation to parasitic infestation with considerable and chronic iron loss is a major factor in the pathogenesis of iron deficiency. It is important to determine on a regional or national basis whether iron deficiency is primarily a nutritional problem, or chronic iron loss due to parasitic infestation, or due to malabsorption. If anemia is due to parasitism, measures necessary to correct sanitation should be strengthened, particularly in areas in which hookworm infestation is prevalent.

Nutritional education programs should be focused at mothers and teenagers to encourage diets which include the eating of vegetables and other foods rich in iron. Several studies have shown that good results can be obtained by the administration of small amounts of iron to schoolchildren even when they are heavily infested with parasites.

7.2. Megaloblastic Anemia

The importance of megaloblastic anemia in the overall nutritional deficiency anemias of adolescence is not certain. In Canada, for example, Nutrition, Canada National Survey (1973) indicated that 10% of their adolescent population had low serum folates, but negligible megaloblastic anemia. In sharp contrast, a study in India reports that out of 132 children with severe anemia 72% had iron deficiency and 10% had nutritional megaloblastic anemia (Betkerur *et al.*, 1971). Further assessment of the overall worldwide importance of megaloblastic anemia in our adolescent population must await further data. The reader is cautioned about the necessity of making a precise diagnosis of the etiology of megaloblastic anemia as outlined in any major hematologic textbook.

8. Pregnancy

In order to more clearly understand nutritional requirements of the pregnant teenager, it is necessary to review briefly the setting in which teenage pregnancy occurs. The most definable landmark of puberty in the teenage girl is menarche. It has been generally agreed, though not well supported by data, that the early menstrual cycles of any girl tend to be anovular. Matsumoto *et al.* (1963) summarized this phenomenon as follows: "The ovarian function attained its maturity not with a leap but passing one after another through various phases following menarche." It appears then that full fertility is not usually attained until perhaps 2 years after menarche. The question does arise, particularly in the younger adolescent girl, whether there is significant competition between the nutritional requirements for growth and those for pregnancy. If this does occur frequently, does it have adverse effects? In teenage pregnancy there are essentially two groups of girls. First, there is the group of those who are still in or have just completed puberty and are age 14 years or less. Some of these girls may still be growing rapidly and may not be fully matured sexually. There is a second group of adolescent girls who are 15 or older and are nearly or completely physically and sexually mature. In both the United States and England the pubescent group constitutes a small minority (7% in Galveston) of pregnant females. In England, such an occurrence is a relatively rare event. Thus, Thomson (1976) has concluded that significant competition between the nutritional requirements for growth and those for pregnancy occurs so rarely that it is not worth considering for nutritional planning on a public health basis. However, there are exceptions to this rule and they need to be handled on an individual basis.

Normally, nutritional requirements increase during pregnancy to support the growth of the fetus, placenta, and maternal tissue. Of particular relevance to the pregnant adolescent are the carefully conducted studies of Calloway and colleagues who evaluated the energy and protein requirements of pregnant adolescents (King *et al.*, 1973). Their data suggested that nitrogen retention was much greater than previously reported. At the same time measures of urinary creatine excretion and total body potassium indicated there was an increase in maternal lean tissue during pregnancy estimated at approximately 3.6 kg, particularly during the last half of pregnancy. Primarily as the result of these studies it was advocated that the allowance for protein in pregnant teenagers should be increased 30 g/day up to 85 g/day through the pregnancy. Estimates of energy requirements indicate that adolescents though very sedentary needed at least 2400–2600 kcal/24 hr. Those teenagers who were more physically active or rapidly growing would require an additional amount of energy, perhaps 50 kcal/kg of pregnant body weight per day (Blackburn and Colloway, 1974). With the increased number of calories other nutrients were also thought to be increased. Thus, on the basis of more recent information it appears that the pregnant teenager, and particularly those who theoretically are still in their own growth phase, do have increased needs for both energy and protein during pregnancy.

9. Obesity

Body weight is made up of a number of compartments and organ systems including the adipose tissue organ. Obesity represents the enlargement of the normally occurring adipose tissue organ until it reaches such a size that we refer to the individual as being obese. Strictly speaking, when a person looks fat he is fat. This increase in the size of the adipose tissue organ in animals, and in all probability in the human, is multietiological, ranging from genetic to hormonal, to environmental, to psychological factors leading to excess size. Therefore, it is important early in our brief discussion of obesity to realize that it is a syndrome. General clinical studies assessing the incidence of obesity have relied upon height and weight relationships and to a lesser extent on more direct measures of adiposity, such as the measurement of subcutaneous fat. Therefore, our estimate of obesity from country to country varies from 6% in the Scandinavian countries up to 15% in adolescent females in the United States (Quaade, 1964). In adolescents there are two forms of obesity. First, the fat child who comes into adolescence and becomes increasingly obese, and second the adolescent who becomes obese. The basic question remains—what is the effect of adolescence upon either the development or the maintenance of obesity? Fat cells develop rapidly late in pregnancy through infancy. During childhood fat cell development plateaus. A final spurt in development of adipose tissue occurs during adolescence. The female adds more fat during adolescence than does the male and at the end of adolescence has twice as much body fat expressed as a percentage of total body weight as does a similar aged male. The evidence also suggests that the number of adipose tissue cells in all probability increases during adolescence. Thus, the size and number of adipose tissue cells become an important ingredient in interpreting the enlargement of the adipose tissue organ such as in obesity. Of most importance is the fact that in studies of weight reduction in teenagers the number of adipose tissue cells remain unaffected and stationary even though body weight and body fat may be lost. Thus, it now appears that the basic problem in the obese adolescent is hypercellularity of the adipose tissue organ. Once this occurs it apparently does not revert back to a more normal state. Therefore, a better understanding and better treatment for the adolescent will revolve around preventing the development of hypercellularity (Knittle, 1972). The metabolic and endocrine responses to the enlarged adipose tissue organ have been well documented, but their meaning is not clear. It now appears that experimentally induced obesity produces the same kind of metabolic and endocrine responses as naturally occurring obesity and, therefore, these changes may not be important from an etiological point of view, but represent the body's adaptive mechanisms to hypercellularity of the adipose tissue organ. Obesity is rarely associated with serious disease in children or adolescents and only occasionally does one need to be concerned about such endocrine disorders as Cushing's disease or other disorders such as Prader–Willi syndrome. In Cushing's disease cessation of growth is usually enough to distinguish it from the normally growing obese teenager with a healthy appetite. Obesity in the adolescent is also generally associated with accelerated growth, such as increases in height and in some cases lean body mass. Their bone age is accelerated and menarche or pubescence also appears to be accelerated. Thus the growth of the obese adolescent is in sharp contrast to hypocalorism where the general growth patterns and pubescence are delayed. The important thing for the health professional to determine is whether obesity is occurring in the presence of unusual eating patterns or whether there is any evidence of hyperphagia. If hyperphagia is clearly demonstrated, it is important to inquire into the family situation and see whether such hyperphagia is a result of unusual family stresses which are being manifested in the teenager's overeating. Frequently there is no such easy answer and one must look for more subtle causes. Treatment of the obese adolescent involves careful attention to whatever psychological factors may be associated with hyperphagia if it is present, encouragement of regular exercise, and a moderate diet that is consistent with adequate growth. Currently there is a great deal of interest in the use of behavioral modification in reducing energy intake. Although there are enthusiastic reports of its inital success for short-term weight loss, the eventual role of behavioral modification in weight control must await long-term evaluation.

What are the effects of over- and undernutrition on growth during adolescence? In the preceding discussion of obesity, it was pointed out that growth is affected. Height, weight, lean body mass (in some cases), and bone age are all accelerated as a function of age (Bruch, 1939). Puberty is reported as occurring earlier (Bruch, 1941). Thus, the effect of obesity is to accelerate the growth process. In contrast, undernutrition is associated with delayed growth and maturation. Dreizen reported on a group of adolescent girls followed longitudinally over two decades in the United States, and compared their growth to a control group. The effect of undernutrition was smaller stature, lower weight, delayed bone age, and delayed puberty during adolescence (Dreizen *et al.*, 1967). Most important, however, is the follow-up data of these girls as young adults. The undernourished girls were as tall and as heavy as their controls. It just took these girls longer to achieve their adult growth when compared to adequately nourished controls. A major and unresolved question on the effect of marginal nutrition on growth remains unresolved. No one will argue that continued and severe deprivation will permanently stunt the growth of animals and humans. What are the effects of early calorie–protein malnutrition on later growth, particularly during adolescence. The recent data suggest that the effects of this nutritional insult despite improved subsequent nutrition do result in measurable retardation of height and weight and a smaller skull at age 15 (Stoch and Smythe, 1976). The ultimate effect on adult proportions will have to await further reports from this important longitudinal study. The effect of undernutrition during the juvenile period on adult growth and longevity is a major and unresolved issue in child rearing. The large number of children exposed to undernutrition emphasizes the importance and urgency in resolving this issue.

10. References

Betkerur, V. N., Shah, D. S., Billar, P. K., and Shah, S. H., 1971, Patterns of anemias in childhood, *Indian Pediatr.* **8**:350.

Blackburn, M. L., and Calloway, D. H., 1974, Energy expenditure of pregnant adolescents, *J. Am. Diet. Assoc.* **65**:24.

Bruch, H., 1939, Obesity in childhood I. Physical growth and development of obese children, *Am. J. Dis. Child.* **58**:457.

Bruch, H., 1941, Obesity in relation to puberty, *J. Pediatr.* **19**:365.

Damon, A., 1968, Secular trend in height and weight within old American families at Harvard, 1870–1965. I. Within twelve four-generation families, *Am. J. Phys. Anthropol.* **29**:45.

Drizen, S., Spirakis, C. N., and Stone, R. E., 1967, A comparison of skeletal growth and maturation in undernourished girls before and after menarche, *J. Pediatr.* **70**:256.

Frisch, R., and Revelle, R., 1969, Variation in body weights and the age of the adolescent growth spurt among Latin American and Asia populations in relation to calorie supplies, *Hum. Biol.* **41**:536.

Frisch, R., and Revelle, R., 1970, Height and weight at menarche and a hypothesis of critical body weights and adolescent events, *Science* **169**:397.

Hegsted, D. D., 1973, Calcium and phosphorus, in: *Modern Nutrition in Health and Disease,* 5th ed. (R. S. Goodhart and M. E. Shils, eds.), Chap. 6, Section A, p. 268, Lea & Febiger, Philadelphia.

Johnston, F. E., Roche, A. F., Schell, L. M., and Wettenhall, N. B., 1975, Critical weight at menarche, *Am. J. Dis. Child.* **129**:19.

King, J. C., Calloway, D. H., and Margen, S., 1973, Nitrogen retention, total body ^{40}K and weight gain in teenage pregnant girls, *J. Nutr.* **103**:772.

Knittle, J. L., 1972, Obesity in childhood: A problem in adipose tissue cellular development, *J. Pediatr.* **81**:1048.

Matsumoto, S., Ozawa, M., Nogami, Y., and Ohashi, H., 1963, Menstrual cycle in puberty, *Gunma J. Med. Sci.* **12**:119.

National Research Council, 1974, *Recommended Dietary Allowances,* 8th ed., National Academy of Sciences, Washington, D. C.

Natvig, H., and Vellar, O. D., 1967, Studies on hemoglobin values in Norway, *Acta Med. Scand.* **182**:183.

Nutrition, Canada National Survey, 1973, Nutrition—A National Priority, Ottawa Department of National Health and Welfare.

Quaade, F., 1964, Prevention of overnutrition, in: *Symposia of the Swedish Nutrition Foundation* (G. Blix, ed.), p. 25, Almqvist & Wiksell, Uppsala.

Stoch, M. B., and Smythe, P. M., 1976, 15-year developmental study on effects of severe undernutrition during infancy on subsequent physical growth and intellectual functioning, *Arch. Dis. Child.* **51**:327.

Tanner, J. M., 1975, Growth and endocrinology of the adolescent, in: *Endocrine and Genetic Diseases of Childhood and Adolescence* (L. I. Gardner, ed.), Saunders, Philadelphia.

Thomson, A. M., 1976, *Nutrient Requirements in Adolescence* (J. I. McKigney and H. N. Munro, eds.), Chap. 15, p. 279, MIT Press, Cambridge, Mass.

Tisserand-Perrier, M., 1953, Etude comparative de certain processus de croissance chez les jumeaux, *J. Genet. Hum.* 2:87.

Wait, B., Blair, R., and Roberts, L. J., 1969, Energy intake of well-nourished children and adolescents, *Am. J. Clin. Nutr.* 22:1383.

Widdowson, E. M., and McCance, R. A., 1936, A study of English diets by the individual method, Part II, Women, *J. Hyg.* 36:293.

World Health Organization, 1968, Nutritional Anaemias, Technical Report Series, No. 405, Geneva.

World Health Organization, 1973, Energy and Protein Requirements, Technical Report Series, No. 522, FAO and WHO, Geneva.

The Adult

Napoleon Wolański

The period of adulthood has for many years been regarded as a phase of stabilization in the development process, which separates the period of "progressive development" from senescence. This conception of adulthood has also been applied to determine nutritional requirements and motor activity. Standards reflecting the physiological requirements have usually been determined according to only the body weight and degree of physical activity.

Many new data indicate that adulthood is also a period of relatively rapid transformations which, though less dramatic than the changes taking place in childhood and adolescence, are as important as those occurring in the aging period. Indeed, some elements of aging are present from the onset of adulthood; it can even be stated that in a way "progressive development" is itself a carrier of aging in the sense of irreversible changes. In fact, in this sense both the progressive and regressive changes have to be regarded by the nutritionist as developmental processes, since they represent transformations of the organism, as compared with the previous state. After the occurrence of these changes, the organism never returns to the previous state. This understanding is very important from the standpoint of nutrition—namely, it indicates that there are permanent requirements in both building and regulating nutrients, and not only of the energy-supplying nutrients. Thus, it is possible to influence by nutrition the developmental changes in the human not only in childhood and adolescence, but also in adulthood. However, the nutritional requirements of the organism are very different during active growth, differentiation, and maturation, as compared with the period of relative equilibrium between the anabolic and catabolic processes. These requirements are even more different at the time when the catabolic processes begin to predominate. To start with, let us consider the nature of adulthood.

Napoleon Wolański ● Department of Human Ecology, Polish Academy of Sciences, Warsaw, Poland.

1. Adulthood—Stabilization or Initial Period of Regression

Depending on the function or property under consideration, the beginning and end of adulthood occur at different periods of an individual's life. In principle, it is assumed that adulthood begins after completion of the skeletal growth and attainment of sexual maturity, together with full reproductive capacity (except for pathological cases). Consequently, the beginning of adulthood usually falls between the ages of 20 and 25 years, differing sometimes in various populations, depending on individual development rate which is related to the genetic properties of the given population and to its living conditions, including nutrition.

There is another completely different criterion for the attainment of adulthood, namely the maximum fitness of the organism or at least a state approaching this maximum. Doubtless, the attainment of the maximum value represents the most clear-cut threshold. However, it is evident that the maximum values (peaks) occur at different ages in an individual, depending on the function or property under consideration. For example, equilibrium sense and dynamic strength already exhibit a maximum at about 10–12 years of life, the trunk muscle static strength at about the 21st year of life, and body height at about the 35th year of life (Fig. 1). This picture was obtained for one particular population. However, in each population—depending on its living conditions, physical activity, nutrition, and other environmental circumstances—maximum values are attained at different ages and in a different sequence. With such a considerable time dispersion, it can be proposed to accept as the beginning of adulthood the age at which the equilibrium sense is best, and as the end of adulthood the age at which the last of all properties completes its progressive development (e.g., the breadth of the hand and its strength, as well as the mandible breadth—at 45–55 years). Although it is relatively easy to specify the final phase of adulthood, it is more difficult to assume the age of 10–12 years to be the beginning of adulthood. Presumably, the conditions of modern civilization are responsible for the fact that the regress in the equilibrium sense begins so early. Consequently, it seems advisable to accept certain intermediate criteria—that is, the period of attainment of complete reproductive efficiency and of union of the epiphyseal cartilage plates. Obviously, when expressed in calendar years, this age is different for various populations and social strata. It is also different when considered from the standpoint of various properties. Namely, at a given time some properties are still in the course of progressive development, whereas others are already regressing (Fig. 2).

It is particularly important that in the case of some properties a second peak is possible. These peaks have so far been demonstrated for the maximum oxygen uptake, minute heart volume, and tidal volume (Pyżuk *et al.*, 1974; Wolański, 1975a; Wolański and Ivanović, 1979). In six urban and rural populations of Poland and Yugoslavia, these peaks occurred at the age of about 45–50 years (Fig. 3). Perhaps the first peak represents the onset of adulthood and the second one, which has not yet been fully identified, its final stage.

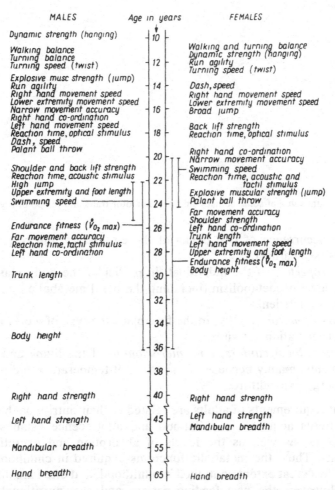

Fig. 1. Age at which the body reaches maximal values of the given traits: the largest dimensions, shortest time of reaction, best sense of equilibrium, greater maximal oxygen uptake, etc.

After the second peak of respiratory and circulatory efficiency, the organism enters into the period of senile involution, which is usually characterized by a drop in adaptability and an increase in the number of genetic information errors, as well as by many important metabolic changes.

According to the present interpretation, adulthood which follows the most intense transformations taking place in adolescence, is a period of considerable metabolic activity as well as of continuing development in the sense of rebuilding of the body tissues and improvement of the functions of the organism. This interpretation of adulthood is indispensable for gaining insight into the specificity of the nutritional requirements of adults. These requirements depend on the following:

1. *Developmental age*, because of the permanent structural rebuilding and functional improvement of the organism, as well as in connection

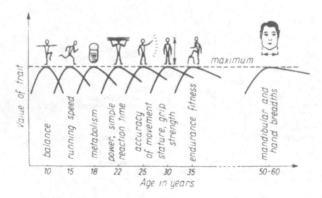

Fig. 2. Age of maximum value of some somatic and psychomotor traits and tempo of regression.

with the occurrence of two peaks of the circulatory and respiratory functions (see earlier).

2. *Genetic properties* of the individual, in particular the body mass and size, the level of metabolism (including the basal metabolic rate), and sensitivity to nutrients.

3. *The nature of motor activity*, in the first place the type of work, and/or sport and recreation activities.

4. *The altitude, local climate, and microclimate* of the living and work environments, mainly because of the effect of temperature and humidity on energy expenditure.

The nutrient requirements of adults are related to their nutritional history. The feeding patterns acquired in childhood and adolescence establish the metabolic pathways, as well as the levels of absorption and excretion of different nutrients. Thus, the metabolic functions acquired in childhood and adolescence are to a great extent continued in adulthood, and often are difficult to overcome. However, the past feeding pattern and past nutritional state cannot always be determined by the current nutritional situation. This is because some end results occur from the compensatory effect of realimentation—that is, from so-called catch-up (biological feedback). Therefore, there are different ways of attaining the "adult" value (Fig. 4). Thus, underfed

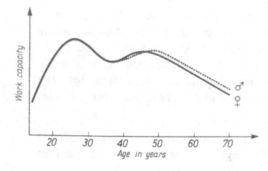

Fig. 3. Two peak values in maximal oxygen uptake, tidal volume, and minute heart volume in men and women, as observed in four Polish and two Yugoslavian populations.

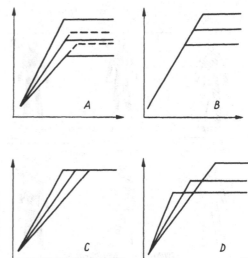

Fig. 4. Theoretical possible patterns of development rate and "inhibition" influencing the definite entity of the feature in adults. Model D is most frequently found in the analysis of phenomena of biological development in the human (see also Wolánski, 1978).

subjects usually exhibit in adolescence a smaller stature as compared with better fed subjects of the same age. However, malnutrition delays not only growth but also sexual maturation and ossification of epiphyseal cartilage plates; consequently, moderately undernourished subjects, though smaller than normal subjects of the same age, grow for a longer time, and may as adults attain an even greater body size than well-fed subjects (model D in Fig. 4). Thus, it has been found that girls with delayed maturity attain at puberty the stature of early maturing girls (Dreizen *et al.*, 1967; Tanner and Eveleth, 1976).

2. Nutrition and Ecosensitivity

The nutrient requirements of adults are related to the energy balance, as judged by the quantity of nutrients ingested, the basal metabolic rate, and degree of motor activity, as well as to the sensitivity of the subject to the different nutrients. Figure 5 shows a model of the latter relationship, which is less well known than the others mentioned.

2.1. Feeding, Ecosensitivity, and Nutritional Status

The question arises as to which kinds of subjects are more and which are less sensitive to nutrients (and to living conditions in general). According to studies on homozygous offspring of parents of positive assortative mating,* the heterozygous subject is physically better developed than homozygous

---

* Opposite to random mating; positive assortative mating if husband and wife are similar, negative if dissimilar.

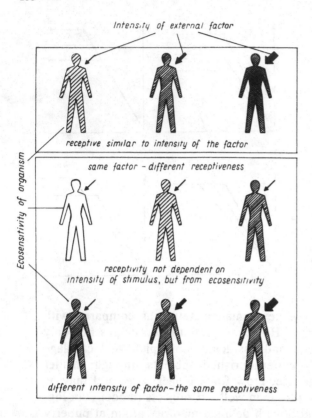

Fig. 5. Intensity of the environmental stimulus and the degree of body sensitivity to it: These are largely independent. As regards strength and duration of the stimulus, various degrees of perception may be observed (middle row); with different intensity of the stimulus, equal degrees of reaction of the organism may be noted (bottom row); finally, the intensity of the perception may correspond to the stimulus strength (upper row).

subjects of the same age in good living conditions (satisfactory nutrition) with respect to the property under consideration. Conversely, under conditions of under- and/or malnutrition, this type of individual exhibits a worse nutritional status than the homozygous subjects of the same age (Fig. 6) (Wolański, 1974a,b). The investigated groups compared with respect to living conditions (including feeding pattern, social status, etc.) showed no differences both in a well-fed population (city of Szczecin, Poland) and in an undernourished one (small individual farm villages). This observation points to the greater sensitivity of heterozygous subjects, as compared with homozygous ones, to a deficiency as well as to an excess of nutrients.

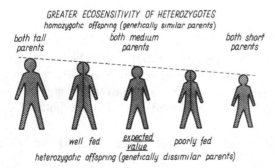

Fig. 6. Expected value of stature of homozygotic offspring, and stature of heterozygotic offspring in well and poorly fed.

In subsequent studies, it was found that these adaptative changes are characteristic of the properties which develop in more than 50% during the postnatal period. The foregoing example concerned stature. At the time of an infant's birth, this measurement is about 30% of the value observed in adults— that is, 70% of its development is after birth (Wolański, 1974c). In the properties that develop mainly during the prenatal period—the nose, face, and head widths, as well as the head circumference, develop in this period in over 50%—the opposite is observed. So these properties are less ecosensitive in heterozygous than in homozygous subjects (Wolański, 1974b,c). Obviously, it is always easier to influence the still developing properties; nevertheless, the difference in sensitivity between homo- and heterozygous subjects seems to be characteristic of adulthood also. In adults the trait values that develop mainly in the postnatal period are usually greater under good than under adverse living conditions, whereas the opposite is true for the properties that develop mainly in the prenatal period. Under conditions of adequate nutrition the properties that develop mainly in the prenatal period exhibit relatively little resistance and great plasticity, and under conditions of poor feeding they are characterized by relatively great resistance and little plasticity. In both cases, they show higher values in the offspring from endogamic than exogamic families (Figs. 7 and 8) (Kasprzak and Wolański, 1975; Wolański, 1975b).

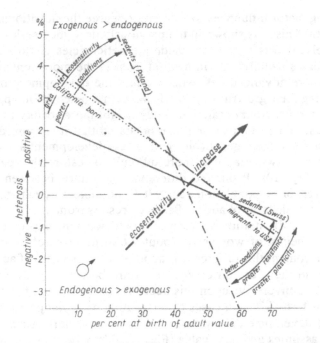

Fig. 7. Percentage of traits value in adult offspring (male) of exogenous couples, as regards offspring of endogenous couples in relation to the value of given traits at birth as a percent of the adult value (schematic picture) in four populations.

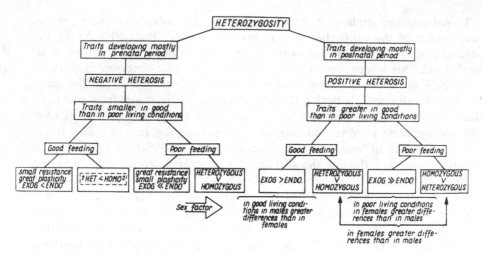

Fig. 8. Heterozygosity and heterosis effects under different living conditions and connected with sex factor.

2.2. Differences in Sex-Dependent Properties as a Criterion of Nutritional Status in Adults

The feeding factor influences to a different degree the nutritional status of men and women. This was shown in the previously described series of studies (Fig. 8). Namely, it was found that moderate deficiencies disturb the developmental processes usually only in men (at least in the more sensitive ones), whereas only more advanced deficiencies affect the development of women, e.g., by inhibiting their growth (Fig. 9). However, realimentation applied after the deficiency periods compensates the body height more readily in men than in women. Thus, in a certain sense it can be stated that it is more difficult to deviate women from their genetically controlled developmental pathway as compared with men, whereas it is more difficult to restore this pathway by realimentation (Fig. 10). From the difference in stature between men and women, observed on a population scale, it is possible to evaluate approximately the nutritional status of adults. Based on results from many populations, the difference in body stature between men and women, expressed as percentage of body height of women, is about 7.6%; in cases of retardation of development in women, this difference is about 9–10%. In cases of acceleration of development in men, the difference in stature between sexes amounts to about 7.8–8.5%. Conversely, upon this acceleration in women the respective difference drops to 6–7.7%. In cases of retardation of development in men or acceleration of development in women, the percentile difference in stature between sexes assumes a similar value (Fig. 10). Namely, upon retardation of development in men the difference is about 7.1–7.4% (Wolański and Kasprzak, 1976).

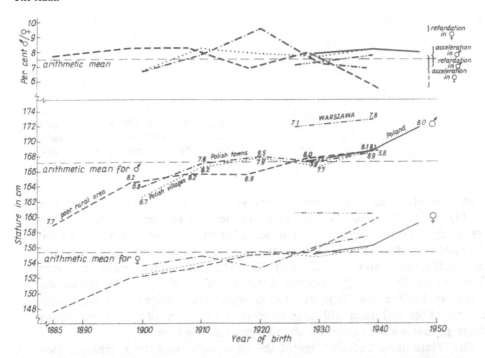

Fig. 9. Stature in adult males and females plotted according to year of birth (1885–1948) from urban and rural areas of Poland, and percentage of differences between the sexes in relation to female stature (mean value 7.6%).

3. Rate of Regressive Changes, and Living Conditions (Including Nutrition)

It has already been mentioned that different body functions attain a maximum level at different ages. It has been shown (Wolański, 1973, 1975b) that

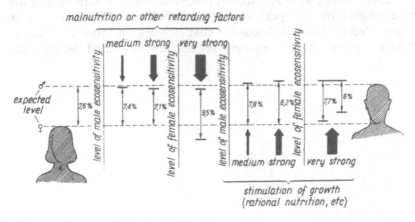

Fig. 10. Schematic picture of changes in stature under various living conditions, and differences between males and females stature.

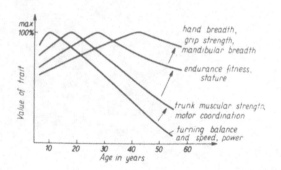

Fig. 11. Age of maximum value of some traits and reverse regression changes. Early peak, rapid regression; late peak, slow regression.

this age predetermines the regression rate of different functions or properties (Fig. 11). When maximum values of properties are attained at about 10–16 years of age, their regression, which begins at the time of attainment of the peak, proceeds at a rate of 5–10% every 10 years. If the peak of the function occurs at the age of about 18–25 years, then the regression proceeds at a rate of 3–6% every 10 years. If it occurs at the age of 30–40 years, the regression rate is only 1–3% every 10 years—for example, body height. The regression of the mandible and hand widths is slowest (0.1–1% per 10 years); the peaks of these properties are observed after 45 years of life (Fig. 1).

The relationship between age at the peak value and the regression rate was determined for different properties in a rural population characterized by an unsatisfactory nutritional status and by moderately heavy physical work (or else by heavy work with transient malnutrition, e.g., during preharvest and harvest). This relationship was verified in other populations with respect to one property, namely the endurance fitness determined from the maximum oxygen uptake (Fig. 12). Observations concerning this property generally confirmed this relationship. In relatively well-fed populations performing heavy physical work and living in a favorable climate (Pieniny, low mountains in south Poland) (Wolański and Pyżuk, 1972), the first peak of maximum oxygen uptake (in fact two peaks were found) was observed after the 35th year of life. For the rural population, the peak was attained before the 35th year of life (Wolański and Pyżuk, 1973). Subsequent studies were performed in farmers employed at the same time as workers in industry (so-called farmers-and-

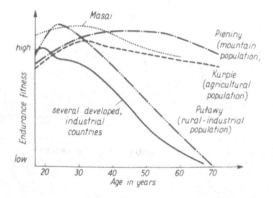

Fig. 12. Age of maximal value of endurance fitness (maximal oxygen uptake) and tempo of regression in different populations.

workers) with high earnings, but without any satisfactory dietary patterns, with a high work load (as farmers and industrial workers), and living under not very favorable environmental conditions in the vicinity of a big plant in Pulawy that produced nitrogen fertilizers. The results indicated that in these subjects the regression of the endurance fitness begins before the 20th year of life. This dependence is confirmed by studies performed in highly industrialized countries (United States, Canada, Sweden) (Shephard, 1969), where the peak of this function is attained during puberty, and its regression is very rapid. This relationship is also confirmed by investigations of African populations such as the Masai group, who live under primitive conditions. These populations exhibit a late peak of this function and consequently its later regression (Di Prampero and Cerretelli, 1969).

These observations stress the part played by nutrition in the regulation of endurance fitness, and the importance of nutrition and physical activity for prolongation of the high fitness period. The relationships between environmental factors and physical activity, on the one hand, and the age at the peak value of a function and the regression rate, on the other hand, are summarized in the scheme presented in Fig. 13. Environmental stimulation, together with motor activity (which requires adequate nutrition), prolongs the development of the different properties or functions. This, in turn, permits later attainment of the maximum value, delays the regression of the given property, and consequently causes its slower involution (Fig. 11). Physical activity, together with adequate nutrition, stimulates development, which results in a higher maximum value; consequently, the regression begins not only later, but also from a higher level of the given property or function (Fig. 13). Thus, a prolonged period of high fitness and vigor is attained for an individual.

4. The Effects of Nutrition in Adults

Variation in the composition of the diet is almost as important in adults as it is in children. This is due to the occurrence of transformations of different

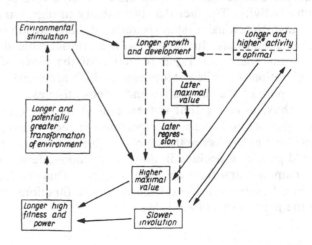

Fig. 13. Expected mechanism of regression connected with environmental and physical activity stimulation.

somatic, physiological, and psychomotor properties, which, as we have shown, continue in adults, as well as to the particular requirements of individuals according to their state of health, professional activities, mode of life, etc.

Deficiencies of the energy-supplying and building nutrients are compensated by smaller energy losses via spontaneous reduction of physical effort, or via utilization of the body's tissues for catabolic processes. In the first place, this utilization affects reserve tissues (Krzywicki *et al.*, 1968); this is an adaptation process causing a decrease in the basal metabolic rate and energy expenditure for work (Fabry, 1969). This is followed by a drop in the ability to undertake physical activity, which also reduces the energy expenditure.

Short-term undernutrition (even as excessively long break between meals) brings about in adults a deterioration of orientation ability, prolongs the reaction time, as well as causes nervous tension and irritability, thus impairing the individual's general fitness and resistance, decreasing the work efficiency, and promoting accidents.

Moderate chronic malnutrition in adults causes an adaptation of the organism which permits maintenance of protein and energy balances. However, this can induce hypovitaminoses, anemia, gastrointestinal diseases, osteoporosis, etc. It can also impair the efficiency of enzymatic systems (especially those of the liver), and thus reduce the ability to detoxify and excrete noxious substances (Pokrovski, 1969).

Inadequate dietary composition, faulty distribution of meals during the day, and especially an excess of food (mainly of sugar and fats) lead to an acceleration of the degenerative processes, obesity, and disturbances in the lipid and carbohydrate metabolism. All this results in a deterioration of the general state of health, and especially to an intensification of atherosclerotic changes.

Adequate nutrition in adulthood is very important for delaying the aging processes. It has been shown that unlimited food consumption in adolescence fails to be followed by an optimum performance of the body in senescence. Lower susceptibility to the degenerative processes of old age results rather from moderation in food consumption since adulthood. Old people ought to limit the caloric value of their diet, and be aware of the great importance of motor activity. The fact that their ability to digest and absorb nutrients from the gastrointestinal system is reduced, fails to lower their requirements of building nutrients. In parallel with a considerable drop in the efficiency of digestion, there is a rapid reduction of the mass of the active tissues and acceleration of senile osteoporosis, which are usually masked by an increase in the pool of the connective and adipose tissues.

Thus, covering (but not exceeding) the protein requirement with high-biological-value protein and meeting the requirements of other building and regulating nutrients are the crucial problems in the nutrition of elderly persons. In old people, especially in those living alone, deficiencies of protein, vitamin C, thiamine, and folic acid are frequent. This reduces the efficiency of the different systems in the organism, whose functions are additionally impaired by the previously suffered diseases.

5. Nutritional Requirements in Adulthood

Because of the increase in mass of adipose tissue with age, observed in the most industrialized countries, the nutritional requirements of adults after the 30th year of life cannot be established on the grounds of their current body weight. After the 25th year of life, there is no further increase in the lean body mass (LBM). Thus, the nutritional requirements ought to be determined according to the body weight measured before the 25th year of life. This body weight has to be expressed relative to the stature, for in the 25th year of life the given person could, in fact, be obese or at least be overweight relative to body height. It is best to determine the fat-free body (FFB) or the lean body mass (LBM), and to refer the nutrient requirements to these values. However, after the 40th year of life LBM decreases; this is obviously indicated by the decrease of muscle strength, frequently with the maintenance of body weight or its increase, but with a rise of the adipose and connective tissues. Thus, with age, the percentage of fat in total body weight increases, whereas the nutritional requirements drop. Therefore, for an evaluation of the nutritional requirements it is necessary to measure first the stature and second the body weight. The resulting values are compared with the regional standard values obtained for presumably well-fed groups of population, with neither food deficiencies nor excessive food intakes.

In establishing nutritional recommendations, the degree of physical activity must be taken into consideration. At least it is necessary to distinguish between rural populations, mostly with moderate physical activity, and urban populations characterized by great differentiation of work load, ranging from light to moderate. In industrialized countries, men usually have light work loads in 50% of cases and moderate work loads in the other 50%. In developing countries, less than one-half of men perform light physical work. The majority of women in the industrialized countries carry out light physical work, and in developing countries moderately heavy work.

Of course, climatic factors have to be taken into account. In hot climates, lower food rations are required because of the lower energy needs. In a cold climate, the energy requirements are higher, and thus the supply of energy-giving and protein foods ought to be higher. The intensified heat production and energy utilization for work increase the energy requirements in a cold climate. In addition, in a hot climate people are probably spontaneously less active physically than in a cold climate.

It can be expected that with further progress toward mechanization and automation in industry, energy requirements will drop. However, in this connection no reduction of the food consumption, but rather an increase in the motor activity during leisure time should be anticipated. A limitation of food consumption ought to be regarded as an undesirable extreme measure.

New data on covering the energy and protein requirements are given by the Joint FAO/WHO Ad Hoc Expert Committee on Energy and Protein Requirements (1973) Report. Satisfaction of the requirements of different amino acids is of special importance. To determine the amino acid pattern and scoring, the following formulas are applied (Energy and Protein Requirements, 1975):

Approximate net utilization figure:

$$\text{Amino acid score} \times \frac{\text{digestibility}}{100}$$

For evaluation of dietary protein need:

$$\text{Safe level of egg and milk protein} \times \frac{100}{\text{amino acid score}} \times \frac{100}{\text{digestibility}}$$

However, it has to be borne in mind that at present in our industrial civilization, contamination (pollution) of food and water may disturb the digestibility of foods. Thus, knowledge of individual requirements and determination of the nutritional value of foods may prove to be insufficient in the face of food contamination.

Present-day diets contain too much fat, but not enough of the indispensable unsaturated fatty acids. Consumers offered high fat dishes ought to consciously limit their intake, for if eaten in excess, these meals promote the development of obesity and atherosclerosis. The percentage of calories derived from fats should not exceed 25–35% of the total daily allowance of calories. As for amino acids, the composition of the ingested fats must be taken into consideration.

Sugar consumption ought to be limited, especially in adults, so that the percentage of calories obtained from carbohydrates be less than 50% of the total calorie allowance. The remaining components of the diet should constitute about half the energy. Milk intake may need to be controlled in adults of those populations with genetic lactose deficiency (McCracken, 1971).

The recommended calorie allowance for adults with moderate activity is now considered to be from 40 (Calloway, 1975) to 46 kcal/kg of body weight (Beaton and Swiss, 1974). If the activity is low, then less food may be taken, but in this case, a higher protein–energy ratio in the diet is required. Minimum energy requirements corresponding to such indispensable activities as washing and dressing amount to about 1.5 BMR (basal metabolic rate) (Payne, 1975).

Cépède (1975) reports the diet composition, at each food level, in "vegetable," i.e., "original," calories of the population's standard diet, as the participation of calories supplied by three food groups: P (animal products such as meat and milk), L (fats), and G (cereals, roots, tubers, sugars, pulses, fruits, etc.):

$$0.01G + 1.56L - 0.86P = 0$$

$$\text{with} \quad P + L + G = 100$$

This author defines the "original," i.e., "vegetable," calories as calories of vegetable origin × 7 (calories of animal origin).

The important question arises whether (in the light of the tendency to reduce the calorie consumption) in adults the requirements of vitamins B_1, B_2, and PP (and perhaps of other nutrients as well) ought to be related to the caloric value of the diet. It seems advisable to maintain the vitamin intakes unchanged, in spite of a reduction of the caloric value of food; however, this problem calls for further study. Perhaps, because of expected future changes in mode of life and the anticipated decrease in the physical activity, there may take place a rise of the requirements of the building and regulating nutrients, e.g., folic acid, vitamin B_{12}. These problems have not yet been fully elucidated.

6. Assessment of Nutritional Status

The problems of assessing nutritional status have been dealt with in many studies (e.g., Jelliffe, 1966; Roche and Falkner, 1974). The essence of the problem, as concerns adults, consists in the difficulty of using body weight for an evaluation of the nutritional status. Plainly, the present tendency to increased fat deposition with age in the population of the developed countries masks the reduction of LBM and renders such an evaluation difficult. Therefore, evaluation according to stature (height) seems to be more reliable. Deficiencies in body height probably testify to dietary deficiency in childhood and/or adolescence. Nevertheless, normal stature in adulthood can be attained even after extreme malnutrition, and, in addition, application of this criterion of nutritional status in influenced by superimposing two phenomena: secular trend and tendency toward the arithmetic mean (Fig. 14). Since these phenomena are superimposed differently in groups of short persons as compared with groups of tall individuals, the resulting evaluation may be faulty. For example, if the progeny of two parents of low stature is taller, then it would be indicated by the mean stature of both parents: this fact accentuates the secular trend pointing to an increase in the mean stature of the populations of the developed countries. On the other hand, if in a group of subjects who are the offspring of two tall parents, there is a tendency to lower stature, then it would follow from the mean stature of both spouses; this is in only some measure compensated by the secular trend.

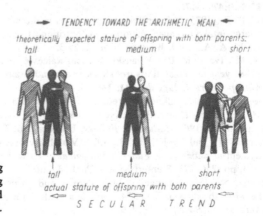

Fig. 14. Expected body height in offspring of two tall and two short parents, according to tendency toward arithmetic mean and modifications connected with secular trend.

In this situation, assessment based on lean body mass or the methodically more difficult fat-free body, and the evaluation of the mineral reserves in the osseous tissue (by evaluation of the ratio of the bone optical density to the cortical thickness) (Wolański, 1967) are, in fact, the only somatic indices of the nutritional status. Difficulties in application of these methods in adults further increase the need for the development of methods of clinical evaluation of nutrient deficiencies, based on the observation of degenerative changes in different organs and of their functions.

Adequate nutrition of adults, thus, is not only decisive in work efficiency, general well-being, and prevention of diseases, but also prolongs the period of full efficiency of the body and delays the aging processes.

7. References

Beaton, G. H., and Swiss, L. D., 1974, Evaluation of the nutritional quality of food supplies: Prediction of "desirable" or "safe" protein: calorie ratios, *Am. J. Clin. Nutr.* **27**:485.

Calloway, D. C., 1975, Nitrogen balance in men with marginal intakes of protein and energy, *J. Nutr.* **105**(7): 914–923.

Cépède, M., 1975, Sociology and nutrition, *Food Nutr.* **1**(2):8–10.

Di Prampero, P. E., and Cerretelli, P., 1969, Maximal muscular power (aerobic and anaerobic) in African natives, *Ergonomics* **12**:51–59.

Dreizen, S., Spirakis, C. N., and Stone, R. E., 1967, A comparison of skeletal growth and maturation in undernourished and well-nourished girls before and after menarche, *J. Pediatr.* **70**:256–264.

Energy and Protein Requirements, 1975, Recommendations by a joint FAO/WHO informal gathering of experts. *Food Nutr.* **1**(2):11–19.

Fabry, P. 1969, *Feeding Pattern and Nutritional Adaptations*, Academia, Prague.

Joint FAO/WHO Ad Hoc Expert Committee on Energy and Protein Requirements, 1973, Report. FAO Nutr. Meeting Rep. Series No. 51, WHO Tech. Rep. Series No. 522.

Kasprzak, E., and Wolański, N., 1975, Ecosensitivity, exogamy-endogamy and heterosis. *Proceedings of International Meeting on Human Ecology*; pp. 461–465, Georgi Publishing, Vienna.

Krzywicki, H. J., Consolazio, C. F., LeRoy, O. M., and Johnson, H. J., 1968, Metabolic aspects of acute starvation. Body composition changes. *Am. J. Clin. Nutr.* **21**:87–97.

McCracken, R. D., 1971, Lactase deficiency: An example of dietary evolution, *Curr. Anthropol.* **12**(4–5):479–517.

Payne, P. R., 1975, Safe protein: calorie ratios in diets. The relative importance of protein and energy intake as causal factors in malnutrition. *Am. J. Clin. Nutr.* **28**:281–286.

Pokrovski, A. A., 1969, *Woprosy Pitanija* **28**:3.

Pyżuk, M., Ivanović, B., Wolański, N., and Raicević, E., 1974, Some respiratory properties of 14–80-year-old inhabitants from seaside and mountain areas in Poland and Yugoslavia, *Glas. Antropol. Drust. Jugoslav.* **11**:O33–45.

Roche, A. F., and Falkner, F. (eds.), 1974, *Nutrition and Malnutrition. Indentification and Measurement*, Plenum Press, New York.

Shephard, R. J., 1969, *Endurance Fitness*, University of Toronto, Toronto.

Tanner, J. M., and Eveleth, P. B., 1975, Variability between populations in growth and development at puberty, in: *Puberty, Biological and Psychosocial Components* (S. R. Berenberg, ed.), pp. 256–273, Stenfert Kroese Publ., Leiden.

Wolański, N., 1967, Changes in bone density and cortical thickness of the second metacarpal between the ages of 3 and 74 years as a method for investigating bone mineral metabolism, *Acta Anat. (Basel)* **67**(1):74–94.

Wolański, N., 1973, Rate of involutional changes in Polish rural populations, *Stud. Hum. Ecol.* **1**:163–166.

Wolański, N., 1974a, Offspring growth and assortative mating of parents, *Stud. Hum. Ecol.* **2**:7–75.

Wolański, N., 1974b, Biological reference systems in the assessment of nutritional status, in: *Nutrition and Malnutrition. Identification and Measurement.* (A. F. Roche and F. Falkner, eds.), pp. 231–269, Plenum Press, New York.

Wolański, N., 1974c, The problem of heterosis in man, in: *Biology of Human Populations* (W. Bernhard and A. Kandler, eds.), pp. 16–30, Fischer, Stuttgart.

Wolański, N., 1975a, Endurance fitness, their development and regress in the light of genetic and ecological studies, *Zaszyty Nauk. Akad. Wychow. Fiz.* **19**:21–46.

Wolański, N., 1975b, Human ecology and contemporary environment of man. XIV Kongres Antropologa Jugoslavije, Zagreb.

Wolański, N., 1978, Secular trend in man: Evidence and factors, *Coll. Anthropol.* **2**:69–86.

Wolański, N., and Ivanović, B., 1979, Physical work capacity of 14–80-year-old inhabitants from seaside and mountain areas in Poland and Jugoslavia, in press.

Wolański, N., and Kasprzak, E., 1976, Stature as a measure of effects of environmental change, *Curr. Anthropol.* **17**(3):548–552.

Wolański, N., and Pyżuk, M., 1972, Morphophysiological characters and physical work capacity in 15–72-year-old inhabitants of low mountains, Pieniny Range, Poland, *Hum. Biol.* **44**(4):595–611.

Wolański, N., and Pyżuk, 1973, Morphophysiological properties in 15–80-year-old inhabitants of rural areas of different mode of life and climatic conditions, *Probl. Uzdrowisk.* **6**:47–62.

Growth Monitoring and Nutritional Assessment

13

Optimal Nutritional Assessment

Stanley M. Garn

1. Introduction

Nutritional surveys and nutritional studies differ very markedly in scope, as in resources, funding, extent of personnel, and facilities for data analysis and the preparation of reports. Many are truly minimal, making use of available personnel and with scanty laboratory support. A few are extensive, involving multiple coordinated field teams, mobile laboratories, and backed by computer analysts, biostatisticians, and professional report writers. There is a tendency to confuse size with success and maximum funding with optimal results. Often, this is not the case.

The optimal survey may not be large, in terms of numbers, but it has a well-defined purpose. Ideally, it is directed to a specific population or population subgroup and with specific nutritional problems in mind. And these may be indicated by previous "spot" surveys, or from available indications of nutritional problems, or from ancillary data, such as infant mortality statistics. Yet the optimal survey must also be large enough, in terms of numbers studied at each age, so that sample sizes (after inevitable losses) will still be large enough to provide useful and meaningful results. The design of the study is therefore the first optimum to consider.

The optimal survey includes demographic and socioeconomic data both in the planning and in the field. It includes enough anthropometric information to appraise size and growth, and sufficient hematologic and biochemical variables to define the nutritional problems actually encountered. It involves personnel both carefully trained and sufficiently motivated, and procedures for testing data quality and observational accuracy. It includes provisions for screening and review of individuals apparently "at risk," and provisions for additional follow-up studies and review. Finally, the optimal study continues

Stanley M. Garn • Center for Human Growth and Development, University of Michigan, Ann Arbor, Michigan.

well after the field phase, through data analysis and completion of useful records.

In many respects, therefore, the optimal study is no more than a sufficient study containing all of the essentials. Surely anthropometric measurements must be taken with care and by well-supervised workers willing to meter their own measurement accuracy. Surely the hematologic measurements must be accorded more care than in the clinical situation, and serum and urinary vitamins determined with comparable care and attention. Surely population-appropriate norms and standards must be employed. Surely radiographic and absorptiometric measurements are appropriate in bone-losing situations, such as protein–calorie malnutrition, starvation, vitamin D deficiency, and, in the other osseous extreme, where fluorosis is common.

It is only because such desiderata have not always been met that we speak now of the "optimal survey." Too often survey teams have been inadequately instructed. Too often the selection of anthropometric, hematologic, or biochemical measurements has been made against a textbook list, ignoring actual field conditions. Ascorbic acid may well be ignored in some tropical situations where electrophoretic separation of hemoglobin fractions may yet be important. In many Westernized areas protein intakes may be less critical to ascertain than fat intakes, and lipid determinations may supersede interest in several of the vitamins. In many surveys fatness has been deduced from weight relative to height, ignoring leg-length variations of major importance.

So the term "optimal" does not refer to the sheer number of measurements but rather their importance and their quality. An optimal survey may involve only a few hundred individuals (at particular risk) or many thousands (to provide population data). An optimal survey is often the second stage, following a spot survey. It may be conducted along with a survey of the local elite, if genetically appropriate, to provide reference data. An optimal survey does not depend on North American or English reference standards alone.

As applied to nutritional surveys and nutritional studies, "optimal" has operational meaning. It means both more than and less than reiteration of previous surveys, either in content or in scope. It refers to the purpose of a study, the design of the study, the problems to be investigated, and the selection of investigations as well as those to be investigated. Perhaps "optimal" is best defined as the opposite of *traditional*. After all, the smallest and the slightest (as caught in the dimensional net) may suffer from chromosomal and genetic problems, not necessarily of nutritional origin!

2. Optimizing Length and Weight

Both body length and gross weight are standard anthropometric measurements to be included in even the minimal nutritional appraisal, and under any circumstances, if only as reference values for caloric intakes. As opportunities afford, both length and weight should be taken with greater care, under constant supervision, and some measure of leg length should also be incorporated into the measuring procedure, since body length alone may not be satisfactory when using standard tables.

Length is commonly measured both as *recumbent length* (crown–heel in infants) and as *standing height* (i.e., stature) in somewhat older subjects. Since standing height may be an unsatisfactory measure below 5–6 years of age, and since it is subject to marked diurnal variations, recumbent length is to be preferred at all ages. Throughout, special care should be given to the measuring device or devices, checking them against standard meter sticks, such as those certified by the Bureau of Standards in the United States. Moreover, personnel should be carefully instructed in the measuring procedure, and their work checked at frequent intervals. In the course of collection, length data should be reviewed, ideally in the form of computer-generated histograms, so as to control reading and recording errors at the source.

Weight is subject to numerous sources of error, beginning with the weighing scales used. These should be checked at regular intervals, using standard weights as supplied by major manufacturers (e.g., Toledo Scale) or standardized weights specially fabricated for the purpose. Moreover, personnel should be instructed how to "balance" the scales at the beginning of every weighing session and in achieving "tare" when receiving blankets or disposable diapers are employed. As with length, accuracy in weighing is improved by careful instruction, replicate measurement, and analysis of data at 1- or 0.5-kg intervals to provide an immediate indication of careless weighing procedures.

With weight, clothing is an obvious and annoying source of error. If nude weight is not practicable (and nude weight after voiding), then there must be correction for clothing worn and a place for the correction on the precoding forms. Standard corrections are not satisfactory for most purposes, especially in adolescents or adults, and especially over the four seasons in cold winter climates. Sample weights for boots, underclothing, shirts, suits, and dresses are then essential if the "corrected" weight is to have accurate meaning.

Between populations there are often large differences in relative leg length, as between Meso-Americans and blacks, Aleut and Nilotes, or where Puebloan Indians and Navajos are to be compared. These differences in leg length complicate survey findings, especially where weight is related to stature as a provisional indication of obesity or leanness. Of course, it is essential that outer fatness be measured separately (see Section 4). But leg length should also be measured, at least in its simplest operational form, i.e., knee height. Then there is less likelihood of deducing "obesity" from a high weight-to-height ratio when the legs are short, or leanness or "underweight" from a low weight-to-height ratio when the legs are long (Figs. 1 and 2).

There are, of course, many other body measurements that can or may be taken. These include arm reach, or "span," and the bi-acromial, bi-iliac, and bi-trochanteric diameters. In general, these three measurements are so correlated with body size and with fatness as to provide little independent information by themselves. Should the opportunities permit, time is better ex-

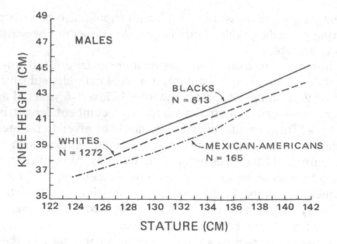

Fig. 1. Leg length relative to stature in blacks, whites, and Mexican-Americans (Chicanos). With major differences in lower-leg length, the same stature norms and weight-for-length norms cannot be used indiscriminantly for all populations (cf. Garn, 1976a).

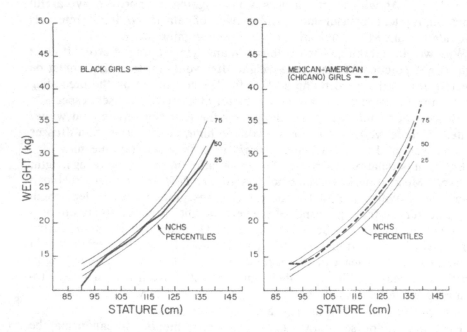

Fig. 2. Weight relative to height plotted for American Negro or "black" boys (left) and Meso-American or "Chicano" boys (right) against the NCHS percentiles as given for the "new" growth charts. With differences in leg length, as in Fig. 1, black children tend to fall in the lower percentiles in weight for height and Chicano children tend to fall in the upper percentiles, so the same growth charts cannot be employed for all racial and ethnic groups.

pended on accuracy of length and weight and on a measure of leg length, supervision of personnel, and data quality (given measuring instruments of quality and accuracy).

3. Optimal Circumferences

A number of circumferences are used, or may be used, in nutritional surveys and clinical assessments. These include circumferences of the head, arm, thorax, and calf. Depending on the age group, sample, purpose, and opportunities, all of these circumferences may be employed, or none of them.

The *head circumference* has at least two purposes and uses. One is to identify developmentally retarded and neurologically impaired individuals, especially in the younger ages. For survey purposes, it is useful to exclude such individuals from those viewed as at risk from a nutritional point of view. Otherwise the subgroup viewed as at risk may be inflated by chromosomal abnormalities, congenital malformation syndromes, and neurologic impairments. A second use of the head circumference is to identify nutritional growth failures, where the chest or arms fail to increase relative to the head, as where early protein–calorie malnutrition (PCM) is common. For some survey purposes, therefore, the head circumference is essential but for others it becomes optional.

Much the same is true of *arm* and *calf circumference*. In areas of acute malnutrition, or where protein–calorie malnutrition is common, "pipestem" arms or legs are of immediate diagnostic importance. One would also include these measures in lower income groupings, or where child neglect is common, or where "macrobiotic" diets exist within a subculture. Yet for most Western populations, and certainly in adolescents and young adults, one may question the separate value of arm and leg circumferences. In the adult female in the United States, most of interpersonal variance in arm circumference is actually variance in subcutaneous fat, more directly and more easily ascertained by fatfold measurements of the arm, back, iliac, and abdominal areas. Indeed, the estimates of upper arm composition, as commonly calculated, may be viewed with suspicion, since they ignore the compressibility of subcutaneous fat, and so overestimate apparent "muscle" (and underestimate fat) as described in Table II in the following section.

Thoracic circumferences, once much used, are now little employed simply because they reflect fatness levels to a most considerable degree. This is perhaps unfortunate with such populations as Turks and Peruvian Indians (who may have large thoracic diameters). Thoracic circumferences may have particular value at high altitudes, well over the 2-km level, as in the Andes, in areas in which atmospheric pollution is high as in inner city belts, near the freeways, and near coal-burning power plants.

For the basic purpose of excluding the neurologically impaired, a simple head circumference is not an optimal addition, but rather an absolute necessity, since the neurologically impaired may constitute a high proportion of infants seemingly at nutritional risk because of their small size and growth delay. Where PCM is common, or in starvation areas, an arm circumference (or a leg circumference) may be a necessity and not just an optimal addition. But for adult females in the Western world and adults in the better nourished countries, in general, circumferences add little that fatfolds do not provide.

4. The Optimal Use of Fatfolds

Fatfolds, commonly but incorrectly called "skinfolds," are an essential part of any nutritional survey where the energy balance is a subject for concern, for fatfolds tell us what length and weight cannot, either separately or together. Fatfolds tell us how fat a person is, both in absolute terms, and (as need be) relative to peers and various standards. Used intelligently and with care, fatfolds can tell us far more, and here we make the shift from minimal to optimal, yet without logistic complications or addition to survey costs.

To begin with, fatfolds measure the *compressed* double fold of fat plus skin, using standardized calipers employing a defined pressure in grams per square millimeter, a defined jaw area, and hence a defined *force*. Because there is a time decay during which the measurement slowly diminishes and then plateaus, there is a waiting period, and personnel must be carefully instructed as to this fact. Moreover, there are situations in which fatfolds should not be taken (among them edema) and there are situations in which the fatfolds are affected by physiologic status (among them pregnancy). Even more than with length and weight, fatfold measurements demand careful instruction and regular replication so as to maximize the value of survey data.

Four or five fatfold sites may be considered, among these the triceps, subscapular, abdominal, and iliac or lowest-rib site. Not every fatfold is equally practicable on the lean and on the obese, both by virtue of modesty and because of the limits of the calipers themselves. So it is useful to incorporate more than one fatfold measurement into the survey scheme. Further, there are some individuals whose outer fat is so rubbery as to make good measurements impractical. This must be explained to survey workers, preferably with an actual demonstration. As we move beyond a single fatfold site and casual measurements, to multiple fatfold sites and carefully supervised measurements we approach optimal fatfold technology (Table I).

Now it is perfectly practical to convert fatfold values (in mm) to fat-weight estimates (in kg) since the age constant regression of weight on fat is nearly linear. The estimate of fat weight (FW) can then be subtracted from total body weight (TBW) to provide an indication of the fat-free weight, since FFW = TBW − FW. Fat weight divided by total body weight yields percent fat (%F) in turn. Such estimates and calculations must be both sex specific and age specific, since outer fat is variously related to fat weight at different ages and in the two sexes. So before comparing fat weight in Bundi and Bostonians or

Table I. *"Lean" and "Obese" Fatfold Values[a] for the Triceps, Subscapular, Abdominal, and Iliac Fatfolds in an American Population*

Age midpoint	Males (mm)								Females (mm)							
	Triceps		Subscapular		Iliac		Abdominal		Triceps		Subscapular		Iliac		Abdominal	
	Lean	Obese	Lean	Obese	Lean	Obese	Lean	Obese	Lean	Obese	Lean	Obese	Lean	Obese	Lean	Obese
1	9.0	16.0	5.0	9.0	3.0	8.0	5.0	10.0	8.0	14.0	5.0	9.0	5.0	11.0	5.0	10.0
2	8.0	13.0	5.0	8.0	3.0	7.0	3.0	7.0	9.0	15.0	5.0	10.0	3.0	11.0	5.0	11.0
3	8.0	13.0	5.0	8.0	3.0	7.0	5.0	8.0	9.0	14.0	5.0	8.0	5.0	10.0	5.0	10.0
4	8.0	13.0	4.0	7.0	3.0	7.0	3.0	7.0	9.0	14.0	5.0	8.0	3.0	10.0	5.0	10.0
5	7.0	13.0	4.0	7.0	3.0	7.0	3.0	8.0	9.0	14.0	4.0	7.0	3.0	8.0	3.0	10.0
6	7.0	12.0	4.0	6.0	2.0	6.0	3.0	8.0	9.0	15.0	5.0	8.0	5.0	10.0	5.0	11.0
7	7.0	13.0	4.0	7.0	3.0	8.0	3.0	8.0	8.0	15.0	4.0	8.0	3.0	12.0	5.0	11.0
8	7.0	13.0	4.0	7.0	3.0	10.0	3.0	8.0	8.0	16.0	4.0	8.0	3.0	13.0	3.0	15.0
9	7.0	16.0	4.0	8.0	3.0	15.0	3.0	12.0	10.0	19.0	5.0	12.0	5.0	18.0	5.0	18.0
10	7.0	15.0	4.0	9.0	3.0	16.0	3.0	16.0	9.0	19.0	5.0	12.0	5.0	18.0	5.0	22.0
11	8.0	21.0	5.0	12.0	5.0	20.0	5.0	21.0	10.0	22.0	5.0	19.0	6.0	25.0	6.0	30.0
12	9.0	22.0	5.0	15.0	5.0	35.0	5.0	32.0	10.0	21.0	6.0	15.0	6.0	22.0	6.0	28.0
13	9.0	22.0	6.0	14.0	5.0	30.0	5.0	28.0	11.0	26.0	6.0	20.0	7.0	30.0	10.0	33.0
14	8.0	17.0	5.0	12.0	5.0	23.0	6.0	26.0	11.0	24.0	7.0	18.0	7.0	22.0	10.0	30.0
15	8.0	20.0	6.0	15.0	7.0	30.0	7.0	30.0	13.0	30.0	8.0	24.0	7.0	30.0	10.0	35.0
16	8.0	20.0	7.0	14.0	7.0	30.0	7.0	31.0	14.0	27.0	8.0	24.0	10.0	32.0	11.0	36.0
17	8.0	23.0	8.0	22.0	10.0	41.0	10.0	45.0	14.0	26.0	8.0	20.0	10.0	26.0	12.0	32.0
18	7.0	17.0	7.0	15.0	6.0	30.0	8.0	33.0	14.0	29.0	9.0	23.0	10.0	26.0	12.0	40.0
20	7.0	22.0	8.0	20.0	8.0	36.0	10.0	40.0	14.0	30.0	8.0	25.0	7.0	31.0	11.0	38.0
30	8.0	22.0	9.0	25.0	11.0	40.0	13.0	42.0	15.0	33.0	8.0	28.0	7.0	35.0	13.0	45.0
40	10.0	22.0	10.0	25.0	10.0	36.0	15.0	40.0	16.0	34.0	10.0	31.0	8.0	40.0	16.0	47.0
50	9.0	22.0	10.0	27.0	11.0	35.0	15.0	42.0	18.0	36.0	11.0	36.0	12.0	45.0	21.0	55.0
60	10.0	23.0	11.0	30.0	12.0	33.0	17.0	42.0	19.0	36.0	13.0	36.0	15.0	47.0	25.0	55.0
70	9.0	22.0	10.0	25.0	10.0	31.0	15.0	40.0	19.0	36.0	12.0	35.0	15.0	45.0	25.0	52.0
80	7.0	20.0	9.0	23.0	6.0	30.0	11.0	33.0	14.0	33.0	8.0	30.0	10.0	45.0	20.0	50.0

[a] Based on 9226 white individuals in the Tecumseh Community Health Study (1962–1965). The statistical definitions of lean and obese correspond to Garn (1972) and Garn and Clark (1976). For the 15th and 85th percentile definitions of "lean" and "obese" see also Seltzer and Mayer (1965). The values in this table are age specific.

in Atitlan and Atlanta there is need for care, and considerable thought, and not just the listing of numbers.

The double fold of fat plus skin is as much as 50% compressible, differing according to age and sex. So fatfolds cannot just be compared, millimeter by millimeter, without consideration of compressibility. Moreover, the compression factor affects such calculations as those relating to the estimated brachial mass and other corrections of circumferences to "eliminate" the amount of fat. Most calculations in the literature ignore fatfold compressibility, and therefore overestimate the muscle mass, and underestimate the mass of fat (Table II).

Under these circumstances, "optimal" fatfold determinations involve (a) more fatfolds, (b) careful selection, (c) training and supervision of personnel, and (d) prospective and retrospective tests of data quality. When these desiderata are accomplished, fatfold–fatfold correlations approach or exceed .9, and show improved correlations with lipid levels, size attainment, and developmental status in childhood and adolescence, with hemoglobin levels, at all ages, and with food intake levels, given constant energy expenditure. More important, fatfold measurements of proven quality far supersede weight or weight for length as indications of fatness, especially in individuals of differing frame sizes or different leg proportions, relative to length. With care, and with imaginative use of the data, fatfolds then become optimal.

5. Optimal Use of Radiographic Information

It is not commonly realized how much information of nutritional relevance (and relevance to nutritional surveys) is contained within a single postero-anterior hand–wrist radiograph. Most workers in applied nutrition are familiar with the use of radiographs to provide bone age or skeletal age estimates, but not with the many other uses, some of them potentially far more important (Fig. 3 and Table III).

A postero-anterior hand–wrist radiograph provides information on "skeletal age," in reference to established North American or British norms (depending on the "Atlas" employed) or simply by counting the number of ossification centers radiographically present. In this way, we can ascertain how retarded or possibly advanced an infant, child, or adolescent is, and thereby we can put anthropometric and other data in proper perspective. Given population differences in stature and the components of stature, there is value in knowing whether a child or an adolescent is (a) skeletally retarded and (b) retarded relative to stature or stem length. Such radiographic and anthropometric information can be extended to such purposes as grouping children of indeterminant age into developmental groupings, for analysis and further study, a real advantage in field situations in which exact age is not known. Such information can also be applied to the separation of those tall for their bone age from those short for their bone age, or in comparing the

Table II. Estimated Arm Muscle Area, Fat Area, and Percent Arm Fat Including Corrections for Fat Compressibility[a]

Age midpoint	Males									Females								
	Uncorrected			20% compressibility			33% compressibility			Uncorrected			20% compressibility			33% compressibility		
	Muscle area	Fat area	% Fat	Muscle area	Fat area	% Fat	Muscle area	Fat area	% Fat	Muscle area	Fat area	% Fat	Muscle area	Fat area	% Fat	Muscle area	Fat area	% Fat
1	1207	645	35	1045	781	43	915	909	50	1084	618	36	957	751	44	837	873	51
2	1284	671	35	1156	816	42	1022	954	49	1241	672	35	1087	815	43	998	948	50
3	1384	692	33	1241	840	41	1079	981	48	1298	692	35	1236	846	43	998	982	50
4	1454	668	31	1306	816	38	1179	955	44	1391	702	33	1236	850	41	1095	988	47
5	1579	650	29	1426	797	35	1290	937	41	1516	730	33	1338	890	40	1187	1038	47
6	1700	625	27	1550	767	33	1405	903	39	1563	742	32	1397	902	40	1256	1052	46
7	1815	670	27	1662	818	33	1506	956	38	1702	764	31	1526	933	38	1354	1092	45
8	1987	706	26	1808	867	32	1660	1021	37	1818	865	32	1632	1057	39	1438	1235	46
9	2074	794	27	1898	975	34	1737	1146	40	1955	936	33	1732	1140	40	1533	1330	47
10	2242	922	29	2035	1127	36	1841	1329	42	2116	1141	34	1869	1386	42	1638	1612	49
11	2408	942	28	2192	1152	34	1982	1351	40	2343	1181	33	2062	1432	41	1815	1673	48
12	2605	1059	28	2363	1298	34	2122	1519	40	2558	1267	33	2256	1548	40	1984	1816	47
13	3013	1026	26	2711	1261	32	2501	1478	38	2711	1423	34	2408	1737	42	2104	2027	49
14	3546	1054	23	3171	1288	29	2938	1518	34	2952	1561	35	2599	1909	43	2258	2236	50
15	3867	1159	23	3573	1413	28	3312	1654	33	3042	1816	37	2574	2213	45	2268	2579	53
16	4184	1080	21	3839	1332	26	3563	1578	31	3197	1807	35	2762	2204	42	2405	2578	49
17	4771	1107	19	4479	1363	23	4163	1615	27	3058	1895	37	2672	2288	45	2328	2677	52
20	5315	1366	21	4971	1684	26	4661	1991	30	3340	1992	37	2923	2419	45	2520	2817	52
30	5802	1645	22	5406	2022	27	5028	2387	32	3606	2326	39	3134	2809	47	2678	3256	55
40	5820	1819	23	5380	2236	29	5003	2627	34	3724	2731	42	3146	3284	51	2616	3801	59
50	5692	1815	23	5292	2230	29	4842	2633	34	3847	2938	43	3257	3529	52	2718	4075	60
60	5445	1601	22	5060	1969	27	4686	2326	32	4132	2956	41	3451	3577	50	2901	4128	58
70	5187	1542	22	4788	1894	27	4493	2237	32	4045	2656	39	3451	3215	47	2948	3748	55
80	4800	1519	23	4473	1864	28	4122	2194	33	3915	2389	37	3387	2893	45	2891	3352	52
90	4116	980	19	3891	1208	23	3642	1428	27	3391	1561	31	3013	1905	38	2822	2237	45

[a] Muscle area = $(\pi/4)[(\text{arm circumference})/\pi - \text{triceps fatfold}]^2$ or $0.785(0.32A - T)^2$. Fat area (FA) = total area (TA) − muscle area (MA). % Fat = (FA/TA) × 100. For references to computational methods see Frisancho (1974).

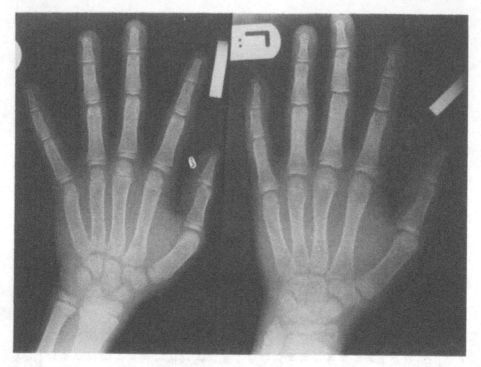

Fig. 3. Two postero-anterior hand radiographs of Nigerian children of closely similar ages but markedly different levels of skeletal development. Note that the older girl shown in the left-hand figure (of 12.33 years) is close to 2 years younger developmentally than the slightly younger girl of 12.25 years shown in the radiograph at the right. Both these girls are developmentally retarded by American and U.K. standards (cf. Greulich and Pyle, 1959; Tanner *et al.*, 1975).

developmental delay in boys as compared with girls. Skeletal age can also be employed to "predict" final stature, and it should be applied to those seemingly most retarded.

Now the conventional anthropometric screenings for small size and light weight and a low weight-to-height ratio inevitably catches in the net two groups. The first group includes those who are indeed nutritionally delayed. The second group includes those who are small or asthenic because of genetic abnormality or chromosomal aberration. The postero-anterior hand–wrist radiograph helps to identify the latter group, especially if metacarpal and phalangeal lengths are measured and the metacarpal–phalangeal profile pattern is drawn in graphic fashion or computer-ascertained and matched. Since some of the small and apparently delayed are dimensionally and developmentally retarded for reasons that are not nutritional, it is useful to identify this fraction, perhaps small in proportion to the total, but large in proportion to those seemingly at risk.

Measurements of bone and medullary widths and calculations of cortical area and percent cortical area (PCA) are especially relevant in chronic and acute malnutrition, protein–calorie malnutrition, and recovery from starvation. In protein–calorie malnutrition, in both children and adults, there may be reduced bone mass relative to skeletal size or volume, i.e., actual loss of bone. The radiographs provide this information. And there are other situations, some nutritional, some genetic, and some both, in which such *radiogrammetric* information is highly relevant. Wheat gluten sensitivity, bone loss associated with antispasmodics, vitamin-D-resistant rickets, and phosphatase-deficiency disease are some examples. Yet in some areas of the world, where the fluoride content of water is high (in excess of 3–4 mg/liter) or where water consumption is excessive, there may be *fluorosis* which the radiographs can help to detect.

The cost of the hand–wrist radiographs may be estimated as $1 per radiograph including processing chemicals but excluding the prorata cost of equipment and personnel. This cost compares favorably with most biochemical determinations, and the radiographic record is not only permanent but useful in many different ways. The radiation cost of the radiographs, 2–5 mR to the skin depending on the equipment, type of film, etc., may be compared with the 1–2 mR daily (background) dosage. Properly taken, with adequate shielding, the gonadal dosage should be far below 0.1 mR for a single postero-anterior hand–wrist film.

The incorporation of hand–wrist radiographs may be viewed as an excessive refinement in small surveys operating out of a suitcase. For new and novel populations or under known nutritional stress, they may be viewed as useful. In large surveys, with laboratory facilities, it is obvious that radiographs should be considered (at least for those at greater nutritional risk). In clinic

Table III. Uses of Hand-Wrist Radiographs in Nutritional Surveys

Use	Comment
1. Determinations of "skeletal age" (Greulich and Pyle, 1959; Tanner *et al.*, 1975; Poznanski, 1976)	Measurement of developmental delay may also be used in stature prediction and as a substitute for chronological age
2. Ossification sequence (Garn *et al.*, 1974b, 1975a; Garn and Bailey, 1977)	Identification of congenital malformation syndromes, such as diastrophic dwarfism
3. Metacarpal–phalangeal lengths and "profiles" (Garn *et al.*, 1972; Poznanski *et al.*, 1972; Poznanski, 1974; Poznanski and Garn, 1974; Garn, 1977)	Identification of congenital malformation syndromes of chromosomal or genetic origin
4. Bone mass and bone density, i.e., bone "quality" (Colbert *et al.*, 1970; Garn, 1970; Dequeker, 1972; Colbert, 1976; Garn *et al.*, 1976c; Mayor *et al.*, 1976)	a. By micrometer-caliper measurement of cortical area and percent cortical area b. By radiographic microdensitometry

and hospital situations, or where bone-losing situations are known or anticipated, such radiographs may be viewed as imperative. And where the skeletal manifestations of nutritional disease are known or suspected, radiographs are then no longer "posh" but a necessary part of the investigative procedure (Fig. 4).

6. Direct Photon Absorptiometry

Direct photon absorptiometry, a filmless noninvasive technique, provides some of the advantages of the radiographic approach but without the need for developing tanks, darkrooms, or hangers. The small isotope source (radioactive iodine or americium) yields an effectively monochromatic radiation which, highly collimated, results in extraordinarily low radiation exposure levels and to a tissue mass only a millimeter wide. In return, direct photon absorptiometry provides indications of (a) bone width, (b) bone mineral, and (c) bone mineral per unit of bone width, i.e., bone "density." Understandably and expectably, bone width so measured corresponds to the radiogrammetric width; bone mass so measured corresponds to cortical area as derived from radiographs; and bone mineral relative to bone volume corresponds to the radiogrammetrically determined percent cortical area.

Since direct photon absorptiometry can be performed with a relatively small instrument of desk-top size, it can be a valuable adjunct to surveys, yielding results at the rate of better than 12 individuals an hour, including time

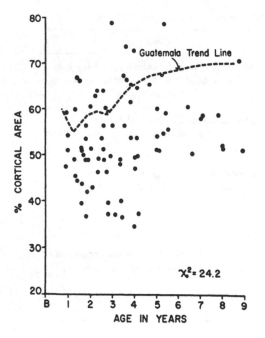

Fig. 4. Bone mass/bone volume (expressed as percent cortical area) in children with protein–calorie malnutrition (PCM). Percent cortical area may be reduced in starvation, juvenile and adult protein–calorie malnutrition, wheat gluten sensitivity, vitamin-D-resistant rickets, chronic renal disease, and adult bone loss (so-called osteoporosis). For details see Garn (1970), Garn and Shaw (1977), Dequeker (1972), and Poznanski (1974). In fluorosis, however, percent cortical area may be increased (cf. Garn, 1973).

for identification, registration, etc. Moreover, the results are instantaneous, in contrast to film-type densitometry which demands both film processing and later analysis employing a recording microdensitometer with computer interface (see Fig. 5 and Table IV).

At the same time, direct photon absorptiometry does not provide all of the advantages of radiographs; it is not a mechanism through which skeletal abnormalities or differences in the size and proportions of individual bones can be ascertained. Direct photon absorptiometry is essentially "blind" and subject to such errors as old fracture lines and bone cysts. Yet given a starvation population and follow-up study, direct photon absorptiometry can be viewed as an optimal technique, as is true for such clinical situations as chronic renal disease, vitamin-D-resistant rickets, hyperparathyroidism, hyperthyroidism, gastrectomy and postgastrectomy states, fluorosis, and renal tubular insufficiency.

7. Optimizing Food Intake Data

Most nutritional investigators are familiar with the various approaches to food intakes, including the market-basket approach, the pot approach, dietary recall, and the 1- and 7-day diary records. Each of these approaches has some value, even if it yields no more than per-capita intakes of calories and protein. *Optimal* dietary information provides enough information to quantify individual caloric, protein, and vitamin intakes for the period corresponding to the survey or study, though longer term information of quality is surely welcome.

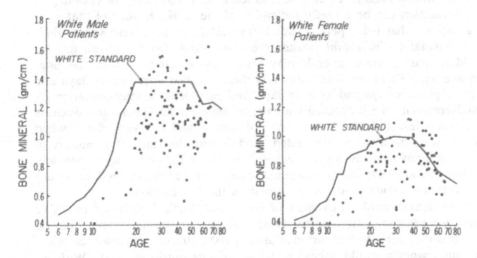

Fig. 5. Reduced bone mass/volume in chronic renal disease patients as measured by direct photon absorptiometry and plotted against the Wisconsin norms. Direct photon absorptiometry, using an ^{125}I source and the Cameron–Norland apparatus, is highly applicable to field studies, yielding direct readings at a rate of 12 subjects per hour or better. For references see Mayor *et al.* (1976).

Table IV. Direct Photon Absorptiometry

Apparatus	Cameron-Norland direct photon absorptiometer (see references)
Radiation source	^{125}I (standard model), americium (modified model)
Elapsed time	60 sec
Readout	Digit reading
Measurements	Bone width Bone mineral Bone mineral (mm²)
Reliability	±3% including positioning errors
Standards	Original Wisconsin norms for whites (Mazees and Cameron, 1973) Unpublished, revised Wisconsin norms (R. B. Mazess, 1975, unpublished)
References	Cameron *et al.* (1968); Jaworski (1976); Mayor *et al.* (1976)

Too many surveys rely upon 1-day dietary records or diaries. Given a large enough subject N at each age (preferably 100 or more for each age–sex group considered) and given care in instruction and recording, the resulting 1-day information can be a useful indication of the intake history of that age–sex *group* for that 1-day period. But 1-day dietary records have severe limitations insofar as *individual* intakes are considered. On any given day an individual man, woman, or child may eat far less or far more than the long-term average. There are feast days and there are fast days, name days and paydays, payless days and leftover days, and each of these may overestimate or underestimate the individual's long-term daily intake. There are sickness days, when less than the average is consumed, and recovery days, when perhaps twice the average is ingested. And despite the vaunted monotony of the diet in many less developed areas, when the food choice is approximately the same day after day, people do get sick and they do become well, crops are short or there is bounty, and there is cash or there is no cash.

A particular, virulent, example of erroneous conclusions based on 1-day dietary records comes from some "analyses" of dietary intake data reported by pregnant women. With reported intakes of 1000, 500, or even fewer calories, some such women would appear to be at obvious nutritional risk. With reported protein intakes of 30 or 15 g (and even less) there might appear to be grave danger to the nutritional status of their conceptuses. Yet these are single-day reports, devoid of indication of what was consumed the day before or the day after. Analyses of the same data also show an excess of high levels of calorie intake, 3000 kcal and more, 120 g of protein and more. With large day-to-day differences in intake, the low values are unlikely examples of long-term intakes. Yet they have been so construed. (One might agonize also about 5000-cal 1-day intakes, though this has not been done!)

One-day dietary intakes are minimal, useful only for large groups, and to provide indication of mean or (better) median intake values. Optimal intake information is surely 7-day in nature, especially as individuals are concerned. Good 7-day intake information, covering the period immediately before the anthropometric, physical, and biochemical examination, is surely to be desired, certainly optimal, but not luxurious (Garn *et al.*, 1976d, 1978).

To be sure, longer term dietary information has much additional value. One would like to know about ascorbic acid intake when fruits are out of season, protein intake before the annual slaughter, and fluoride intake during long, hot, dry, and thirsty summers. During the rainy season in the Kalahari, with decreasing body fat and body weights, the diet is surely different. Caloric intakes may maximize during the Christmas holidays, and fat intakes too, but with traditional alcoholic beverages then a major source of energy but a minimal source of nutrients. Truly optimal dietary information would sample each season, for each individual studied, with a yearly and year-end summary then available. Then the 7-day period immediately prior to the anthropometric and medical examination could be placed in proper perspective, along with information on seasonal changes in weight and body composition.

Such a perspective would be optimal. What we so often achieve is less than minimal, and with the dietary information (by virtue of simple expediency) far out of phase with the various anthropometric and biochemical determinations. At the very least reported intakes and measured levels of serum and urinary vitamins need to cover the same general period of time, and with reasonable temporal depth. Too often they do not.

8. Optimizing Hematologic Determinations

Although hemoglobin concentrations (Hgb) and packed-cell volumes (Hct) have been a regular part of most nutritional surveys, there has been rather little attempt to optimize their value. For the most part these common hematological measures have been taken with no more than usual clinical care, interpreted against usual clinical standards, then reporting the proportion of individuals deemed "deficient" or "low."

It should be obvious that a drop of blood squeezed out of a heel prick may not yield adequate hematological indications. It should also be obvious that hemoglobin determinations read to the nearest 0.5 g/100 ml are too coarse, readout accuracy then being half a standard deviation (SD)! And though hemoglobin and hematocrit values on the same individuals are highly correlated ($r \simeq .9$), individual discrepancies (as indicated by the McHc ratio) are of considerable importance. So the first steps in optimizing the value of the hematologic determinations reside in the drawing of blood, care in handling the blood sample, the determinations themselves, and computing the McHc ratio.

The question of norms or standards deserves considerable thought, beyond the simple "ranges" in common clinical use. Here it may be useful to set the lower limit at the 5th percentile, from truly large-scale "normal" studies, but with attention to the corresponding upper limit (i.e., the 95th percentile) as well. Low hematologic values are associated with many nutritional dificiencies, including simple malnutrition, but high hemoglobins and hematocrits, at sea level, may be indicative of excessive iron intake, respiratory disorders, or atmospheric pollution. Furthermore, since hemoglobins and hematocrits are elevated in the obese, adjusted cutoff values may be employed for them.

During adolescence Hgb and Hct rise rapidly in the male, and in exact accordance with maturity status. So in boys aged 13–17, it is necessary to employ maturity criteria in addition to age *per se*. Otherwise there may be a spurious excess of "low" values, when some of the subjects are sexually immature and others quite mature. Moreover, the physiologic decrease in Hgb and Hct during pregnancy demands careful attention to pregnancy status. Otherwise there may be an excess of apparently low values if the pregnant and the nonpregnant are irretrievably mixed in the data base.

In populations in which abnormal hemoglobins and/or G6PDD are common, it may be necessary to ascertain sicklers, thallasemics, the G6PDD homozygotes, etc., and to separate their hematologic values from the total sample. And, of course, cases of chronic renal disease (CRD) should be excluded from this apparently normal population as a matter of course. Furthermore, individuals of largely African descent tend to lower hemoglobin and hematocrit levels by 0.75–1.00 g/100 ml. This rather large hematologic difference obtains at all socioeconomic levels, at all educational levels, in athletes and in infants, during pregnancy, and even in chronic renal disease (CRD) and in nephrectomized individuals. So there is need for differential hemoglobin and hematocrit standards for both blacks and whites, otherwise the proportion of blacks viewed as "deficient" or "low" will be spuriously large (Table V).

Of course, the clinician, seeing only one patient at a time, may find it difficult to put all of these desiderata into use, except perhaps the generalizations pertaining to both blacks and adolescents. The clinician employs the hematologic workup to identify disease, and may care little about the 5th and 15th percentiles for hemoglobins or hematocrits. But in the nutritional survey, with hundreds or even thousands of participants, optimal use of hematologic data demands more than routine care in determination and interpretation. Neither hemoglobin alone nor a hematocrit alone is enough, and there should be follow-up studies of those who are exceptionally low or exceptionally high or with a discrepant McHc ratio. Follow-up studies should also include the transferrins and other measures of iron saturation, and provisions for supplementation as well as therapy.

The textbook "normal" range of human hemoglobins (excepting those in the newborn) approximates that in the horse, cow, sheep, and mouse. The

Table V. Black–White Differences in Hemoglobin and Hematocrit Levels

General, and income-corrected	Whites exceed blacks by ~1.0 g/100 ml (Garn *et al.,* 1974a, 1975b; cf. Johnson and Abraham, 1979)
Infancy	White infants exceed black infants even when supplemented (Garn *et al.,* 1975c)
Pregnancy	White pregnant women exceed black pregnant women, income-corrected (Garn *et al.,* 1967a, 1977)
Chronic renal disease (CRD)	White CRD patients exceed black CRD patients, correcting for disease status and therapy (Garn *et al.,* 1976b)

optimal use of survey data in the human goes beyond these simple values, asking whether hematologic status could be improved by nutritional means. There is evidence that hemoglobins and hematocrits at the higher end of the normal range are associated with more rapid growth and larger size attainment.

As a matter of policy, the hemoglobin and hematocrit norms used should be derived from adequately large samples, carefully age-adjusted, with separate values for blacks and whites (Table V), and defining *deficient* as the 5th percentile and *low* as the 15th percentile as suggested in Table VI.

9. Serum and Urinary Vitamins

It has been common practice to include determinations of serum and urinary vitamins in most nutrition surveys without attention to dietary information, the time of day, or the composition of foods recently consumed. This practice has been justified by the facts that intake data are rarely available in analyzed form during the clinical examinations, and that serum and urinary vitamin levels show distressingly low correlations with intake levels calculated from food-frequency recall or from 1-day dietary histories. To laboratory workers, moreover, the colorimeter or spectrophotometer is far more accurate than human memory, under the best of conditions.

But under optimal conditions, intake information should certainly be a guide as to what vitamins are likely to be underingested, and preliminary intake data may be used to provide such a guide, under field conditions. In tropical areas, where fruits constitute a large part of the diet, ascorbic acid may not be a vitamin of concern, except perhaps in the affluent with Westernized food habits. Vitamin A should not be a problem where papayas and yams are in abundance. Surely even market-basket dietary information can help to identify possible vitamin lacks and so determine what serum and urinary vitamins to measure during the survey.

North Americans tend to consume their citrus or tomato juice as part of

Table VI. Suggested Values for "Deficient" [a] *and "Low"* [b] *Hemoglobin Concentrations (g/100 ml)* [c]

| | Whites | | | | Blacks | | | |
| | Males | | Females | | Males | | Females | |
Age	Deficient	Low	Deficient	Low	Deficient	Low	Deficient	Low
1	—	—	—	—	—	—	—	—
2	9.3	10.8	10.2	10.7	9.8	10.4	6.3	9.9
3	10.0	11.2	10.5	11.0	10.0	10.6	7.5	10.5
4	10.4	11.2	11.0	11.4	10.4	10.8	9.4	10.6
5	10.9	11.4	11.2	11.6	10.5	10.8	10.2	10.8
6	11.2	11.8	11.2	11.8	10.6	11.0	10.4	11.3
7	11.3	12.0	11.4	11.9	10.6	11.3	10.5	11.3
8	11.4	12.0	11.5	12.0	10.7	11.4	10.6	11.1
9	11.5	12.0	11.5	12.0	10.7	11.4	10.6	10.9
10	11.5	12.0	11.6	12.1	10.7	11.5	10.7	11.0
11	11.6	12.1	11.5	12.2	10.8	11.5	10.7	11.3
12	11.7	12.2	11.6	12.3	10.8	11.5	10.8	11.4
13	11.9	12.6	11.6	12.3	10.9	11.5	10.8	11.3
14	12.1	12.9	11.8	12.1	11.0	11.8	10.8	11.4
15	12.1	13.2	11.7	12.1	11.7	12.3	10.8	11.3
16	12.4	13.5	11.6	12.2	12.2	12.8	10.8	11.2
17	12.6	14.0	11.5	12.1	12.6	13.1	10.8	11.5
20	13.5	14.3	11.4	12.3	12.8	13.2	10.7	11.8
30	13.5	14.3	11.3	12.3	12.6	13.3	10.4	11.5
40	13.2	14.1	11.4	12.2	12.1	13.0	10.2	11.3
50	13.0	13.9	11.5	12.4	12.0	12.8	10.1	11.4
60	12.8	13.6	11.6	12.5	12.1	12.5	10.0	11.5
70	12.3	13.2	11.7	12.5	12.0	12.3	10.0	11.3

[a] 5th percentile.
[b] 15th percentile.
[c] Interpolated from unpublished, smoothed, income-restricted Ten-State Nutrition Survey data (S. M. Garn and P. E. Cole, 1977, unpublished). For comparison see Johnson and Abraham (1979).

their morning meal, so it may make a difference whether the samples are taken at 9 A.M. or 9 P.M. Moreover, they may take vitamin supplements, especially if they anticipate queries as to their dietary practices. It makes quite a difference whether subjects have taken a vitamin supplement shortly before the physical (medical) examination, but it is a rare survey that pays attention to that fact. (A few vitamin supplements can raise the mean level of most vitamins considerably.)

Many water-soluble vitamins are expressed relative to urinary creatinine, as a correction for (or compensation for) the level of dilution. In this way the "casual" urine sample, which dilutes vitamin levels and creatinine alike, is corrected for the amount of water recently consumed. However, urinary creatinines are affected by dietary practices, being higher where meat intake is high, and lower when the intake of flesh is low. Even the 24-hr urinary creatinine may vary considerably, depending on the diet, and such variations or differences may affect apparent vitamin levels by as much as 20%.

In the optimal survey, therefore, these problems bear thinking through. Surely the vitamins looked for (in blood or urine) should have some relation to vitamins viewed as inadequately ingested, in market-basket or intake survey. Surely ascorbic acid should not be measured after the daily dollop of tomato juice or in individuals consuming an additional 1000 mg/day. Surely people on a high flesh dietary must be separated from those with little intake of flesh, if urinary creatinines are the reference standard. Surely vitamin levels should be analyzed separately for the poor and for those more affluent (Fig. 6). The optimal survey demands that these problems will be thought through and resolved.

10. Screening the At-Risk Group

International, national, and local nutrition surveys too often limit themselves to ascertaining the proportion of infants, children, and adults apparently at risk with respect to nutrition and without follow-up of individuals so identified. With a large survey team, a typically tight schedule, and both equipment and personnel to move from place to place, such approaches are, perhaps, inevitable. But the practice of being "in like Flynn" severely restricts the value of any survey and the value of nutritional findings that can be made.

Of individuals defined as at risk by virtue of small size, limited weight, or developmental delay, a relatively large proportion may be chromosomal abnormalities, malabsorption states, single-gene substitutions, and the progeny of the small. Indeed, if the 5th percentile is used as the cutoff for length or weight, and if 1–2% of all live-born infants evidence karyotypic abnormalities,

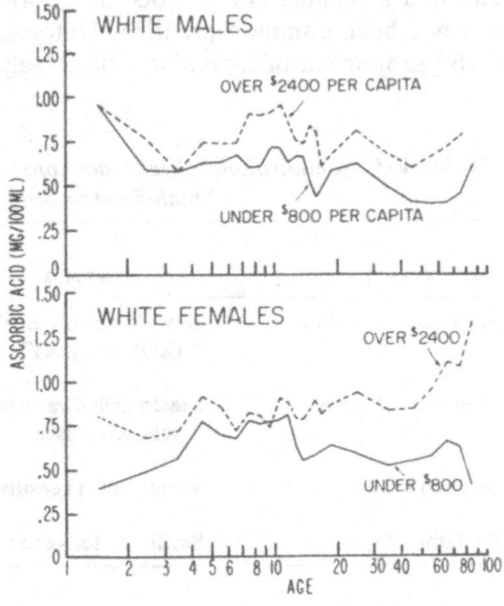

Fig. 6. Effect of economic status on ascorbic acid levels. This plot, based on over 5800 determinations, exemplifies the need to consider socioeconomic status (SES) in nutritional studies and shows the need for preliminary information at the planning stage. In tropical areas, where the poorer individuals consume a less acculturated diet, the income-related vitamin C curves shown here may well be reversed.

then a relatively large proportion of apparently at-risk individuals may have other than nutritional problems. Moreover, since two short parents will generally have short children, this simple fact (plus the chromosomal and genetic abnormalities just mentioned) may well place the *majority* of individuals caught by the anthropometric net in the nonnutritional category (Table VII).

From this standpoint alone, it is imperative that individuals provisionally defined as at risk by anthropometric, biochemical, or clinical criteria be given a second and considered look. By calling them back for restudy, some proportion will immediately be eliminated as the result of measuring and recording errors or as simple identification errors. Another proportion, possibly as high as 50%, in some circumstances may prove to be chromosomal reduplications, chromosomal deletions, or examples of various types of dwarfism or malabsorption syndromes. Still another proportion will, upon investigation, prove to be the sons and daughters of the genetically small or, alternatively, small-for-term infants whose size even as late as the seventh year is ordinarily restricted.

Reviewing those at apparent risk with respect to hemoglobins and hematocrits some proportion again will prove to be measuring or recording error. Differential blood analysis will also identify some as exhibiting blood dyscrasias, renal complications, or bacterial infections—placing them at risk indeed, but not necessarily at nutritional risk. Some individuals apparently low in serum and urinary vitamins will prove to be the result of measuring or recording error or metabolic abnormality with little correspondence between water-soluble vitamins ingested and those recovered.

Such screening constitutes a service to the subjects and an improvement in the quality and meaning of the data. Actually, such attention and consideration should be built into all nutritional surveys and studies. If the process of recall and screening is described as "optimal" here, it is merely because this has not been common practice. Unfortunately, it is all too common to report the proportion of individuals putatively at risk without confirming the

Table VII. Nonnutritional Size Reductions Likely to Be Concentrated in Small-for-Age Subjects

Description	Examples	Frequency per 1000 live births
Chromosomal abnormalities	Down's syndrome, Turner's (XO), many XYY	10
Single-gene substitutions	Diastrophic dwarfism, achondroplasias	2
Malabsorption syndromes	Wheat gluten sensitivity	1
Small-for-term infants	Small size for gestation length	25–50

measurements or being sure that the numbers truly reflect nutritional problems and not chromosomal or genetic disease, or simple familial shortness.

11. Optimal, Necessary Follow-Up Studies

As indicated earlier, the majority of nutritional surveys have been designed for maximum speed in data collecting, without opportunity to "screen" individuals apparently at nutritional risk, and without provisions for follow-up studies. Reported findings, therefore, are at best approximate, both exaggerating the proportion seemingly at risk and underestimating the prevalence of other clinical, biochemical, and anthropometric indications. In order to optimize the value of the investigations, there is need to review and repeat studies on some fraction of the subjects surveyed and there is need for a variety of follow-up studies as well.

Surely those confirmed cases of inadequate growth, consistently low hemoglobins and hematocrits, or repeatedly low serum and urinary vitamin levels merit nutritional supplementation, if need be, nutritional therapy. Indeed, the ability to improve growth rates and health status should be viewed as one of the purposes of any nutritional survey, and budgetary provisions should be established at the onset. It is of little practical value to ascertain that 23% of children are at "nutritional risk" as an isolated statistic, if nothing is then done about it. Follow-up, supplementation, and therapeutic programs are therefore both optimal additions and moral imperatives.

Many nutritional surveys have been underfunded with respect to data analysis, and with final reports demanded in an impracticably short time. Results have therefore been given as simple tabulations, usually with respect to arbitrary cutoff values and with much valuable data remaining unanalyzed and then lost. Even worse, some data have been made available to organizations not involved in the original survey, and motivated to make sensational (if unfounded) discoveries. As a single example, 1-day dietary records have been treated as if they were 7-day dietary records, ignoring the fact that 1-day records tend to exaggerate the proportion of apparently low and apparently high dietary intakes (Garn *et al.*, 1976d, 1978).

So the word "optimal," as applied to nutritional surveys and nutritional studies, must include screening of apparently at-risk individuals, some of whose difficulties may be of chromosomal or genetic origin. It must also include follow-up studies, in part to make sure that children and adults designated borderline at the time of the survey do not decline into a less satisfactory category. It must include provisions for nutritional supplementation and nutritional therapy, in part to test and confirm the nutritional hypotheses. And it must include provisions for adequate data reduction and adequate data analysis to make sure that the findings are commensurate with the investigative effort (Table VIII).

Once again the term "optimal" merits definition. Once again the term

Table VIII. Steps in Screening and Follow-Up of Those at Apparent Nutritional Risk

1. *Ascertain* individuals at apparent nutritional risk by anthropometric and other criteria, using a broader than minimal "net" to correct for measuring and other errors

2. *Screen* this group, by repeat examinations, to exclude measuring errors and to exclude those with congenital malformation syndromes and the progeny of the dimensionally small

3. *Follow-up* those remaining, instituting supplementation and/or therapy, noting response to supplementation or therapy

"optimal" does not mean overfunded, overintensive, or posh, but more nearly adequate in data gathering, data reduction, and data analysis.

12. Optimal Steps in Data Recording, Reduction, and Analysis

As compared with serial, longitudinal growth studies (or in aging studies of the ongoing variety) where the participants are all well known to the investigators, and vice versa, most nutrition surveys involve a seemingly endless progression of subjects and endless opportunities for error in identification and data recording. With numbers hastily written down, and by a variety of workers, honest errors multiply like yeast organisms. A birth date may be 1–11–77 in the American system and 11–1–77 in the Continental system, quite a difference to the age assessment of an infant! When it comes to the precoding forms (for punch card and tape transcription), sixes may look like fives, nines and one look like sevens, and zeros (noughts) and sixes may appear much alike. Probably 3% of values ultimately recorded in most surveys are in error for the reasons outlined in Table IX. In the optimal survey these recording errors are reduced to a minimum.

Often, each subject is issued a numbered identification card to be retained through the steps of a survey. But identification cards and subjects can become separated, or exchanged. The mother may retain the cards for her children, and soon her identity and her child's identity become confused. (Reviewing radiographs from many surveys, it is remarkable how often an obviously adult female hand is associated with the I.D. of a 2-year-old, and vice versa.) Or siblings may exchange cards, so that (in the radiographs) the older appears to be the younger, and vice versa. In the optimal survey a Polaroid (instantaneous) photograph is stapled to each identification card, and successive workers see to it that photograph and subject match (Garn, 1976b).

There are defined "codes" for no information, no specimen, inadequate identification, etc. Usually these are 9-9-9, 8-8-8, 9-8-8 codes, and the like. Too often, in the steps of data reduction and data analysis, there are problems with these codes. A 9-9-9 may be recorded as 999 mg, 999 mm, etc. Even in the steps of data cleaning, errors may creep in. Working with Ten-State, Pre-School Nutrition Survey, and NINCDS data, we have encountered seemingly

999-kg women, 998 mg of riboflavin, zero birth weights, and even negative birth weights and birth lengths! We have encountered children aged minus 6 years, if the calculated data were to be trusted.

There are other problems with age, in nutritional surveys, when age is variously given as (a) past whole year, (b) next whole year, (c) nearest whole year, or (d) exact age. There are problems when an age interval is identified as the past age, the next age, the exact age, etc., making useful comparisons less than possible.

In the optimal survey, great care is used in recording numbers and in their transcription. Participants are photographed and identified at each step of the survey. Codes are rigidly observed and rigidly employed in data reduction and data analysis. The data are "cleaned" both by computer program and by "eyeball" review of individual printouts, both for single measurements and for bivariant distributions, in computer plots. Just a few 9-9-9, 9-9-8, 8-8-8, and similar codes computer-treated as actual values can reap havoc with data otherwise meticulously handled. It is amazing what a few card-punching errors of the decimal-point variety can accomplish. In the optimal survey great care is taken in data cleaning to exclude such improbabilities, as a matter of course.

13. Optimal Assessment

The fact remains that the term "optimal" as applied to a nutritional survey or a nutritional study is not defined simply in terms of funding, personnel, laboratory facilities, or wealth of statistical assistance. Rather, an optimal nutritional survey may be defined in terms of planning, design, training and motivation of personnel, and the objectives (both investigative and human).

Table IX. Common Sources of Error in Recorded and Coded Survey Data

Type of error	Causes
Identification	Wrong identification number, wrong identification card (see Garn, 1976b)
Measurement	Readout errors, especially key digits
Transcription	Misreading of look-alike numbers (1 and 7, 5 and 6), punching errors, decimal-point errors
Coding	Use of wrong code, especially for "no information"
Card-reading	Inadequate attention to codes, improper coding, wrong card format, etc.

With regards to anthropometric, clinical, or biochemical and hematologic data, less can often be more, but one more essential measurement may often supersede a number of measurements of lesser immediate value.

There is little point in duplicating all the measures used in some previous nutrition survey, simply to duplicate them. A study of starvation survivors is vastly different from a survey of a Westernized city, and a locale where obesity is common scarcely demands the same measures where protein–calorie malnutrition is prevalent. An arm circumference may be quite invaluable in one context and superfluous in another, radiographs (and direct photon absorptiometry) may be essential or redundant, and the classic signs of vitamin deficiencies may be a mere exercise or an absolute necessity to record.

I stress personnel training, supervision, and motivation, to maximize data quality. This is both optimal and minimal. I stress the selection of key problems on the basis of prior information. This is both optimal and obvious. I see the need to make hemoglobin and hematocrit determinations knowing the prevalence of hemoglobinopathies and available dietary iron. This is elementary. I emphasize the need to screen the listing of those apparently at nutritional risk to eliminate chromosomal and genetic abnormalities from the 5–15% of those so designated. Actually, this should always be done. I urge follow-up studies on those seemingly at risk. This is optimal, in that it is uncommon, but desirable and to be viewed as the norm.

Why compare longer legged blacks and shorter-legged Chicanos (or Eskimos or Filipinos) with Boston or Hanes weight-for-length developmental norms? Surely this strategy is less than optimal. Why emulate every other survey if parasitic infection is the pervasive health and nutritional problem? Why use 1-day dietary records if the problem is individualistic, to identify low intake primaparas or poverty-level children? Why accept the usual, hasty simple tabulations (simply to get a report written) ignoring all the possibilities in massive data extensively collected?

To some, still, "optimal" may be identical with redundant, superfluous, luxurious, or overmanned. To some, centered upon the nutritional problems of children, it may appear excessive to include adults or pregnant women or the aged. Properly viewed and properly defined, the optimal study is economical and neither oversupplied with human and technical resources nor deficient in what needs to be done. It may be optimal, compared to what has been the practice in the past, but not wasteful in being either too little or too much.

14. References

Cameron, J. R., Mazess, R. B., and Sorenson, J. A., 1968, Precision and accuracy of bone mineral determination by direct photon absorptiometry, *Invest. Radiol.* 3:141–150.

Colbert, C., 1976, *Radiographic Absorptiometry*, 2nd ed., Radiological Research Laboratory, Wright State University, Dayton.

Colbert, C., Mazess, R. B., and Schmidt, P. B., 1970, Bone mineral determination *in vitro* by radiographic photodensitometry and direct photon absorptiometry, *Invest. Radiol.* 5:336–340.

Dequeker, J., 1972, *Bone Loss in Normal and Pathological Conditions,* Leuven University Press, Belgium.

Frisancho, A. R., 1974, Triceps skinfold and upper arm muscle size norms for assessment of nutritional status, *Am. J. Clin. Nutr.* 27:1052-1058.

Garn, S. M., 1970, *The Earlier Gain and Later Loss of Cortical Bone,* Thomas, Springfield, Ill.

Garn, S. M., 1972, The measurement of obesity, *Ecol. Food Nutr.* 1:333-335.

Garn, S. M., 1973, Fluoride nutrition and bone loss, in: *Mineral Nutrition Today,* pp. 107-115, 1972 Miles Symposium, Bellevue, Ontario.

Garn, S. M., 1976a, The anthropometric assessment of nutritional status, in: *Proceedings of the Third National Nutrition Workshop for Nutritionists on University Affiliated Facilities* (M. A. H. Smith, ed.), pp. 3-21, University of Tennessee Center for the Health Sciences, Memphis, Tenn.

Garn, S. M., 1976b, The use of photographs in nutritional surveys, *Ecol. Food Nutr.* 5:115.

Garn, S. M., 1977, Patterning in ontogeny, taxonomy, phylogeny, and dysmorphogenesis, in: *Colloquia in Anthropology,* Vol. 1 (R. K. Wetherington, ed.), pp. 83-106, The Fort Burgwin Research Center, New Mexico.

Garn, S. M., and Bailey, S. M., 1977, Genetics of maturational timing, in: *Human Growth: A Comprehensive Treatise,* Vol. 1 (F. Falkner and J. M. Tanner, eds.), pp. 307-330, Plenum Press, New York.

Garn, S. M., and Clark, D. C., 1976, Trends in fatness and the origins of obesity, *Pediatrics* 57:443-456.

Garn, S. M., and Shaw, H. A., 1977, Extending the Trotter model on bone gain and bone loss, *Yearbook Phys. Anthropol.* 20:45-56.

Garn, S. M., Hertzog, K. P., Poznanski, A. K., and Nagy, J. M., 1972, Metacarpophalangeal length in the evaluation of skeletal malformation, *Radiology* 105:375-381.

Garn, S. M., Smith, N. J., and Clark, D. C., 1974a, Race differences in hemoglobin levels, *Ecol. Food Nutr.* 3:299-301.

Garn, S. M., Poznanski, A. K., and Larson, K. E., 1974b, Race differences and sex differences in ossification order, *Am. J. Phys. Anthropol.* 41:481.

Garn, S. M., Poznanski, A. K., and Larson, K. E., 1975a, Magnitude of sex differences in dichotomous ossification sequences of the hand and wrist, *Am. J. Phys. Anthropol.* 42:85-90.

Garn, S. M., Smith, N. J., and Clark, D. C., 1975b, The magnitude and the implications of apparent race differences in hemoglobin values, *Am. J. Clin. Nutr.* 28:563-568,

Garn, S. M., Smith, N. J., and Clark, D. C., 1975c, Lifelong differences in hemoglobin levels between Blacks and Whites, *J. Natl. Med. Assoc.* 67:91-96.

Garn, S. M., Shaw, H. A., and McCabe, K., 1976a, Black-white hemoglobin differences during pregnancy, *Ecol. Food Nutr.* 5:99-100.

Garn, S. M., Stinson, S., and Mayor, G. H., 1976b, The race difference in hemoglobin level in chronic renal failure patients, *Am. J. Clin. Nutr.* 29:240-241.

Garn, S. M., Poznanski, A. K., and Larson, K. E., 1976c, Metacarpal lengths, cortical diameters and areas from the 10-State Nutrition Survey, including: estimated skeletal weights, weight, and stature for Whites, Blacks, and Mexican-Americans, in: *Proceedings of the First Workshop on Bone Morphometry* (Z. F. G. Jaworski, ed.), pp. 367-391. University of Ottawa Press, Ottawa.

Garn, S. M. Larkin, F. A., and Cole, P. E., 1976d, The problem with one-day dietary intakes, *Ecol. Food Nutr.* 5:245-247.

Garn, S. M., Shaw, H. A., Guire, K. E., and McCabe, K., 1977, Apportioning Black-White hemoglobin and hematocrit differences during pregnancy, *Am. J. Clin. Nutr.* 30:461-462.

Garn, S. M., Larkin, F. A., and Cole, P. E., 1978, The real problem with one-day diet records, *Am. J. Clin. Nutr.* 31:1114-1116.

Greulich, W. W., and Pyle, S. I., 1959, *Radiographic Atlas of Skeletal Development of the Hand and Wrist* 2nd ed., Stanford University Press, Stanford, Calif.

Jaworski, Z. F. G., 1976, *Proceedings of the First Workshop on Bone Morphometry,* University of Ottawa Press, Ottawa.

Johnson, C. L., and Abraham, S., 1979, Hemoglobin and selected iron-related findings of persons 1–74 years of age: United States, 1971–74, in *Advancedata from Vital & Health Statistics of the National Center for Health Statistics,* no. 46, U.S. Department of Health, Education, and Welfare, Hyattsville, Md.

Mayor, G. H., Sanchez, T. V., and Garn, S. M., 1976, Determining bone mineral status in renal patients—The use of photon absorptiometry, *Dialysis Transplant.* 5:36–39, 51.

Mazess, R. B., and Cameron, J. R., 1973, Bone mineral content in normal U.S. Whites, *International Conference on Bone Mineral Measurements* (R. B. Mazess, ed.) DHEW (NIH) 75-683, Dept. of Health, Education and Welfare, Washington, D.C.

Poznanski, A. K., 1974, *The Hand in Radiologic Diagnosis,* Saunders, Philadelphia.

Poznanski, A. K., 1976, *Practical Approaches to Pediatric Radiology,* Yearbook, Chicago.

Poznanski, A. K., and Garn, S. M., 1974, Skeletal measurement in the evaluation of congenital malformation syndromes, *Skeletal Dysplasias* 10:125–138.

Poznanski, A. K., Garn, S. M., Nagy, J. M., and Gall, J. C., Jr., 1972, Metacarpophalangeal pattern profiles in the evaluation of skeletal malformations, *Radiology* 104:1–11.

Seltzer, C. C., and Mayer, J., 1965, A simple criterion of obesity, *Postgrad. Med.* 38:A101–A107.

Tanner, J. M., Whitehouse, R. H., Marshall, W. A., Healy, M. J. R., and Goldstein, H., 1975, *Assessment of Skeletal Maturity and Prediction of Adult Height (TW2 Method),* Academic Press, New York.

Reference Data

Charlotte G. Neumann

Selection of reference data and levels for assessing normal physical growth continues to cause disagreement and at times heated debate. The main issues center about the meaning of normalcy; use of local versus international standards; choice of elite versus representative local reference populations; relative importance of environmental versus genetic factors; clinical validity of classification systems; and choice of parameters most valid for assessment of nutritional status (Jelliffe and Gurney, 1974). Also there is a dearth of reference data for newborns for arm circumference and fatfolds in growing children.

1. Uses For Reference Data

Anthropometric measurements are most useful in the assessment of nutritional and health status (Jelliffe and Gurney, 1974; Yarbrough *et al.*, 1974). Appropriately developed standards can serve as a reference against which to measure change in nutrition and health in a given population, to evaluate results of intervention programs, and to estimate the deviation from the genetic potential for physical growth (International Union of Nutritional Sciences, IUNS, 1972).

On an individual basis, reference data should provide a frame of reference for selecting the person who falls outside a predetermined level of growth performance for further scrutiny or investigation for over- or undernutrition. More specifically, this refers to protein–calorie malnutrition (PCM) or obesity. A level must be set above or below which an individual is considered to be at risk. Growth charts based on reference data are useful in the longitudinal monitoring of an individual's growth and selecting those at risk for developing nutritional problems (National Center for Health Statistics, NCHS, 1979).

On a community-wide or population basis, reference data for growth are used to define the extent and severity of malnutrition or overnutrition (Water-

Charlotte G. Neumann • Division of Population, Family, and International Health, School of Public Health and Department of Pediatrics, University of California, Los Angeles, California.

low, 1976). In early childhood and among physiologically vulnerable adults (e.g., pregnant or lactating women) the prevalence and incidence of PCM is of particular importance (Jelliffe, 1966). While overnutrition in adulthood is of major concern in the affluent nations (Taitz, 1971; Shukla *et al.*, 1972; Neumann and Alpaugh, 1976), overnutrition in infancy and early childhood has become a recent concern not only in the affluent nations but in developing countries as well (Jelliffe and Jelliffe, 1975).

Use of universal reference data allows international comparisons of health and nutritional status and presentation of hard data in comprehensible form so as to stimulate necessary action and further the aims of public health measures (Waterlow, 1976). In surveillance, reference data can be used to observe trends over time, or in program evaluation as a baseline to measure the effects of intervention (Joint FAO/Unicef/WHO, 1976). Thus, the most common uses of reference data are to identify those individuals who fall outside acceptable limits, either to be treated or followed up, to identify at-risk populations for whom public health measures need to be taken, and to provide a nutritional profile of a given community or several communities for purposes of comparison.

2. Meaning of Reference Data

Reference data for growth are not necessarily optimal, normal, ideal, desirable, or to be considered a target, nor does a reference population imply a healthier population relative to physical growth attainment (Wolanski, 1974; Jelliffe and Gurney, 1974). However, "normal" in operational terms must be equated with health. The concept of "reference" data is preferable to the term "normal" or "ideal" as most standards are not readily obtainable for populations which are being assessed (Waterlow, 1976).

The problem of overnutrition in a reference population is illustrative of this. Obesity is increasing among younger age groups including infancy, not only in the industrial nations from which reference standards are apt to come but also from urban and elite populations in the developing countries. The pushing of solid foods early in infancy, the use of concentrated feedings and overfeeding due to the use of bottle-feeding and processed foods, and abandonment of breast feeding are all contributory (Jelliffe and Jelliffe, 1975; Taitz, 1971). Doubling of birth weight occurs at a younger age, as early as 2–3 months (Neumann and Alpaugh, 1976). In a current, collaborative study of 300 Kenyan urban infants, about a quarter were found to be clinically obese at the 6-month follow-up due to the addition of artificial milk, sweetened juices, and sweetened solids to the regimen of breast milk (Neumann, 1977). In developing countries, as the way of life changes, particularly in urban areas and among elite families, one needs to be on the alert for potential dangers and harmful effects and practices being copied from industrialized nations (Jelliffe and Gurney, 1974).

The health-related effects of overgrowth and undergrowth such as mental and neurologic development, immunocompetence, and long-term sequelae need further elucidation. The assumption that "larger is better" with greater growth equated with better health is questionable (Cheek, 1968; Wolanski, 1974). Is a large, rapidly growing, early maturing child supposedly a healthier child and biologically better off? Long-term sequelae may be diabetes, obesity, or one in five succumbing to atherosclerosis (Kannel and Dawber, 1972). Children who are tall and mature early have not been shown to have any real biologic advantage. In terms of natural selection there may even be survival value of smaller children. Malcolm, (1974) noted that the mortality rate was highest in those toddlers who were growing most rapidly and had presumably greater nutritional requirements. As birth weights increase the larger neonate may be in difficulty when born to women with constricted birth canals, either due to acquired causes such as circumcision or malnutrition or genetic causes, as found among Masai, Watusi, and some other East African tribes (Gebbie, 1974).

Some animal work points to decreased survival with overnutrition (McCance, 1964; Halec, 1962). Obese infants have been shown to have an increased incidence of lower respiratory infections (Tracey *et al.*, 1971; Hutchison-Smith, 1970). There is a need for greater clarity and understanding of the significance of bigness in relation to health and survival. Another ill effect in selecting undesirably high standards of growth based on data from measurements of overfed people is that the standard may prove unobtainable in developing countries and may make the nutritional picture appear impossibly bleak.

Therefore, the yet unresolved problem is how to ensure that measurements are not made on overnourished individuals and to define very precisely how optimal growth or growing populations would be sought from a variety of genetic pools in representative parts of the world (Jelliffe and Gurney, 1974). This would require, for example, that dietary or feeding practices of at least a subsample of the reference population be investigated and that some documentation of health status be made. F. Falkner [1977 (personal communication to D. B. Jelliffe)] and others (Fomon, 1974) advocate that for the infant age group up to 6–8 months, the growth patterns of basically healthy, breast-fed infants should be used for reference purposes as these infants have the best chance for not being overfed (Jelliffe and Jelliffe, 1975; Neumann and Alpaugh, 1976).

3. Criteria for Ideal Reference Data

So great is the need for ideal reference data, that IUNS (1972) convened a task force for the purpose of making recommendations for the construction of sound international reference data, particularly for preschool children, based on a well-nourished, not overnourished, reasonably cared for population with data by sexes presented separately. Paraphrasing the task force's recommendations (IUNS, 1972), the population must be large, representative, well described, well defined, and with a minimum of 100–200 children per age group. Data must be periodically updated so as to be reasonably contemporary. Measurement techniques must be well described and well executed, with validation of measuring techniques. Standardization and uniformity of techniques and measurements are necessary to ensure comparability of results among different countries and different surveys carried out at various times and places. Knowledge of precise chronologic age is preferable to age-independent standards, though these may become essential in lesser developed countries where age is not always precisely known. The measurements selected must reflect nutritionally vulnerable labile body tissue such as arm circumference and body weight. Also reference levels which are not age dependent have to be resorted to. Standards must relate to health status and should indicate the level below which an individual or group is at risk for PCM and an upper level above which an individual is at risk for overnutrition.

Choosing populations deliberately from the elite groups with reasonably adequate health care and nutrition may be racially unrepresentative of an entire country. However, it would dramatize the degree to which the growth potential of a given group is unfulfilled and increase the probability of detecting significant undernutrition (Goldstein, 1974). It would be helpful, but probably not possible, if the reference population were of the same ethnic mixture as far as the population to be studied were concerned. Finally, representation of reference data needs to be practical, useful, and presented in as simple and understandable a way as possible, not too finely categorized, and with clinical reality for each category.

4. International Reference Data

There is a great need for a single set of international reference data for growth particularly in preschool children based on a well-nourished, but not overnourished, well-cared-for population (IUNS, 1972; Waterlow, 1976; Jelliffe, 1966; McLaren, 1976). Few countries, particularly the lesser developed ones, can afford sufficiently large, well-executed, statistically valid studies on representative populations. Internationally available reference data can be used as the basis for comparing one community with another, assuming measurements were made in a uniform way (IUNS, 1972). Confusion results when community data are compared with data from other areas but using different reference data, derived even by different methods.

Historically, the reference data derived from the Harvard growth curves (Stuart and Stevenson, 1975) and English data of Tanner *et al.* (1966) have served as international reference standards. The Harvard standards have been used by WHO since 1966 (Jelliffe, 1966) and have had the advantage of being internationally available. The more contemporary National Center for Health Statistics (NCHS) (1976) standards, described later, may replace the latter two. Reference data for arm circumference and skinfold obtained specifically on the population on whom height, weight, and head circumference measurements are obtained are badly needed in place of the patchwork standards now available using different populations for different measurements.

Limitations in the use and application of international reference data certainly exist and as desirable as they may seem, some caution must be applied (Walker and Richardson, 1973). The "nutritional ideal" may have to be modified, even radically, for lesser developed countries as rapid changes occur in the way of life for many people, and potentially harmful or dangerous practices, particularly nutritional, impose themselves in these countries (Jelliffe and Gurney, 1974). The most obvious criticisms are directed toward possible genetic differences between the reference population and the population under study.

The WHO monograph by Jelliffe (1966) attempted to produce a common anthropometric language and present internationally useful reference data, although these were based on various populations at varying times and by different methods and measuring equipment (Jelliffe and Gurney, 1974). Zerfas (1977) and many others feel that these should not be replaced unless reference data for international use can meet the following criteria: The reference data should have sufficient following and acceptance; the reference data should be able to measure in another population the level near which or under which the risk of undernutrition exists; the lesser developed countries should benefit from using the reference as a target for desirable growth to realize its full potential relevant to health and survival as health care and nutrition improve; and finally the reference data should provide more significant information on nutritional status than is currently available.

5. Locally Constructed Reference Standards

Ideally, the construction of local reference standards representative of the same ethnic, genetic mix of the population being evaluated, would obviate problems with genetic differences between reference population and study population. The dilemma in preparing country-specific standards is that if one chooses an elite population which is reasonably well nourished and well cared for medically, the numbers are usually inadequate and may be unrepresentative of the country or racially different and no better in this regard than international standards (Waterlow, 1976). If one draws upon a sample representative of the entire country in a developing country, then children with poor medical care, illness, and moderate and mild malnutrition are included in the reference standards (Wolanski, 1974). For example, when the All Indian Council of Medical Research standards (ICMR, 1972) were applied to Indian children, a much lower percentage were classified as malnourished compared to the higher percentages when the Harvard standards were used (Neumann *et al.*, 1969;

Naik *et al.*, 1975). A high percentage of children with subclinical malnutrition and poor care and health are included in the All-India Standards with no attempt to screen these out.

Nonetheless, a number of children from well-defined elite groups may be obtained from private nursery schools and day care centers, as well as children of university faculty or government officials. Where a national nutritional board or university nutrition center exists, it may be a much easier task to produce local reference standards. However, more often than not, great difficulties are encountered in preparation of local standards due to the lack of money and trained manpower (Goldstein, 1974).

To be useful, according to IUNS (1972), local standards should identify the median and distribution of anthropometric measurements in the country concerned. The reference population should have a cultural or ethnic background similar to that of the total population or have the same proportions present as in the general population. If very widely divergent groups are present within a country, standards may need to be constructed separately. Local standards should be monitored every few years to detect secular changes.

6. Genetic versus Environmental Influence on and the Use of Reference Data

What is the justification for applying international reference data, usually derived from industrialized countries, to developing countries which are overtly different from an ethnic, racial, cultural, climatic point of view? Physical growth is an expression of multiple interacting forces, environmental, nutritional, genetic, infection, and disease experience, and it is often impossible to clearly define these components (Jelliffe and Gurney, 1974). The implication of genetic influence is that these govern the full potential for growth given favorable nutritional and environmental circumstances. Genetic differences become very apparent in comparing adult body build in certain groups, as in the often referred to example of pygmies versus the Watusi or Masai (McLaren, 1976; Jelliffe, 1966). Also genetic variation probably expresses itself in fat distribution, as in triceps skinfold measurements and steatopygia seen in Africans (Robson, 1964). Racial and genetic differences become more marked in older children of school age and separate standards may be necessary for the very short and the very tall groups.

The use of standards based on parental stature (average of mother and father's height in the fourth decade) has been suggested by Garn and Rohmann (1966) and Tanner *et al.* (1970). Use of parent-specific height standards have been determined on white and black children and other racial groups in California among a middle class membership, of comparable economic standards, in a large prepaid health care plan (Wingerd *et al.*, 1971). It was found that mean heights for the groups diverged after 1 year and that black children were the tallest and racially mixed children the shortest. Positive correlation between mid-parent and child height increased with age, with black children taller for parent height than white children. Construction and use of a slide rule was introduced to determine parent-specific height percentiles for each child by sex and race. The discrepancy may reflect not merely differing growth patterns but, more importantly, that black parents had been relatively deprived compared to the white parents and that their children, because of better circumstances, were able to reach their growth potential. It would be desirable to know the degree to which growth potential of various groups is fulfilled. Standards based on elite populations of the same genetic, racial, or ethnic mix and background as the study population should be of help in this regard.

The current situation is that there are few well-constructed "local" national standards available to fit all the criteria for satisfactory reference data. Thus, there is justification for using general international reference data for growth, although they may be derived from North American and European children. Young children of developing countries, particularly in the first 2 years of life who are of the elite group, well fed, and free of disease and infection, show growth curves strikingly similar to those derived from the United States, Scandinavia, Holland, and England (Habicht, 1974). Evidence for this has been seen in anthropometric studies from such widely scattered places as India (Neumann *et al.*, 1969), Nigeria (Janes, 1974), Ghana (Neumann *et al.*, 1975), Ehtiopia (Eksmyr, 1969), and Malaysia (McKay, 1969). Recent results of an international growth project (Tanner, 1971) indicate that genotype differences in growth scarcely exist between African and European children.

Watson and Lowry (1975), studying 7000 American children ageg 6–10 years, compared the growth of black and white children. No significant differences were noted for height, but a greater variation in weight was observed. For black girls, there appeared to be an earlier prepubescent spurt than in white girls. Nonetheless the authors concluded that there were no significant differences and that it was not necessary to publish separate standards for the two races. Wingerd *et al.* (1971), studying 5000 children in their first 2 years of life, 66% black, 23% white, with the remainder racially mixed (all Kaiser Health Plan members), found similar growth curves for height, length, and head circumference for all groups. Of interest is the fact that these data obtained between 1959 and 1967, agreed more with contemporary late 1960 English standards of Tanner *et al.* (1966) than with the older U.S. Harvard standards obtained in the late 1930s (Stuart and Stevenson, 1975).

Habicht (1974) points out that any racial or genetic effect on mean preschool growth is small compared to environmental effects. Therefore, he feels it is justifiable to use a uniform elite-based set of data as an international reference standard. Habicht's conclusions are based on studies in a variety of lesser developed countries throughout the world that show that the differences in growth of preschool children associated with social class and environment are many times greater than those attributable to ethnic, genetic, or racial factors alone. However, it must be cautioned that the distribution about the mean may be much broader in developing countries and some caution will have to be used in the interpretation. For example, in certain areas, elite children may be healthy, well nourished, and function normally but have a height that is 95% and weight that is 85% of the international standard. Their levels may have to be considered normal. Birth weight distribution also appears to be more closely related to socioeconomic status than to genetic differences (Rosa and Turshen, 1970; Latham and Robson, 1966; Naeye *et al.*, 1971). Interpretations using international reference data applied to a given population, therefore, should be made with caution and common sense.

7. Other Factors Affecting Reference Data

For the first 2 years of life, male–female differences in growth, although small, are definite (Foman, 1974). Females are slightly lighter and shorter than males in the first few years of life. This growth difference between sexes is slight in industrial nations but in developing countries, such as areas in Southeast Asia, India, and Pakistan, it is well recognized that there exists inferior treatment toward female infants and children as reflected by poorer growth performance by females and in female death rates exceeding the death rates of males as high as 2 to 1 (Gordon *et al.*, 1965). Male–female differences in fatfolds are quite striking, particularly in the prepubertal period and in all ensuing periods. Head circumference, early in life, shows no significant variation based on genetic or racial factors but there does appear to be a slight increase in male as compared to female head circumference (Nellhaus, 1968). Therefore, male–female differences must not only be accounted for in standards but must be interpreted cautiously in terms of environmental factors.

Birth order and number of siblings are potent environmental factors influencing nutrition and infection experience of the child (Wallace, 1973; Dingle *et al.*, 1964): These account for differences in growth and are almost never accounted for in reference standards (Goldstein, 1974). For example, weight gain in early childhood decreases with increasing birth order as does general nutritional status. There is also an increase in incidence and severity of infection with increased birth order and number of siblings.

8. "Best" Available Height and Weight Reference Data for International Comparisons

What is the availability of contemporary, well-constructed, universal reference data for international comparisons? The reference standards of Tanner constructed in the late 1960s are based on a large sample and fully described (Tanner *et al.*, 1966) but represent a fairly homogeneous population. A comprehensive set of standards was assembled on Dutch children by Van Wieringen (1972) with several thousand children represented in each age group but date back to 1966, again in a fairly homogeneous group of children. In response to dissatisfaction with existing reference standards in the United States, the National Center for Health Statistics (NCHS), the Fels Research Institute, and the Center for Disease Control (CDC) collaborated in developing modern reference standards according to the recommendations of the National Academy of Science for assessing growth of contemporary children and adolescents in the United States (NCHS, 1976). These are based on a large, nationally representative probability sample, using guidelines of a group of experts on physical growth, pediatrics, and clinical nutrition. In contrast, most previous internationally used reference standards, the widely used Harvard and Iowa standards, were derived from a small sample size from biased populations of limited SES and ethnic characteristics, mainly middle class white children, and constructed in the 1930s and early 1940s (Stuart and Stevenson, 1975). The new growth charts have been prepared using data from the following sources: birth to 36 months, Fels Research Institute; 2–6 years, CDC Health and Nutrition Survey; 6–12 years, CDC Health and Nutrition Survey Cycle II; 12–18 years, CDC Health and Nutrition Survey Cycle III. Percentile values between the 25th and 75th percentiles are taken to represent normal growth. Values at or below the 5th percentile and at or above the 95th percentile for weight, length, or height are taken to represent undernutrition and overnutrition, respectively, and indicate risk for ill health compared to the rest of the population. It is of interest to compare the Harvard curves with that of the NAS curves (see Fig. 1). The percentile curves for body weight for associated length or height are considered age independent until the onset of puberty (Fig. 2).

9. Reference Data for Skinfold (Subcutaneous Fat)

As concern with overnutrition grows, even in developing countries, the need for better skinfold reference data becomes apparent. Reference data, particularly for young children, are not generally available in the United States at this time although the Ten-State Nutrition Survey (TSNS) data are published

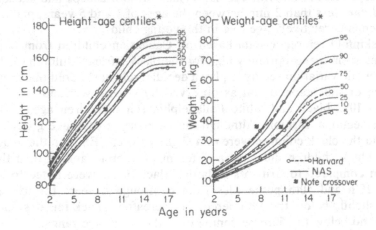

Fig. 1. Comparison of Harvard and NAS growth curves for girls (Zerfas, 1977). *CDC data.

in percentile distributions (Garn and Clark, 1976) and NCHS will have reference data published in the near future. European reference data on skinfolds in young children do exist. Karlberg *et al.* (1968) have published data on young Swedish children for the first 3 years of life by sex and by 3-month intervals and then 6-month intervals after 1 year of age. Tanner and Whitehouse (1962) compiled extensive data in the 1960s on triceps and subscapular skinfolds on British children (London City Council Study) aged 5–16 years based on 1000 children of each sex at each year of age. Data on children aged 2–5 years are based on longitudinal growth studies of the Child Study Center and Harpenden Growth Studies on approximately 100 girls and 100 boys. Data for children aged 1 month to 2 years were taken from a longitudinal growth study in Brussels, Belgium using the same technique (Tanner and Whitehouse, 1961). From the data for males and females triceps and subscapular skinfolds are constructed covering values for the 3rd to 97th percentile.

Pett and Ogelvie (1956) produced excellent reference data for Canadian children and adults. Cross-sectional data on third generation French Canadian Montreal children were gathered between 1969 and 1970 on 2722 boys aged 6–17 years and an equal number of girls aged 6–16 years, with 60–169 children per 6-month age intervals (Jenicek and Demirjian, 1972). Upper, middle, and lower socioeconomic classes were included. Values were all higher than the previous Pett and Ogelvie data, with high subscapular skinfold and triceps skinfolds resembling Tanner's data with an overall trend in increasing fatness.

Keet *et al.* (1970), in examining the value of skinfolds in assessing suboptimal nutrition in children, showed the mean percentages of standard values in well-fed children were near 100% of standard values for weight, height, head circumference, subscapular skinfold, and mid-upper arm circumference. However, for triceps skinfold the mean value of 93.4% of standard was noted. Weight correlated significantly with skinfold and even more highly with mid-upper arm circumference (MUAC). Triceps skinfolds of the normal groups were above the 10th percentile and in the group with PCM well below the 3rd percentile except in children with edema. Values of both triceps and subscapular skinfold vary less than 2 mm between the ages of 1 and 5 years, making these measurements relatively age free in the young child.

Triceps skinfold reference data based on Caucasian children from temperate climates seem inappropriately high for the black African children (Robson, 1964). Genetic variation seems to play a decided role in skinfold thickness and patterning of subcutaneous fat as observed by several workers. Malina (1966) studied 1092 black and white Philadelphia schoolchildren aged 6–12 years in cross-sectional fashion. Although the numbers in each age grouping were small and the black children were of a slightly lower SES than the white children, the black children did have a different percentile ranking than the white children compared to British standards. Black males were noted to be between the 25th and 50th percentiles from 6 through 8 years for triceps skinfold, but slightly below the 50th percentile thereafter. Black females fluctuated above and below the 50th percentile over the entire age range.

A number of studies on African children are relevant. Robson (1964) studied skinfold thickness in normal East African adolescents receiving an adequate diet of over 3000 cal/day, but living in the tropics and performing hard manual labor. Mean triceps skinfold thickness was less than in English boys of 11–15 years of poor nutritional status but subscapular skinfold thickness was only slightly less than in English boys of good nutritional status. Differences in fat distribution in African boys whether due to genetic differences or environmental factors such as physical activity or different body insulation requirements appear definite.

Data on Ghanaian children aged 6 months to 6 years show a similar discrepancy (Neumann, 1974). The triceps skinfold measurements in 200 Ghanaian children correlated poorly with weight for age and arm circumference, not only in the severe and moderately malnourished group, but even in the normal groups (see Table I). Skinfold standards were derived from Hammond (Jelliffe, 1966).

Data on 647 Honduran children (Frisancho and Garn, 1971) showed that relationships between triceps skinfold and stature, within the range of subcutaneous fat observed, were poor. Only those with triceps skinfolds above the 95th percentile and those below the 5th percentile differed markedly in stature. Of normal healthy children, 40–45% had values of triceps skinfolds that were 66% of standard compared to the English (Tanner and Whitehouse, 1962) and Swedish (Karlberg *et al.*, 1968) standards. Thus, the degree of fatness is not always an effective criterion of nutritional status unless dealing with extremes.

Cross-sectional observations on rural, Punjabi Indian children aged 6 months to 5 years, in whom height, weight, arm circumference, and triceps skinfold were simultaneously obtained, show a wide discrepancy between classification of PCM using Harvard standards for height and weight and classification of PCM according to triceps skinfolds using Tanner's skinfold standards (Neumann *et al.*, 1969). In apparently healthy Punjabi children in the group with normal height and weight 43.9% of the triceps skinfold measurements were 60% or less than the Tanner standards (Tanner and Whitehouse,

Table I. Discrepancy in Categories of Percent below Standard of "Normal" Children (0–60 mo) for Weight vs. Triceps Skinfold for Age[a]

	Weight/age[b] (%)	Triceps skinfold[c] (%)
Above standard (std.)	17.1	12.1
100–91% of std.	34.1	0
90–81%	51.2	4.4
80–71%	0	9.7
70–61%	0	29.8
60% below	0	43.9

[a] From Neumann *et al.* (1975).
[b] Harvard standards 50th percentile values.
[c] Hammonds standards as found in Jelliffe (1966).

Table II. Discrepancy in Percentages of Children (Punjabi–Rural) in Categories below Standard for Weight and Triceps Skinfold for Age[a]

Age (mo)	Above std.[b] Wt.	TSF	100–91% std. Wt.	TSF	90–71% std. Wt.	TSF	70–61% std. Wt.	TSF	60–51% std. Wt.	TSF	<50% std. Wt.	TSF
0–6 (N = 60)	16.7	13.3	23.3	15.0	46.7	28.3	8.3	13.3	1.7	11.6	3.3	18.5
7–12 (N = 84)	4.8	2.3	8.3	3.5	61.9	15.5	15.5	10.7	7.1	21.4	2.4	46.4
13–24 (N = 171)	0	1.7	4.1	1.9	19.3	8.7	32.7	9.9	25.7	22.8	4.7	55.5
25–40 (N = 74)	0	4.1	2.3	0	0	17.5	0	21.5	12.2	14.4	1.4	47.2

[a] From Neumann *et al.* (1969).
[b] Standard weight for age = 50th percentile, Harvard values (Jelliffe, 1966); standard TSF for age = 50th percentile, Hammond values (Jelliffe, 1966).

1962) (see Table II). Robson *et al.* (1971) measured triceps and subscapular skinfolds in studies of 1389 Dominican children of African ancestry aged 1 month to 11 years who appeared healthy and well nourished and attended crêches and private schools. Significant differences were found between the English standards and the mean values in these Dominican children. Using a grading system of Hammond (1955), the nutritional status of the Dominican children would be rated as "normal using subscapular skinfolds, and poor using triceps skinfold." The 5-year-old boys when compared to more local standards, using data of Montoye *et al.* (1965), would have a subscapular skinfold thickness above the 50th percentile and a triceps skinfold below the 20th percentile.

Hammond's data (1955) has been adopted and published as a reference standard for mean triceps skinfold for age (Jelliffe, 1966). However, the mean values are on the average of 6.4% higher for boys and 9.1% higher for girls than in Hammond's original data (Robson *et al.*, 1971). The application of these triceps skinfold standards to Dominican children, as with African, Honduran, Punjabi, and American black children, places them in a category that would be lower than their appropriate nutritional status.

Furthermore, the above triceps skinfold thickness standards have been used as the basis of the standard of reference for mid-arm muscle circumference (MAMC) which is calculated from mid-upper arm circumference (MUAC) and triceps skinfold (TSF) (Jelliffe, 1966), thus giving inflated values for MAMC for perhaps many areas of the developing world.

Seltzer and Mayer's (1967) widely used criteria for identification of obesity from childhood to adulthood use triceps skinfold data obtained on a Caucasian population. These authors feel that the triceps is superior yet it appears that triceps skinfolds are larger for whites than for blacks at most ages. Caution is therefore needed in applying these standards to non-Caucasian populations.

In summary, differences between racial groups and between those living in tropical and those in temperate zones seem most marked for triceps skinfolds. Since this measurement is used as an index for overnutrition and obesity and to a lesser extent for undernutrition, separate race-specific or ecology-specific standards seem necessary.

10. Arm Circumference Reference Data

Mid-upper arm circumference, because of its simple and rapid measurement technique, ease of teaching to unskilled staff, and excellent correlation with weight for age and weight for height, is being used increasingly as a public health index of PCM (Jelliffe and Jelliffe, 1969). Decreased arm circumference may represent a diminution of muscle mass and/or subcutaneous tissue and an increase may represent increased fat and/or muscle mass. In the WHO monograph by Jelliffe (1966) the standards are based on healthy Polish children (Wolański, 1961).* Although Wolański's data show erratic rises and falls, there is a very slow increase in the MUAC during the second to fourth year, thus making arm circumference useful when age is not precisely known. The differences between sexes are small, and for survey use combined male and female values appear to be adequate.

Wolański's data approximate data on healthy British children quite closely (Jelliffe and Jelliffe, 1969). There is a rapid increase in the first 6–8 months followed by a slow increase until 4 or 5 years of age. The standard deviation between healthy, well-nourished children is small, especially for older children. Because of high correlation with weight for age or weight for height, it has been suggested that MUAC replace weight in survey work.

Both Burgess and Burgess (1969) and El Lozy (1969) felt that Wolański's curves should be smoothed out rather than left in its erratic form. The modification consisted of smoothing out the curve by using midpoint age group values only with a slight variation produced in the actual figures. Arm circumference measurements from various countries (England, Poland, and Sweden), although not identical, are similar (Jelliffe and Jelliffe, 1969).

The situation for reference data in adults is that there are very meager and outdated ones currently available. In the WHO monograph by Jelliffe (1966) the standards are derived from the U.S. garment industry data used in pattern construction.

There have been relatively few arm circumference studies among non-Caucasian, elite children. Ninety-eight percent of an elite group of children in Ethiopia fit Wolański's standards (Eksmyr, 1969). Studies on well-nourished children in Ghana, the Punjab, Malaysia, and Guyana all found arm circumference to be lower than the Caucasian reference data (Neumann *et al.*, 1969; McKay, 1969; Ashcroft *et al.*, 1968). In the absence of more contemporary or universal standards, the modified midyear data of Wolański for arm circumference can be used. Measurements may be presented as means plus or minus

---

* Referred to by Jelliffe (1966) and Burgess and Burgess (1969).

standard deviations (SD), percentiles, or 10% levels below standard. Wolański's values are for 1–2 years, 16 cm; 2–3 years, 16.25 cm; 3–4 years, 16.5 cm; and 4–5 years, 17.6 cm (Jelliffe and Jelliffe, 1969). Other arbitrary "normal" levels which seem appropriate in light of local experience and need are, of course, valid.

11. Reference Data for Head Circumference

Head circumference (HC) alone is not helpful in assessing nutritional status of an individual or a group. This measurement is most commonly used to monitor brain growth, for HC, at least in the first year of life, correlates significantly with brain cell number (Winick and Rosso, 1969). For the individual child it is a good screening method for detecting microcephaly, hydrocephaly, or macrocephaly. Long-standing malnutrition of early onset, particularly in the first few months of life, or intrauterine growth retardation may result in decreased cell number and a diminution of HC of about 2 SD below the mean (Winick, 1968; Lubchenco *et al.*, 1966; Stoch and Smythe, 1963). Head circumference in the absence of severe and early onset PCM may provide an estimate of age when age is not known in a child under 2 years.

Currently available reference data have been published as part of the new NCHS Growth Charts and Standards based on a large, representative sample of children aged 0–36 months (NCHS, 1976) (see Fig. 2). The methodology and techniques used were uniform and carefully standardized. The HC data were obtained from the same children on whom height, length, and weight measurements were obtained. Values above the 95th percentile or below the 5th percentile are indications for close scrutiny or further evaluation. Similar standards were constructed by Karlberg *et al.* (1968) also for the 0- to 36-month age group.

The Nellhaus HC standards (Nellhaus, 1968) for boys and girls from birth to 18 years of age is a composite standard based on reports from the world literature since 1948. The normal range is defined as an HC lying within 2 SD above and below the mean for age and sex, although a single measurement could be misleading, giving no clue as to rate of head growth. Although Nellhaus ascertained from the published reports that the subjects studied were all free of neurologic disease or retardation, that they were full term at birth, and that measurements were carried out in a similar fashion, there is no real assurance that these assumptions are valid. No significant racial, national, or geographic differences in HC were found from the various reports used. The main advantages of the Nellhaus standards are the large age span covered and the global "representativeness" of the data.

Variations and increase in skull size may be due to genetic variation or may occur in pathologic conditions. Sickle cell or Cooley's anemia may cause extramedullary hematopoiesis with skull enlargement (Jelliffe, 1966) or rickets may cause frontal bossing of the skull (Fomon, 1974). Cultural practices such as binding of the head in infancy may cause severe molding and make HC measurements difficult to interpret (Jelliffe, 1966).

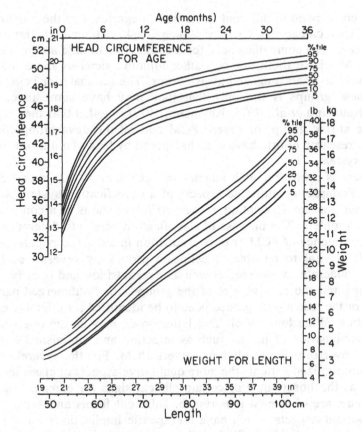

Fig. 2. Example of new NAS or NCHS growth curves. The data is for girls, from birth to 36 months.

In summary, for purposes of nutritional assessment, a slight diminution in expected HC (about 2 SD below the mean) may indicate severe PCM with onset early in life (Winick and Rosso, 1969).

12. Classification of PCM and Overnutrition

A uniform classification system for describing PCM and overnutrition in terms of severity, approximate duration, and predominate type is still far from agreed upon (McLaren, 1976). This chaotic situation is exemplified by the current terminology of various workers: PCM (protein–calorie malnutrition), PEM (protein–energy malnutrition), EPM (energy–protein malnutrition), PJM (protein–joule malnutrition), JPM (joule–protein malnutrition), and the numerous classification systems.

The problems of classification are intimately tied up with choice of reference data, designation of upper and lower levels of normality, clinical correlates which correspond to different numerical categories, and the prognostic capabilities of each category in terms of etiology, risk, outcome, and intervention needed. There is a compelling need for identification of mild and moderate degrees of PCM where there are no pathognomonic signs as there are for severe PCM such as marasmus and kwashiorkor. The rationale for being able to identify these groups is not only that they may have some functional impairment (Neumann *et al.*, 1975; Keilmann, 1977) but also that they represent a sizable at-risk group for severe PCM and all the devastating consequences. Also research efforts have been hampered by lack of a uniform PCM classification system.

Any system is arbitrary, with adaptation necessary to suit a particular purpose. For example, quantitative aspects of a classification system would be most relevant for the description of the nutritional status of a community in terms of the problem. The purpose of classification here is to determine the incidence or prevalence of PCM or of overnutrition in a defined population or community. It is vital to be able to delineate groups by severity such as normal, mild, moderate, or severe, as well as overnutrition and obesity. The marginal groups and the relative size of the groups, the "submerged part of the iceberg," or the true at-risk groups need to be identified for effective early preventive action (Waterlow, 1976). This latter group may be the one which, with proper combination of insults such as infection and diminished intake, can be pushed over into frank, obvious, severe PCM. For the severely malnourished or hospitalized patients the more qualitative aspects of classification systems such as the observance of certain clinical signs for categorizing severely ill patients are of utmost importance. Different forms and severity of PCM are important aspects which have therapeutic implications for the severely malnourished individual. Clinical signs such as hair and skin changes and edema, a pathognomonic feature of kwashiorkor, are combined with anthropometric measurements such as weight, arm circumference, and weight for height. In the presence of edema anthropometry alone would give inflated values. Although it is generally agreed upon that full-blown kwashiorkor and marasmus present the extremes of the full spectrum of PCM there is usually a predominant picture which may give clues as to etiology and prognosis, such as the relative balance of protein versus calories in the diet though this is not wholly clear (Waterlow, 1976).

13. Commonly Used Classification Systems

13.1. Weight for Age

The most popular method of classification of PCM and overnutrition and the most easily understood by the general public is based on weight for age. The reference data used most commonly have been the Harvard (Stuart and Stevenson, 1975), and Tanner standards (Tanner *et al.*, 1966), which give similar results. Weight represents nutritionally labile tissue such as muscle and fat and reflects acute changes of recent onset. The presence of edema may make weight misleadingly high. Also the exact age or good approximation of age must be known. Another problem with this classification system is that no estimate of duration of the malnutrition can be made except to say that marasmus appears to have a slower onset and course of development than kwashiorkor. Furthermore, children with very low weight for age do not necessarily have marasmus but may have stunting or nutritional dwarfism, and their low body mass is appropriate for the low height. Also low-birth-weight infants will give low percents of standard. Weight for age is expressed in percentiles, standard deviation(s) above or below the mean, or 10% categories below standard (Jelliffe, 1966; Fomon, 1974; NCHS, 1976).

Similarly, weight for age above standard can be categorized as overnutrition and obesity. In the problem of the child with excessive weight for age, it is not clear if the child has merely increased adipose tissue and is therefore truly obese or has an increase in lean body mass such as muscle and skeletal tissue (Forbes, 1964). Thus, weight for age can be misleading in assessing overnutrition.

The most commonly used weight for age classification systems have been devised by Gomez *et al.* (1956) and Jelliffe (1966) using the Harvard Growth Standards as a reference (see Table III). In the Gomez classification PCM is divided into first, second, and third degrees corresponding, respectively, to 90–75%, 75–60%, and less than 60% of the Harvard weight for age standards. Jelliffe's classification expresses weight for age in 10% levels below the Harvard weight for age standards: 1 SD of mean weight for age equals about 10% of standard weight; 90–60% is designated as mild to moderate PCM; and below 60% is characterized as severe PCM. In both of these systems no attempt is made in the "severe" category to distinguish between marasmus and kwashiorkor.

13.2. Combination of Weight for Age and Presence of Edema

The addition of the absence or presence of edema to weight for age classification has helped distinguish marasmus from kwashiorkor in severe PCM. The Wellcome Trust and Bengoa (1970) classification systems attempt to do this (see Table IV). The Wellcome Trust working group defined the following criteria for different syndromes: Children whose weights fall between

Table III. Most Commonly Used Classifications of PCM by Weight for Age[a]

			Protein–calorie malnutrition (%)		
Author	Standard	Normal (%)	Mild	Moderate	Severe
Jelliffe, 1966	Harvard (1975)	110–90	90–81 First level	80–61 Second and third levels	<60 Fourth level
Gomez, 1956	Harvard (1975)	>90	90–75 First level	75–61 Second level	≤60 Third level

[a] Weight as percentage of standard—Harvard standard 50th percentile value.

Table IV. Most Commonly Used Classifications of PCM by Weight for Age and
Presence or Absence of Edema[a]

| Author | | Malnutrition (%) | | | Severe (%) |
	Standard	Normal	Mild	Moderate	
Bengoa, 1970	Harvard	>90	89–75	74–60	<60, and all cases with edema regardless of weight
Wellcome Trust Working Party, 1970	Harvard	>90	80	80–60, with edema = undernourished, no edema = undernourished	<60 std. with edema = marasmic kwashiorkor, edema = marasmus

[a] Weight as percentage of standard—Harvard standard 50th percentile value.

60 and 80% of standard weight for age with edema are diagnosed as having kwashiorkor and without edema are diagnosed as being undernourished; those whose weights fall below 60% of standard with edema are diagnosed as having marasmic kwashiorkor and those without edema are diagnosed as having marasmus. Garrow similarly suggested that children who are over 60% standard weight for age and have edema be called kwashiorkor; children who are under 60% of standard weight for age without edema be called marasmus; and children whose weights are below 60% of standard and have edema be called marasmic kwashiorkor. McLaren's (1976) classification, in addition to this, also relies on serum albumin which greatly complicates matters under field conditions. Bengoa (1970) included all children with edema regardless of weight deficit in the third degree PCM group in Gomez's classification.

Basically, then, all of these recommend a combination of the presence or absence of edema and percent below standard weight for age or body weight deficit in delineating the main PCM syndrome. The criticisms leveled at these methods are that, "there is confusion as to the type and degree of PCM present and that there is loss of the 'spectral concept' of PCM" (McLaren and Read, 1972).

13.3. Weight for Height or Length

An important modification for classification purposes has been the introduction of weight for height or length which are excellent indicators of current nutritional status and may be considered virtually age independent (Waterlow, 1976) (see Table V). Baldwin (1925) over 50 years ago first called attention to the value of weight for height as a sensitive indicator of nutritional status. Jelliffe (1966) introduced the use of weight for height and length in nutritional assessment and then almost simultaneously Waterlow (1972) and McLaren and Read (1972) promoted its use as well. The Wellcome Trust Working Party (1970) in trying to define "nutritional dwarfing" also tried to use weight for height but only in a qualitative manner.

Table V. Most Commonly Used Classifications of Overnutrition and PCM by Weight and Height Relationships

Authors	Indices and standards	Normal	Overnutrition	Malnutrition–PCM		
				Mild	Moderate	Severe
McLaren and Read, 1972	Weight/height/age[a] index Harvard std.[b]	110–90%	>110%	90–85%	85–75%	<75%, no edema = marasmus, with edima = kwashiorkor
Waterlow, 1972, 1976	Weight/height[c] (degree of wasting) Harvard std.[b]	110–90%	>110%	90–80%	80–70%	<70%
Kanawati and McLaren, 1970	Height for age[d] (degree of stunting) Harvard std.[b]	≤95%	—	95–90%	90–85%	85%
New NCHS, 1976	Weight/height[c] percentiles (below puberty) NCHS	25th–75th percentile	90th–95th percentile, overweight		10th–5th percentile	Under 5th percentile
Zerfas, 1977	Weight/height[c]		>95th percentile 85–80% obese			<80%

[a] Observed weight as percentage of ideal weight for a given height or length for age.
[b] Harvard standards—50th percentile values.
[c] Percentage of standard weight for height or length
[d] Percentage of standard height or length for age.
[e] Potential or suggested—needs to be field tested.

The major advantages of using weight for length or height is that these, for all practical purposes, may be considered nearly race and age independent except as puberty is approached and at the extremes of weight for height values (Waterlow, 1972, 1976). Although McLaren (1976) strongly feels that the weight of a given child at a given height or length does depend on age, particularly during the first year of life, the relationship may be considered age independent between 1 and 5 years. Ehrenberg (1968) and Kpedekpo (1971) and Dugdale (1971) confirm this age-independent relationship between weight and stature.

Another advantage of using the weight for length or height relationship is that it can help distinguish which child whose weight for age is below 60% of standard has nutritional dwarfing and has an appropriate weight for stature or which child has marasmus and a very low weight for stature. The method will not tell if a child is appropriately short or tall for age.

Weight for height may be reported as percentiles (NCHS, 1976) or as 10% levels below the Harvard standards (Jelliffe, 1966). The lower limits of "normal" have been far from agreed upon, with 90–95% of standard suggested as the cutoff point for normal depending on the judgment and interpretation of the local circumstances. Waterlow (1976) suggests that mild, moderate, and severe PCM be represented by 90–80%, 80–70%, and below 70% of the Harvard standards, respectively. Eighty percent of standard is considered to represent moderate malnutrition. There actually are few data on weight for length or height levels in children with severe PCM.

It is useful to understand how weight for height standards were derived. Jelliffe (1966) plotted a graph of 50th percentile weight values against the 50th percentile length or height values at each age and then calculated values corresponding to 90, 80, 70, and 60% of the Harvard standards. The recently published standards (NCHS, 1976), which have been recommended as a replacement for the Harvard standards, calculated mean weight at a given height at intervals of 3 cm. These weight and height or length measurements were simultaneously obtained on a given child. In practice, there is little difference found in the median values of weight for height between the Harvard and the NCHS standards except in the young infant between 1 and 3 months (Zerfas, 1977).

13.4. Weight for Height Combined with Height for Age

A further refinement was introduced into anthropometric thinking by Waterlow (1972) and Seoane and Latham (1971) who suggested that by using height for age in conjunction with weight for height one could classify PCM according to severity and also gain some idea of duration. Waterlow (1972) suggests the term "stunting or retardation" to "dwarfing" or "chronic or previous PCM" (Seoane and Latham, 1971) for deficits in height for age, and "wasting" instead of "acute PCM" (Seoane and Latham, 1971) for deficits in weight for height. The presence or absence of edema can be added to either classification. The systems will then pick out the following broad categories of PCM (Waterlow, 1976): (a) normal, (b) wasted only (acute malnutrition only), (c) wasted and stunted (acute and chronic or past malnutrition), and (d) stunted only (nutritional dwarfism or recovered past PCM).

The distribution of heights for a given age may show considerable genetic variation from one population to another, and caution, common sense, and judgment must prevail in the application of the Harvard or other similar standards. In regards to height for age, suggested classifications are presented in Table V. In using the Harvard standards, values which are 97% of standard are equal to 3rd percentile values, or 1 SD below standard. Data from developing countries indicate that 1 SD is closer to 95% of the Harvard standard (Waterlow and Rutishauser, 1974) and this value rather than 97% of standard be taken as the cutoff point for mild retardation of height. Long-term severe undernutrition or stunting is probably represented by 85% of the Harvard standard. Waterlow (1972) suggests slightly lower cutoff points for classification of severity of stunting into three grades (see Table V).

The use of development quotients (DQ) index has been introduced by Graham (1968) not only for weight but for height. The weight or height age is divided by the chronologic age, age being measured from conception instead of from birth. Although this may be a more sensitive indicator of height or weight retardation, the obvious problems of estimating age from conception exists and greatly hampers the use of this method except possibly for research purposes.

14. Other Methods of Classification of PCM and Overnutrition

14.1. Arm Circumference

The mid-upper arm circumference (MUAC), described earlier, is used as a rapid classification method in assessing nutritional status (Jelliffe and Jelliffe, 1969). Under difficult field conditions this method may well replace those that rely on weight, height, or length measures and knowledge of age. MUAC correlates highly with percent standard weight for age (Shakir, 1972) and with weight for height (Rutishauser, 1969). MUAC has not been used to any great extent for assessing overnutrition but may well become a simple screening method in this regard. MUAC is also used to calculate the mid-arm muscle circumference (MAMC) useful in assessing protein lack relative to lack of calories (Jelliffe, 1966).

Suggested cutoff points for separating normal individuals from those with PCM have varied. Jelliffe and Jelliffe (1969) and Robinow and Jelliffe (1969) first suggested using 85% of Wolański's year constant standards (see Table VI which are equivalent to 2 SD below the standard value or the 3rd percentile which corresponds to 80% of the Harvard standard weight for age (Jelliffe and Jelliffe, 1969). Other workers, notably Kondakis (1969) and Karlberg *et al.* (1968), based on their experiences, feel that 80% of the standard is much more reasonable and realistic. For 1–5-year-olds these authors consider 14 cm as the lower cutoff for normal nutritional status and 12.5 cm as the upper limit for defining PCM. Two-year-olds with kwashiorkor but without gross edema have been noted to have MUAC values in the range of 10.2–12.4 cm and those with marasmus to have MUAC values in the range of 8.2–10.4 cm, corresponding in turn to 70 and 60% below standard (Jelliffe and Jelliffe, 1969).

14.2. QUAC Stick: Arm Circumference for Height

Arm circumference for height has added little additional information beyond the use of MUAC but has become popular as an age-independent measurement in areas of acute food shortage in the 1- to 12-year-old age group (Arnhold, 1969; Waterlow, 1976).

In emergency (Arnhold, 1969) and under nonemergency situations (Jelliffe and Jelliffe, 1969) the QUAC stick has a demonstrated usefulness as a rapid, simple nutritional assessment tool. The method was based on the assumption that a child with an average height for the community should have an ideal MUAC. Initial standards were based on Wolański's median MUAC values for height on rural Nigerian children from birth to 5 years (Morley *et al.*, 1968) and estimates for older children based on the 10th percentile heights for London children (Tanner *et al.*, 1966). There is no clear agreement as to the appropriate cutoff point but 85–80% MUAC for height have been suggested as the upper limits for PCM. A percentile classification was used by Sommer and Lowenstein (1975) in Bangladesh and showed that the 50th percentile values are approximately 85% standard and 9th percentile are 75% of QUAC-stick standards. In the United States MUAC for height standards were prepared based on the Ten-State Nutrition Survey data (Frisancho, 1974) but have shortcomings of relating elite percentiles to nonelite populations. Although useful in identifying children below a certain level of nutritional status, the QUAC stick is not appropriate for comprehensive classification of PCM.

14.3. Skinfolds

Skinfold measurements, in general, have been used much more extensively in the classification of overnutrition and obesity than for undernutrition. Based on Karlberg's data on triceps and subscapular skinfold data for children 0–36 months (Karlberg *et al.*, 1968), values exceeding the mean by over 2 SD indicate obesity (Fomon, 1974). For age groups from 3 through 17 years Fomon recommends that 90th percentile skinfold values are an indication of overnutrition, as does Tanner (1962), using his own standards, and that 95th percentile values or greater indicate definite obesity (Fomon, 1974). The reference data Fomon refers to are Karlberg *et al.* (1968) and the U.S. Health Evaluation Surveys Cycle II and III (NCHS, 1974). Garn and Clark (1976), with data on 42,000 individuals from ages 1 to 80 years in the Ten-State Nutrition Survey, used the 15th through 85th percentiles to define the categories of lean, medium, and obese, respectively, with separate standards for males and females, black and white. Seltzer and Mayer (1967) published obesity standards for Caucasian

Americans aged 5–50 years indicating minimum TSF levels for obesity which are greater than 1 SD above mean TSF values.

In the classification of undernutrition, although the interpretation of TSF measurements may present problems (as discussed earlier) in severe PCM, in the absence of edema, skinfolds below 6 cm in the 1- to 5-year age group indicate gross fat deficiency. Values as low as 2 and 3 cm are seen in marasmus, with most values seen below the 3rd percentile of standard (Keet *et al.*, 1970; Neumann *et al.*, 1969). Triceps skinfold values are essential for the calculation of mid-arm muscle circumference (MAMC) useful in evaluating the relative deficits of fat versus muscle and calorie versus protein deficiency (Gurney and Jelliffe, 1973). Since there is little agreement on cutoff points for infants and children, especially of various genetic backgrounds, results for skinfolds can be best expressed in percentiles and the reference data specifically identified.

14.4. Sequential Nutritional Diagnosis

A useful concept presented in flowchart form by Zerfas (1977) is that of "sequential nutritional diagnosis" (see Fig. 3). This can assist in the interpretation of the anthropometric measurements usually obtained in assessment. In addition to height or length and weight it may include MUAC, skinfold, arm fat area, and MAMC in distinguishing various forms of undernutrition or overnutrition.

In summary, classifications systems should be able to express severity, type, and duration of PCM and overnutrition. Age-independent methods must be available where needed. In general, composite indices using nonanthropometric data and the more complex methods using logs, squares, cubes, roots, or equations require extra work and added complications under difficult field conditions and may have very little relevance to biologic reality. These may become a block to understanding and the communication of information.

Whatever system is adapted, it must be mutually agreed upon and widely accepted. The classification must be sufficiently simple and understandable and offer a guide to practical action or management. For example, which individual or group is to receive supplementary feedings, intervention, or generally improved care. The system needs to be sufficiently sensitive for early detection of problems. The accuracy of a classification system is reflected

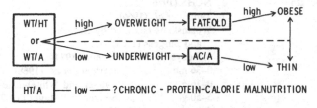

Fig. 3. Sequential nutritional diagnosis (Zerfas, 1977). WT/HT = weight for height; HT/A = height for age; WT/A = weight for age; AC/A = arm circumference for age.

by the extent to which chidlren are misdiagnosed and put into the wrong category. This in turn depends on the ratio between the precision with which a parameter is estimated and the interval between categories which expresses meaningful groupings. A summary of "best" available reference data is presented in Table VI.

15. Conclusion

A review has been undertaken of principal reference data and classification systems available for a number of anthropometric measurements used in the assessment of protein–calorie malnutrition and overnutrition. These are reviewed in the context of their usefulness, limitations, and future needs.

Classification is the mechanism for transforming raw anthropometric data into categories that can be translated into action intervention where needed. It can define the extent, severity, and type of problem, the at-risk or potential problem group, and give a sense of priority for action. It is well to realize that the arbitrary nature of cutoff points for normalcy and the appropriateness of the reference data and classification system used require great caution in the interpretation of conclusions reached about the nutritional status of a given population. Expected outcomes from the application of various standards and classifications will differ. At the crux of the whole matter is the prevailing need for common sense, of local conditions where a given standard and classification system is being applied, and the expected outcomes from the application of different standards and classifications.

Although there is no substitute for local reference data being collected on elite groups from a variety of genetic and ecologic settings around the world, a uniform single reference standard and classification system which can deal with undernutrition and overnutrition is thought to be highly desirable. The newly published NCHS (1976) growth standards carry great promise as serving as the "new improved" international standard, currently for weight, stature, and head circumference, and shortly for arm circumference and skinfolds as

Table VI. Summary of "Best Available" Reference Data for International Use

Weight for age and height or length for age	Harvard (1946)	
	Tanner *et al.* (1966)	
	NCHS (1976)	
Weight for height or length	NCHS (1976)	
	Tanner *et al.* (1966)	
Arm circumference	Wolański as modified by Burgess and Burgess (1969)	
Skinfold	0–8 days	Fomon (1974)
	0–36 mo	Karlberg *et al.* (1968)
Triceps	0–16½ yr	Tanner and Whitehouse (1962)
	5–50 yr	Seltzer and Mayer (1967)
Subscapular	All ages	Hammond (1955)
	2–21 yr	Pett and Ogelvie (1957)

well. Waterlow *et al.* (1977) and WHO (1978) have recommended that these replace the time-honored and historically important Harvard standards which have served the world so well. The valuable contributions of those whose works are cited repeatedly in this chapter have furthered the rational use of anthropometry in nutritional assessment. The introduction of a common anthropometric language, feasible, reproducible methods of measurement, and the presentation of the then best available standards for the interpretation and presentation of anthropometric data by Jelliffe (1966) went far in making order out of chaos.

ACKNOWLEDGMENT. The author wishes to thank Dr. Alfred J. Zerfas, who very generously shared resources, materials, and suggestions and allowed the reproduction of Figs. 1 and 3.

16. References

Arnhold, R., 1969, The QUAC stick: A field measure used by the Quaker Service Team, Nigeria, *J. Trop. Pediatr.* 15:243 (Monograph No. 8).

Ashcroft, M. T., Bell, R., and Nicholson, C. C., 1968, Anthropometric measurement of Guyanese school children of African and East Indian racial origins, *Trop. Geogr. Med.* 20:159.

Baldwin, B. T., 1925, Weight–height–age standards in metric units for American-born children, *Am. J. Phys. Anthropol.* 8:1.

Bengoa, J. M., 1970, Recent trends in the public health aspects of protein–calorie malnutrition, *WHO Chron.* 24:552.

Burgess, H. J. L., and Burgess, A. P., 1969, A modified standard for mid-upper arm circumference in young children, *J. Trop. Pediatr.* 15:189 (Monograph No. 8).

Cheek, D. B., 1968, Conclusions and future implications, in: *Human Growth* (D. B. Cheek, ed.), p. 616, Lea & Febiger, Philadelphia.

Dingle, J. H., Badger, G. F., and Jorden, W. S., 1964, *Illness in the Home: A Study of 25,000 Illnesses in a Group of Cleveland Families,* Press of Western Reserve University, Cleveland.

Dugdale, E. E., 1971, An age independent anthropometric index of nutritional status, *Am. J. Clin. Nutr.* 24:174.

Ehrenberg, A. S. C., 1968, The elements of law-like relationships, *J. Roy. Stat. Soc.* 131:288, Series A.

Eksmyr, R., 1969, The upper arm circumference in privileged Ethiopian children, *J. Trop. Pediatr.* 15:195 (Monograph No. 8).

El Lozy, M., 1969, A modification of Wolański's standards for the arm circumference, *J. Trop. Pediatr.* 15:193 (Monograph No. 8).

Fomon, S. J., 1974, *Infant Nutrition*, 2nd ed., Saunders, Philadelphia.

Forbes, G. B., 1964, Lean body mass and fat in obese children, *Pediatrics* 34:308.

Frisancho, A. R., 1974, Triceps skinfold and upper arm muscle size: Norms for assessment of nutritional status, *Am. J. Clin. Nutr.* 27:1052.

Frisancho, A. R., and Garn, S. M., 1971, Skin-fold thickness and muscle size: Implications for developmental status and nutritional evaluation of children from Honduras, *Am. J. Clin. Nutr.* 24:541.

Garn, S., and Clark, D. C., 1976, Trends in fatness and the origins of obesity, Ad Hoc Committee to Review the Ten-State Nutrition Survey, *Pediatrics* 57:443.

Garn, S. M., and Rohmann, C. G., 1966, Interaction of nutrition and genetics in timing of growth, *Pediatr. Clin. North. Am.* 13:353.

Gebbie, D. A. M., 1974, Obstetrics and gynecology, in: *Health and Disease in Kenya* (L. C.

Vogel, A. S. Muller, R. S. Odingo, Z. Onyango, and E. De Geus, eds.) pp. 485–498, East African Literature Bureau, Nairobi.

Goldstein, H., 1974, Some statistical considerations on the use of anthropometry to assess nutritional status, in: *Nutrition and Malnutrition* (H. F. Roche and F. Falkner, eds.), pp. 221–230, Plenum Press, New York.

Gomez, F., Galvan, R. R., Cravioto, J., and Frank, S., 1955, Malnutrition in infancy and childhood with special reference to kwashiorkor, in: *Advances in Pediatrics,* Vol. 7 (S. Levine, ed.), p. 131, Yearbook, New York.

Gomez, F., Ramos-Galvan, R., Frenk, S., Cravioto, J. M., Chavez, R., and Vasquez, J., 1956, Mortality in third degree malnutrition, *J. Trop. Pediatr.* **2**:77.

Gordon, J. E., Singh, S., and Wyon, J. B., 1965, Causes of death at different ages, by sex, and by season, in a rural population of the Punjab, 1957–1959: A field study, *Indian J. Med. Res.* **53**:9.

Graham, G. G., 1968, The later growth of malnourished infants. Effects of age and subsequent diet, in: *Calorie Deficiencies and Protein Deficiencies* (R. A. McCance and E. M. Widdowson, eds.), pp. 306–316, Churchill, London.

Gurney, J. M., and Jelliffe, D. B., 1973, Arm anthropometry in nutritional assessment: Nomogram for rapid calculations of muscle-circumference and cross-sectional muscle and fat areas, *Am. J. Clin. Nutr.* **26**:912.

Habicht, J. P., 1974, Height and weight standards for preschool children: How relevant are ethnic differences in growth potential? *Lancet* **ii**:611.

Halec, Å., 1962, Studies of protein reserves: The relation between protein intake and resistance to protein deprivation, *Am. J. Clin. Nutr.* **11**:574.

Hammond, W. H., 1955, Measurement and interpretation of subcutaneous fat with norms for children and young adult males, *Br. J. Prev. Soc. Med.* **9**:152.

Hutchison-Smith, B., 1970, Respiratory infections in obese and infants of normal weight, *Med. Off.* **123**:237.

Indian Council of Medical Research (ICMR), 1972, in: *Growth and Physical Development of Indian Infants and Children,* Technical Report Series, No. 18.

International Union of Nutritional Sciences, 1972, The creation of growth standards, a committee report, *Am. J. Clin. Nutr.* **25**:218.

Janes, M. D., 1974, Physical growth of Yoruba children, *Trop. Geogr. Med.* **26**:389.

Jelliffee, D. B., 1966, *Assessment of Nutritional Status of the Community,* Monograph Vol. 53, WHO, Geneva.

Jelliffe, D. B., and Jelliffe, E. F. P., 1975, Fat babies: Prevalence, perils and prevention, *Environ. Child. Health,* Monograph No. 41:124–159.

Jelliffe, E. F. P., and Gurney, M., 1974, Definition of the problem, in: *Nutrition and Malnutrition* (A. F. Roche and F. Falkner, eds.), pp. 1–14, Plenum Press, New York.

Jelliffe, E. F. P., and Jelliffe, D. B., 1969, Current conclusions, in: The arm circumference as a public health index of protein–calorie malnutrition of early childhood, *J. Trop. Pediatr.* **15**:253–260 (Monograph No. 8).

Jenicek, M., and Demirjian, A., 1972, Triceps and subscapular skin-fold thickness in French-Canadian school-age children in Montreal, *Am. J. Clin. Nutr.* **25**:576.

Joint FAO, UNICEF, WHO Expert Committee, 1976, *Methodology of Nutritional Surveillance,* Technical Report Series, No. 53, pp. 20–60, WHO, Geneva.

Kanawati, A. A., and McLaren, D. S., 1970, Assessment of marginal malnutrition, *Nature* **228**:573.

Kannel, W. B., and Dawber, T. R., 1972, Atherosclerosis: A pediatric problem, *J. Pediatr.* **80**:544.

Karlberg, P., Engström, I., Lichtenstein, H., and Svennberg, I., 1968, The development of children in a Swedish urban community. A prospective longitudinal study, III. Physical growth during the first three years of life, *Acta Paediatr. Scand. Suppl.* **187**:48.

Keet, P., Hansen, J. D. K. L., and Truswell, A. S., 1970, Are skinfold measurements of value in the assessment of suboptimal nutrition in young children? *Pediatrics* **45**:965.

Keilmann, A. A., 1977, Weight fluctuations after immunization in a rural preschool child community, *Am. J. Clin. Nutr.* **30**:592.

Kondakis, X. G., 1969, Field surveys in north Greece and Dodoma, Tanzania, *J. Trop. Pediatr.* **15**:201 (Monograph No. 8).

Kpedekpo, G. M. K., 1971, Pre-school children in Ghana: The use of prior information, *J. Roy. Stat. Soc.* **134**:372, Series A.

Latham, M. C., and Robson, J. R. K., 1966, Birth weight and prematurity in Tanzania, *Trans. Roy. Soc. Trop. Med. Hyg.* **60**:791.

Lubchenco, L. O., Hansman, C. H., and Boyd, E., 1966, Intrauterine growth in length and head circumference as estimated from live births at gestational ages from 26 to 42 weeks, *Pediatrics* **37**:403.

Malcolm, L. A., 1974, Ecological factors relating to child growth and nutritional status, in: *Nutrition and Malnutrition* (A. F. Roche and F. Falkner, eds.), pp. 329–352, Plenum Press, New York.

Malina, R. M., 1966, Patterns of development in skin-folds of Negro and White Philadelphia children, *Hum. Biol.* **38**:89.

McCance, F. R., 1964, Some effects of malnutrition, *J. Pediatr.* **65**:1008.

McKay, D. A., 1969, Experience with the mid-arm circumference as a nutritional indicator in field surveys in Malaysia, *J. Trop. Pediatr.* **15**:213 (Monograph No. 8).

McLaren, D. S., 1976, Protein energy malnutrition (PEM), in: *Textbook of Paediatric Nutrition* (D. S. McLaren and D. Burman, eds.), pp. 105–117, Churchill–Livingstone, London.

McLaren, D. S., and Read, W. W. C., 1972, Classification of nutritional status in early childhood, *Lancet* **2**:146.

Montoye, H. J., Epstein, F. H., and Kjelsberg, M., 1965, The measurement of body fat: A study in a total community, *Am. J. Clin. Nutr.* **16**:417.

Morley, D. C., Woodland, M., Martin, W. I., and Allen, I., 1968, Heights and weights of West African village children from birth to the age of five, *West Afr. Med. J.* **17**:8.

Naeye, R. L., Diener, M. M., Harcke, H. T., Jr., and Blanc, W. A., 1971, Relationship of poverty, race to birth weight and organ and cell structure in the human, *Pediatr. Res.* **5**:17.

Naik, P. A., Zopf, T. E., Kakar, D. N., Singh, M., and Sandhu, S., 1975, Primary school children in rural Punjab: Nutritional and anthropometric profile, *Indian Pediatr.* **12**:1083.

National Center for Health Statistics (NCHS), 1976, Growth charts, *Monthly Vital Stat. Rep.* **25**:Suppl. No. 3.

Nellhaus, G., 1968, Head circumference from birth to 18 years. Practical composite: International and interracial graphs, *Pediatrics* **41**:106.

Neumann, C. G., 1977, Collaborative current study, Ministry of Health, Nairobi City Council, Nairobi Medical School, University of California (UCLA), unpublished results.

Neumann, C. G., and Alpaugh, M., 1976, Birth weight doubling: A fresh look, *Pediatrics* **57**:469.

Neumann, C. G., Shanker, H., and Uberoi, I. S., 1969, Nutritional and anthropometric profile of young rural Punjabi children, *Indian J. Med. Res.* **57**:1122.

Neumann, C. G., Lawlor, G. J., Stiehm, E. R., Swendseid, M., Newton, C., Amman, A., and Jacob, M., 1975, Immunologic responses in malnourished children, *Am. J. Clin. Nutr.* **28**:89.

Pett, L. B., and Ogelvie, G. G., 1956, The report on Canadian average weights, heights and skinfolds, *Can. Bull. Nutr.* **5**:1.

Robinow, M., and Jelliffe, D. B., 1969, The use of arm circumference in a field survey of early childhood malnutrition in Busaga, Uganda, *J. Trop. Pediatr.* **15**:217 (Monograph No. 8).

Robson, J. R. K., 1964, Skin-fold thickness in apparently normal African adolescents, *J. Trop. Med. Hyg.* **67**:209.

Robson, J. R. K., Bazin, M., and Soderstrom, B. S., 1971, Ethnic differences in skin-fold thickness, *Am. J. Clin. Nutr.* **24**:864.

Rosa, F. W., and Turshen, M., 1970, Fetal nutrition, *Bull. WHO* **43**:785.

Rutishauser, I. H. E., 1965, Heights and weights of middle-class Baganda children, *Lancet* **2**:565.

Rutishauser, I. H. E., 1969, Correlations of the circumference of the mid-upper arm with weight and weight-for-height in three groups in Uganda, *J. Trop. Pediatr.* **15**:196.

Seltzer, C. C., and Mayer, J., 1967, Greater reliability of the triceps skinfold over the subscapular skinfold as an index of obesity, *Am. J. Clin. Nutr.* **20**:950.

Seonane, N., and Latham, M. C., 1971, Nutritional anthropometry in the identification of malnutrition in childhood, *J. Trop. Pediatr. Environ. Child. Health* **17**:98.

Shakir, A., 1972, Pattern of protein–calorie malnutrition in young children attending an outpatient clinic in Baghdad, *Lancet* **2**:143.

Shukla, A., Forsyth, H. A., Anderson, C. M., and Marwah, S. M., 1972, Infantile overnutrition in the first year of life: A field study in Dudley, Worcestershire, *Br. Med. J.* **4**:507.

Sommer, A., and Lowenstein, M. S., 1975, Nutritional status mortality: A prospective validation of the Quacstick, *Am. J. Clin. Nutr.* **28**:287.

Stoch, M. B., and Smythe, P. M., 1963, Does undernutrition during infancy inhibit brain growth and subsequent intellectual development? *Arch. Dis. Child.* **38**:546.

Stuart, H. C., and Stevenson, S. S., 1975, Physical growth and development, in: *Textbook of Pediatrics*, 10th ed. (W. E. Nelson, V. C. Vaughan, and R. J. McKay, eds.), pp. 13–51, Saunders, Philadelphia.

Taitz, L. S., 1971, Infantile overnutrition among artificially fed infants in the Sheffield region, *Br. Med. J.* **1**:315.

Tanner, S. M., 1976, Population differences in baby, size, shape, and growth rate, *Arch. Dis. Child.* **51**:1.

Tanner, J. M., and Whitehouse, R. H., 1962, Standards for subcutaneous fat in British children, *Br. Med. J.* **1**:446.

Tanner, J. M., Whitehouse, R. H., and Takaishi, M., 1966, Standards from birth to maturity for heights, weights, height velocity and weight velocity: British children in 1965, *Arch. Dis. Child.* **41**:454, 613.

Tanner, J. M., Goldstein, H., and Whitehouse, R. H., 1970, Morley standards of children's height ages 2 to 9 years allowing for height of the parents, *Arch. Dis. Child.* **45**:755.

Ten-State Nutrition Survey (TSNS), 1972, Department of Health, Education, and Welfare, Document HO, HSM 72-8130.

Tracey, V. V., De, N. C., and Harper, J. R., 1971, Obesity and respiratory infections in infants and young children, *Br. Med. J.* **1**:16.

Van Wieringen, J. C., 1972, Height and weight surveys in the Netherlands, as quoted in Waterlow (1976).

Walker, A. R. P., and Richardson, B. D., 1973, International and local growth standards, *Am. J. Clin. Nutr.* **26**:897.

Wallace, H. M., 1973, The relationship of family planning to pediatrics and child health, in: *Maternal and Child Health Practices* (H. Wallace, E. Gold, and E. F. Lis, eds.), pp. 795–801, Thomas, Springfield, Ill.

Waterlow, J. C., 1972, Classification and definition of protein–calorie malnutrition, *Br. Med. J.* **3**:566.

Waterlow, J. C., 1976, Classification and definition of protein–energy malnutrition, in: *Nutrition in Preventive Medicine* (G. H. Beaton and J. M. Bengoa, eds.), pp. 530–555, WHO, Geneva.

Waterlow, J. C. Buzina, R., Keller, W., Lane, J. M., Nichaman, M. Z., and Tanner, J. M., 1977, The presentation and use of height and weight data for comparing the nutritional status of groups of children under the age of 10 years, *Bull. WHO* **55**:489.

Waterlow, J. C., and Rutishauser, I. H. E., 1974, Malnutrition in man, in: *Early Malnutrition and Mental Development* (J. Cravioto, ed.), p. 13, Almquist & Wiskell, Uppsala.

Watson, E. H., and Lowrey, G. H., 1975, *Growth and Development of Children*, 6th ed., Yearbook, Chicago.

Wellcome Trust Working Party, 1970, Classification of infantile malnutrition, *Lancet* **2**:302.

Wingerd, J., Schoen, E. J., and Solomon, I. L., 1971, Growth standards in the first two years of life based on measurements of White and Black children in a prepaid health care program, *Pediatrics* **47**:818.

Winick, M., 1968, Changes in nucleic acid and protein content of the human brain during growth, *Pediatr. Res.* **2**:352.

Winick, M., and Rosso, P., 1969, Head circumference and cellular growth of the brain in normal and marasmic children, *J. Pediatr.* **74**:774.

WHO, 1978, Personal communication.

Wolanski, N. L., 1974, Biological reference systems in the assessment of nutritional status, in: *Nutrition and Malnutrition* (A. F. Roche and F. Falkner, eds.), pp. 231–269, Plenum Press, New York.

Yarbrough, C., Habicht, J. P., Martorell, R., and Klein, R. E., 1974, Anthropometry as an index of nutritional status, in: *Nutrition and Malnutrition* (A. F. Roche and F. Falkner, eds.), pp. 15–26, Plenum Press, New York.

Zerfas, A., 1977, Liberia National Nutrition Survey Report, UCLA Nutrition Assessment Unit, School of Public Health, Los Angeles, unpublished report.

Clinic Assessment

David Morley

1. Introduction

"For us, clinical assessment of nutritional state is difficult. We do not know the age of our children, and we accept that a plastic strip round the arm can give a more accurate assessment of nutritional state than a pediatrician."

"Our health workers find it difficult to offer effective preventive care along with curative care of the presenting condition."

"Records of previous illness or clinic attendance are not usually available. We have to rely on the mother's memory, and she has little conception of the passage of time."

"Home visiting without continuing records of the priority group, the small children, is inefficient. A meaningful two-way dialogue with the parents on their small children's health is difficult."

These statements would be accurate in the majority of clinics, health centers, and small hospitals, the only places where four out of five of the children on our planet can hope to get effective health care. This chapter does not consider the approach of the academic anthropometrist to growth, but rather lays emphasis on what can be practically achieved for the majority of the children of the world.

The first objective of the growth chart to be described is that of maintaining adequate growth. However, in those units where these records are being used effectively, none of the four statements at the beginning of the chapter would be true.

The record being described here was first developed in West Africa. The need for such a record became apparent during one of the first and only longitudinal studies of young children growing up in a village environment. The specific development that occurred at this time was the innovation of a

David Morley • Institute of Child Health, University of London, London, England.

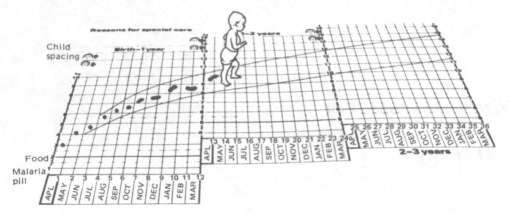

Fig. 1. A simple growth chart. The use of the calendar system to record the passage of time has made this particularly appropriate. Upper line: 50th centile boys. Lower line: 3rd centile girls (International Children's Centre Study, U.K. Children).

calendar system for recording the passage of time (Fig. 1). Practical experience at village level showed that this was the critical difference which made weighing of several hundred children and plotting the weights on charts feasible by village workers with no more than a few years of primary education and a few sessions of teaching and continuing supervision of their work.

The advantage of such charts as been appreciated by innovators across the world, and an analysis of 100 different charts, all containing this basic innovation, has just been completed (Woodland and Kelly, 1979). There is now a very considerable literature on these charts (Morley, 1973). In this chapter the major developments in their use are summarized from a forthcoming book (Morley, 1979).

2. A Home-Based Record

Any traveler who calls in to clinics and small hospitals around the developing world will soon discover that adequate record systems rarely exist. For a number of reasons, record systems of more than approximately 10,000 individuals cannot be kept without specialized and expensive training, which is rarely available in Third World countries. As this difficulty is being appreciated, more workers are turning to a natural alternative—that is, giving the record of the child, suitably protected in a polythene envelope, to the mother herself. The first reaction of most health workers to such a change is that the mothers will lose the records. This is not borne out in practical experience. Once the mothers appreciate the value of the record to their child, it is treasured. In the clinic situation, records are not "lost" but only "mislaid." One study in Botswana (Stephens, 1976) showed that in one clinic only 1% of the mothers failed to bring back their records, whereas the clinic could not find 18%.

The present drive by the international agencies and governments toward developing primary health care systems as a priority has underlined the need for records such as this. They will be particularly necessary as governments accept the widespread need for immunization programs, and some record of the immunization state of every child in the community becomes essential.

3. The Growth Chart in Undernutrition

A study of growth curves on weight charts shows that a pattern of curve can be identified, associated with differing conditions. In marasmus, growth fails from the early months of life (Fig. 2). However, before the condition of marasmus is recognized, there has frequently been a failure for a year or even 18 months. Marasmus is only the end of the spectrum of failure of growth, and the majority of children growing up in the rural areas and slums of developing countries show periods of inadequate growth.

Over the next decade, failure of growth in the first year, particularly when it follows a failure of growth before birth, leading to a small-for-date baby, will receive increasing political importance, as communities become more

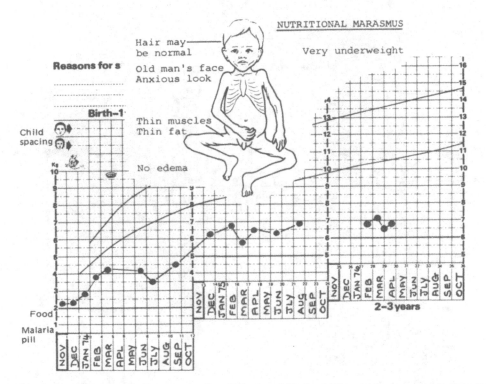

Fig. 2. Marasmus usually starts in the first 3 months of life, but is frequently not recognized until many months later. Upper line: 50th centile boys. Lower line: 3rd centile girls (International Children's Centre Study, U.K. Children).

aware of the effect of such growth on the intellectual development of their children. Politicians will perforce have to place more emphasis on the health care, adequate nutrition, and stimulating environment which are now accepted as essential if the intellectual potential of the individual is to be achieved later in life.

A wide variety of presentations are seen from marasmus through marasmic kwashiorkor to the "pure" kwashiorkor. Even the latter, however, has usually but not always shown a prolonged period of poor growth (Fig. 3). Occasionally kwashiorkor may arise in quite well-grown children, either because there has been a shift over a number of months to a diet with a very low protein content or more commonly because of an illness such as severe measles, leading to a dramatic loss of weight over a few weeks, and precipitating the child into kwashiorkor.

The growth curve in rickets may also be typical, arising perhaps in a child who has shown poor growth, and then possibly due to a change in environment the growth is suddenly accelerated, and the stores or supplies of vitamin D are not adequate for this increased growth. Similarly, in xerophthalmia there is usually a period of inadequate growth lasting many months. The appreciation

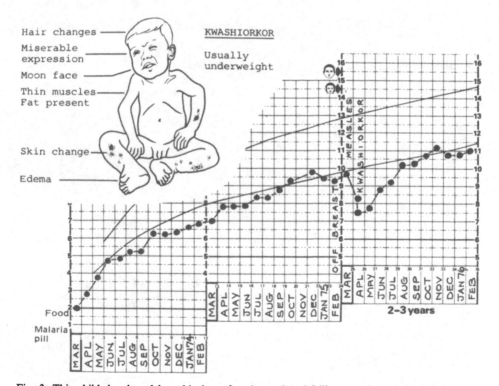

Fig. 3. This child developed kwashiorkor after 4 months of failing to gain weight, followed by measles which precipitated the illness. Upper line: 50th centile boys. Lower line: 3rd centile girls (International Children's Centre Study, U.K. Children).

of the importance of this failure of growth has recently led to extending the concept of nutrition rehabilitation as a method of treating xerophthalmia.

The growth curve is also specific in a number of other illnesses. Measles has already been mentioned. In the child who survives the onslaught of whooping cough in the first 3 months of life, there is often a prolonged period of poor growth. In urinary tract infection, primary tuberculosis, congenital heart disease, and handicapping conditions, a poor gain in weight may be the earliest symptom.

4. The Growth Chart in Recovery

Not only inadequate nutrition but the majority of other chronic illnesses of childhood present with poor growth, among other symptoms. Improvement of growth may also be one of the early signs of recovery, and is an inspiration both to the health worker and the mother to persist with the improved diet or therapeutic measure.

A wider recognition of the "protein disaster" (McLaren, 1974) and the realization of the problems of bulk in the diet and that the majority of children may not be able to obtain adequate energy without the use of foods with a high energy content (Rutishauser, 1976) will now make it possible for nutritionists and health workers to give more appropriate advice to mothers, and expect a more dramatic response. This has recently been highlighted by a study in Papua New Guinea (Binns, 1977) in which a newly introduced crop, palm oil, used for cooking, was fed to children and led to a dramatic and significant improvement in their growth. This improvement depended on the amount of oil consumed, and Binns calculated that if the oil had been made available at 4 months and throughout childhood the children would have remained along a growth curve which would have been considered normal in Western countries. This study was undertaken in a sweet potato staple area, and yet plasma proteins showed no change, although no additional protein had been given with the oil.

5. The Growth Chart, Breast Feeding, and Birth Interval

The tragic spread of bottle feeding in Third World countries has now fortunately been recognized by health workers and governments, and many are attempting to prevent further spread of this disastrous innovation of Western technology. Papua New Guinea has lately banned the import of feeding bottles and teats. Breast feeding is now appreciated even more for its nutritional value as a source of protein, and also because its calories are of such low bulk and so easily absorbed. The many ways in which breast milk protects the young child against infection are also now better understood. We also better comprehend the value to the child and the mother of mother–child bonding and the part played by breast feeding and the skin contact that it involves.

Breast feeding has recently been recognized as one part of the reproductive process, and this concept can be well expressed on the growth chart. The suckling infant, through stimulus to the areola and the resultant prolactin production, is now recognized as the major contraceptive mechanism operating in Third World countries. There are many reasons, therefore, for the health worker to be concerned with the support and encouragement of breast feeding. At the same time she must discuss required contraceptive methods with the mother. In many rural societies mothers do not conceive in the first 9 months or even year after the birth of a child, and in such areas introducing contraceptive devices in this period may be counterproductive. The weight chart may be used to identify the vulnerable month (Morley, 1973), that is, the month in which 5% of mothers in any given community can be expected to conceive again. This is the time when after suitable discussion with health workers the mother may accept an appropriate family planning method so that she can prolong the interval between her children. An excellent recent study (Martin, 1978) has shown that in Chinese children in Singapore, the interval between births, particularly that before the birth of the index child, is one of the most significant factors in child's growth at 9 years in terms of height and weight. In addition, and perhaps more importantly, the child with the long birth interval has an advantage in psychometric tests and the teacher's assessment as to whether the child is above average in attainment.

6. Growth Charts and the At-Risk Child

The simple concept that children who are most in need should receive additional services is seldom achieved in practice. The growth chart is a mechanism by which the children at greater risk can be identified. By displaying these risk factors is a prominent place on the chart the health worker can be repeatedly reminded that such children should at all times receive additional care. For the 5-10% of high risk children in the community, an additional paper weight chart can be maintained at the local health center or subcenter. The weight of the child is recorded at each attendance, and the staff and those responsible for supervising the center can quickly identify whether the child is responding to the additional care, and satisfy themselves that the staff are regularly in touch with the child.

7. New Weighing Methods

The emphasis on simple growth charts has led to a reexamination of the methods used in weighing children. Except for the infant in the first month of life, an accuracy of 100 g is adequate for all weighing; for this level of accuracy a hanging scale has been found to be economical, reliable, and resistant to even the most adverse environments. The child can be suspended from this in trousers, a locally made chair, or a number of other containers, whichever is

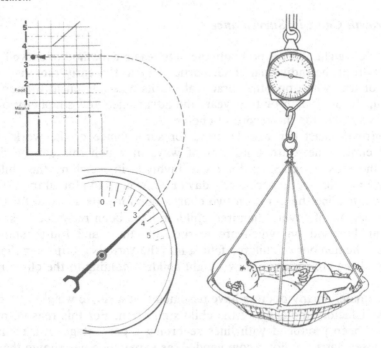

Fig. 4. The method of weighing children which is simple and easy. The markings on the scale are similar to those on the chart. The "bowl" for the baby is made from a metal or bamboo ring and netting.

suitable to the local community (Fig. 4). As will also be seen in this figure, the marks on the chart are similar to those on the face of the scales. This simplifies the plotting of the weight considerably, which can be done without knowledge of the decimal system.

The use of this low cost type of weighing machine* has taken the weighing of children to the small subcenter and even the home, and has for the first time made it possible for the less well-qualified health worker to have an ongoing assessment of nutrition. Unfortunately the concept of a growth curve and its significance is one that may not be achieved during the present educational system in many developing countries. For this reason, a number of visual aids, including transparencies, a large transparent chart for the overhead projector, a flannelgraph, and precut duplicating stencils, are all available.†

Probably the major limitation in the spread of these charts has been due to a misunderstanding of the need to teach the symbolism involved with a growth curve. However, it has been shown that an understanding of these growth curves can be achieved by illiterate people.

* Available with this special face from CMS Weighing Equipment Ltd., 18 Camden High Street, London NW1 OJH.
† Teaching Aids at Low Cost (TALC), Institute of Child Health, 30 Guilford Street, London WCIN 1EH.

8. The Growth Chart in Surveillance

A simple weight chart is probably the only feasible method of surveillance of the growth of the individual child, particularly in the early months of the first year of life, when no other practical means exist. Even at a later age, certainly up to the fourth or fifth year, the advantages and simplicity of the weight curve make this the record of choice.

The growth chart can also be used for surveillance of the work of an individual clinic. There are a number of ways in which this can be done. Probably the most practical is for those involved in weighing the children every 3 or 6 months over a period of a day or a few days to plot all the weights of children attending the clinic on two charts, one for girls and one for boys. When the weights of several hundred children have been recorded, a special overlay can be used on which are marked the third and fourth standard deviations. The number of children falling into the various groups can then be counted, and the proportion of low weight children coming to the clinic monitored.

Unfortunately many doctors have assumed that a single weight of a child can be a satisfactory indicator of the child's nutrition. For this reason, many charts have been produced with lines recording a percentage of a standard weight. These charts are not recommended, as experience has shown that the health worker then pays too much attention to whether a child is above or below a line and is not sufficiently concerned with the ongoing weight curve which is the best assessment of the child's growth. However, from an epidemiological standpoint, the proportion of children of low weight in a community is a satisfactory measure.

At the national level, the charts may also be used as the most sensitive indicator of the nutrition of the country. This is achieved by using a recording system (Fig. 5) so that each time the child is registered, the mark is different depending on whether his weight is above or below the lower line on the growth chart. If among the children attending clinics across the country, the number below and above this line is then sent each month to district, regional, and central headquarters, all will have a sensitive indication of any changes in the nutrition of the most vulnerable age group in the country (Cole-King, 1975).

9. Summary

This chapter has attempted to show how through a simple growth chart the understanding of nutrition and growth can be brought down to village level and the health worker given a tool which, if adequately understood and used, can revolutionize the understanding of the problems she meets. The chart is considered an essential part of that microplan of services (King, 1978) through which the less fortunate four-fifths of the world's children in developing countries can hope for health care, adequate nutrition, and the stimulating and

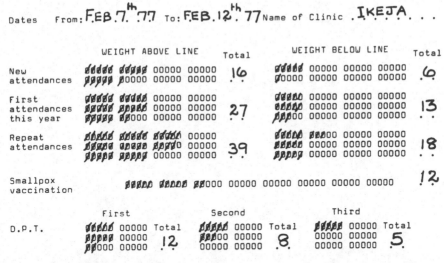

Fig. 5. Under fives' clinic attendances. Registration of children attending a clinic should separate those whose weight falls above or below the lower line. Over a period of time this will give a very sensitive index of the nutrition of those attending.

loving environment so essential to the full development of the human and his intellect.

10. References

Binns, C. W., 1976, Proceedings 12th Annual Symposium Medical Society of Papua New Guinea, Lae.

Cole-King, S., 1975, Under-fives clinic in Malawi: The development of a national programme, *J. Trop. Pediatr. Environ. Child Health* 21:183.

King, M., 1978, Health microplanning in the developing countries: A systematic approach, *J. Int. Health Services* 8(4), in press.

Martin, E. C., 1978, Birth intervals and development of nine-year-olds in Singapore, *IPPF Med. Bull.* 12(1):3.

McLaren, D., 1974, The great protein fiasco, *Lancet* 2:93.

Morley, D. C., 1973, *Paediatric Priorities in the Developing World*, Butterworths, London.

Morley, D. C., 1979, *See How They Grow*, Macmillan, London, in press.

Rutishauser, I. H. E., 1974, Factors affecting the intake of energy and protein by Ugandan pre-school children, *Ecol. Food Nutr.* 3:213.

Stephens, B., 1976, A Health Information System for Outpatient Services Based on Family-Retained Health Records, University of Botswana, Lesotho and Swaziland, duplicated report.

Woodland, M., and Kelly, M., 1979, *J. Trop. Pediatr. Environ. Child Health*, in press.

Anthropometric Field Methods: General

Alfred J. Zerfas

Anthropometric field methods refer to the use of selected body measurements in a field situation which can be used to describe the nutritional status of communities. Such techniques are especially valuable in developing countries, particularly in young children, and the present account deals largely with these circumstances.

There are two important components to the field definition of such communities. First, the study should comprise a well-defined population. Subjects may be examined in their home, community, or in a situation which can act as a collecting point, such as a local school or clinic. The population may also be hospital attenders or institution based. Second, such studies should be concerned with the practical outcome of the method—in other words, the information should be used for the benefit of the population examined. This will often be related directly to the population from which the sample was taken, but may include a research study based on some fact revealed by the field survey. Conversely, a community study undertaken purely for research can rightly be regarded as an academic scientific exercise, if there is no practical value for the population assessed.

Community assessment may also have a negative implication. For example, if nutritional status is found to be satisfactory, essential resources may not be required for nutritional improvement and can be used for other purposes.

1. Relevance of Anthropometry to Nutritional Status

Body measurements reflect the end results of genetic and environmental influences on body size or tissue bulk (Jelliffe, 1966). Environmental factors include those related to nutrition, such as diet and infections, and those not directly related, such as physical activity or patterning of body build according to age or sex.

Alfred J. Zerfas • Division of Population, Family and International Health, School of Public Health, University of California, Los Angeles, California.

In developing countries particularly, infections and poor diet are among the strongest factors influencing growth and nutritional status of the preschool child (Habicht *et al.*, 1974). In many traditional societies, this close relationship may be greatest and most apparent in the second year of life during the transitional period from breast milk to adult food (Gordon *et al.*, 1967; Jelliffe, 1968a). In some urban areas, where breast feeding often ceases during early infancy, the period of maximal ill effects may be in the first year of life. In all these situations, results from simple body measurements, such as weight according to age, may be used to identify the growth failure and undernutrition resulting from a deficiency of calories and/or protein utilized by the child.

In view of the strong relationship between nutrition and growth of young children, this group is often selected for anthropometric assessment. Moreover, young children are a well-recognized group for risk of undernutrition and ill health (Jelliffe and Jelliffe, 1973).

An obsession with highly selective at-risk groups alone may be counterproductive. High prevalence rates of low body measurements and undernutrition dramatize a need in these groups, but often supply no basis for comparison with the rest of the population. Moreover, responsibility for nutritional improvement in a community or nation may go hand in hand with limitations imposed by the age groups assessed. For example, in many countries this responsibility may be almost entirely within the maternal and child health department of the Ministry of Health.

Where feasible, anthropometry should be used to describe a varied number of age groups within the population, bearing in mind the need for different interpretations of the results. Neonatal measurements will reflect maternal health and nutrition during pregnancy (Rosa and Turshen, 1970); those of infants, an interplay of fetal, natal, and postnatal influences (Fomon, 1974); those of young children, the interaction of nutrition and infection. The degrees and forms of physical activity and distribution of body build will influence the interpretation of measurements in older children. Thus, increased exercise in this age group may be an important factor related to a low arm circumference or fatfold measure. This is only one of several important nonnatritional factors, such as timing of adolescent growth spurt influencing growth and body size of the older child (Tanner, 1962).

In developing countries, skeletal measurements of children largely reflect past nutritional status, but the interpretation is influenced by many factors, including the age of the child, the timing and intensity of the episode of malnutrition, and the recovery pattern after the period of undernutrition (Graham and Adriansen, 1972; Ashworth, 1969). For example, short stature in a 5-year-old may result from single or multiple episodes of nutritional deficiency during any previous period. The measure alone would not indicate when the episode occurred. There is evidence, however, that head circumference size may reflect nutritional deficiency that occurred at an earlier age than does stature (Yarbrough *et al.*, 1974).

Many of these problems arise as a result of measurements being taken on a single occasion. This supplies limited information about past growth and nutritional history.

Longitudinal anthropometric assessment of children can give more information than cross-sectional assessment, and should be done whenever feasible, even though the measurements used may be simpler and less varied (Billewicz, 1974; Morley, 1970). A combination of both methods is often justified and can yield complementary information.

It should be noted that even with careful and regularly repeated assessment, nutritional status may change dramatically in young children as a result of acute infection, and some anthropometric measurements, notably weight, decrease rapidly with dehydration.

In addition, anthropometric measurements can be used as some of the numerous indicators which suggest the children who are at risk of infections and further deterioration in nutritional status (WHO, 1976a).

2. Surveys, Screening, and Surveillance

Anthropometry, together with other methods (e.g., clinical signs, biochemical tests), may be used in three main types of nutritional assessment activity—surveys, screening, and surveillance (WHO, 1976b).

A *survey* is a procedure that collects information about individuals or groups in a given setting, usually at one period of time (cross-sectional survey). Usually a sample is taken of the total population to be studied. Any decision to act with appropriate programs is invariably delayed, usually until long after the field phase. The analysis of the data, the writing of the report and recommendations, and the subsequent effect on planning, policymaking, and allocation of resources is often time-consuming and complicated. A survey performed and interpreted by representatives of the actual community or population studied, or at least from within a similar geographical or administrative area, is often by far the most efficient procedure, when feasible (King *et al.*, 1972). This enables insights to be obtained that are not possible for outsiders and it involves the community in planning and implementing programs geared to their real priorities.

The purpose of *screening*, on the other hand, is to identify individuals or groups, so that an immediate decision can be made to refer malnourished individuals or to act on the general results (Cochrane and Holland, 1971). This decision may involve further anthropometric and other measurements to clarify and expand the nutritional diagnosis, close follow-up for identified at-risk individuals, or the implementation of appropriate action, either preventive or therapeutic.

The question of referral needs to consider the added expense and time required from the point of view of the patient and family, as well as the service. Thus, the anthropometric criteria for screening decisions depend on the numbers likely to be at risk and the actual or potential resources available. Resources are almost always fixed, or uncommonly, may be modified according to need. Two examples are presented.

1. Obesity. Triceps fatfold measurements may be used to determine obesity (American Academy of Pediatrics, 1968.) Over 10% of children in the Ten-State Nutrition Survey had obesity as defined by triceps fatfold thickness exceeding the 95th percentile (Garn and Clark, 1976). These results demonstrate the scope of the problem and the need for public health measures. On the other hand, for individual children with obesity defined in this way, adequate resources would be unlikely to cater for the large numbers screened at that level. The numbers for referral or attention may be reduced by raising the level to the 97th or 99th percentile, so that only the "most obese" by anthropometric criteria are included, or by including nonanthropometric factors, such as clinical features or health status.

2. Body or arm wasting. Degrees of wasting as shown by anthropometry (weight according to height, arm circumference) can be identified by categories, such as mild, moderate, and severe. In the Nigerian Civil War, body measurements were used in this way to determine the type of intervention required—mild for preventive feeding programs, moderate for outpatient therapy, and severe for hospital treatment (Aall, 1970).

Results from the anthropometric information from growth charts can encompass survey, screening, and surveillance (Morley, 1973). When collated for individual or combined clinics, cross-sectional survey data are available for any one point in time. When collected for periods over time for the same child, longitudinal survey data are available, equivalent to a surveillance procedure. This procedure is also apparent when repeated cross-sectional data over time are available from individual or combined clinics, so that rates of change may be estimated.

Screening occurs when a child's measure is considered above or below a certain set value (usually weight) for age and a decision is made at the time regarding if and what action is to be taken. This decision is far more valid when the immediate result obtained can be compared with previous growth measurements, to indicate the child's own progress.

With respect to the at-risk child (IUNS, 1977) determined at least in part by growth status or body measurement: screening will identify such a child, a cross-sectional survey will indicate the proportion of at-risk children by age and locality for any one time, and surveillance will assess change over time.

3. Approach to Field Methodology

Several aspects of field methodology have been reviewed elsewhere (Jelliffe, 1966; ICNND, 1963; Weiner and Lourie, 1969, WHO, 1970; Behar, 1975) and this section is, therefore, limited to certain important features. Field procedures are influenced by the extent, depth, and duration of information desired and the purpose for which it is to be used, the required precision of the result, and the known or expected constraints. Long before the field phase begins, careful planning is required to determine the scope, content, and feasibility of the assessment, indeed whether it is really necessary at all.

The following sections comprise a check list of questions which should be considered.

3.1. Goals

What will be done with the results? Who will use these results? If policymakers, do they understand their meaning and implications? If senior technical advisors (often from different ministries and agencies), have they been consulted and their needs fully reviewed? How should the results be presented; is more than one presentation required according to audience?

Due consideration should be given to whatever nutrition plans, programs, and projects have been instituted, so that results may be incorporated within the system (Berg, 1973). For example, if results are to provide baseline information, can the assessment be repeated later as a part of evaluation procedures? Thus, a simple measure such as arm circumference may need to be included initially for this reason.

Is the study truly to perform a service and provide essential, valid information, or for vague or less immediately practical research reasons, or a compromise of both? Will the results be used to "sell" the need to consider nutrition in health and development programs? This is particularly relevant to anthropometry where the classifications and definitions used are not universally agreed upon (McLaren and Burman, 1976).

3.2. Objectives

What specific questions will the results answer? Should the term protein-calorie malnutrition (PCM) be used, or substitutive descriptive terms, such as wasting and stunting? Are definitions and selection of reference values in agreement with the views of local senior technicians? Should a selected group of upper socioeconomic status also be assessed within the population, so that "optimal growth" results can be used for comparative purposes?

If comparisons are required between this assessment and other studies, including elsewhere in the world, are the same reference values being used or have efforts been made to ensure comparability?

3.3. Scope of Work

How will the population to be investigated (population universe) be selected and identified? Is stratification required according to administrative, ecological, or cultural areas? What target groups will be measured and where—in clinics, community meeting places, homes?

Will the sample be selected from census results, preregistration, household listing, or other methods? Will sampling be multistage or simple, systematic or cluster? Is the selection of the sample random, by convenience, haphazard, or purposive (according to known representativeness)? What is the optimal sample size (Kish, 1965; Cochran, 1963)?

What provision will be made to keep the number of absentees selected in the sample to the minimum? Will follow-up visits be required?

Is formal clearance necessary for the study? What identification will the surveyors have in order to facilitate their work—and ensure their acceptability? Will protocol be respected at political, administrative, and cultural levels?

What is the duration of the study—for preparation, training, field trials, the field phase, writing the report, and the follow-up? Is the study cross-sectional, longitudinal, or a combination of both?

What anthropometric measures will be selected, and what will be the purpose of each? How difficult is age determination with respect to the possible use of age-related measures, such as weight for age (Jelliffe and Jelliffe, 1971).

What equipment will be required? Will this be locally made or purchased? What considerations will be made for portability, and for provision for loss or breakage?

How will measurement procedures ensure quality control? What checks will be instituted during the training and in the field? Have due allowances been made for initial and continuation motivation and fatigue of the measurers?

3.4. Constraints

The following constraints should be reviewed with respect to the scope and quality of work: Physical, such as the distance to be traveled and the terrain; political, such as local acceptance to survey procedures (fatfold); cultural, such as concepts of modesty in relation to examination of older children and adults of both sexes; time; personnel; transport; equipment if bulky; and, above all else, cost.

Cost is considered in more detail. It is often difficult to estimate hidden or unexpected opportunity costs. Time-related costs obviously depend on the duration of the study, and include personnel: travel, per diem, salaries; transport: vehicle maintenance, insurance, hire, depreciation, and gasoline; supplies and equipment: repair and replacement, hire, and depreciation. Costs are not related to the duration of the survey include those for preparation and purchase.

Time-related costs will depend on survey content, travel, rest periods, and delays. An estimate of total duration of the survey may be derived from the following equation, if approximately the same amount of time is spent at each sample site:

$$D = \frac{s \cdot d}{t}$$

In this equation, D is the total duration of the survey, s the number of sample sites to be visited, d the duration in each site, and t the total number of teams on the survey.

Table I. Survey Duration by Site in Hours According to Population Density and Procedure[a]

| Procedures[b] | Population density | | | | | |
| | Dense | | Average | | Sparse | |
	Number of hours	%	Number of hours	%	Number of hours	%
Travel						
Between sites	1		5		15	
Within sites	2		5		10	
Subtotal	3	18	10	37	25	57
Nontravel						
Local preparation	2		3		4	
Interview	6		6		6	
Review data	3		3		3	
Rest period	3		4		6	
Subtotal	14	82	16	63	19	43
Total per site	17	100	26	100	44	100
Number of sites	30		75		15	
Total per survey (hr)	510		1950		660	
Total per survey (days)	64		244		82	

[a] Hours divided by 8 (number of working hours per day).
[b] To the final total must be added duration from one area of the country to the next, and contingency (often related to vehicle breakdowns and delays).

Table I gives a detailed account of the estimated duration of the different component activities in a recent national nutrition survey in Africa. In each site, a group of 30 children under 5 years of age and their mothers had anthropometric measurements, as well as other tests. It is obvious that the proportion of time for travel rose and for the interview fell with increasing population sparsity in the area. This might suggest that an increase in duration and depth of the interview in sparsely populated areas would be an efficient procedure, considering the time taken to reach the survey subjects. This would depend on a sufficient number being selected in these areas. Added training time would be required for the extra information. Alternatively, the extra time for the interview may be best allocated to validating and checking procedures. Also, in sparsely populated areas, the education of the interviewee may be relatively poor, and age, in particular, difficult to determine.

Total time-related costs may be reduced by decreasing the total number of sites or the duration of time spent in relation to each site (particularly travel). Increasing the number of teams is limited by training problems, their field support, and supervision. In addition, more than ten teams introduces problems of observer variability between teams. This is particularly important if the results from different parts of the country are to be compared.

Judicious team placement, in making sure that the different populations to be described are measured by all teams, will minimize the problem. This may require increased transport needs in order to travel "caravan style" from one part of the country to the next.

The choice usually has to represent a suitable balance of resources and time with feasibility and validity. Usually resources and personnel are the limiting factors and should be given the highest priority in the planning, training, and implementation phases. Time is often closely related to costs and resources. There may be added reasons for time being a critical factor, such as the onset of the rainy season limiting the duration of the survey. Where survey results are required quickly, the duration should be viewed in its total perspective—that is, from planning to the writing of the report, not just the field duration.

The approximate cost of certain anthropometric instruments is illustrated in Table II for comparative purposes. Costs vary according to several factors, including place of purchase and suitability of the instrument for certain age ranges. Low cost instruments for the measurement of fatfolds are yet to be developed (IUNS, 1972).

A graduated stick for height can be attached to a wall for height measurement or on a table for length, providing there is a suitable fixed end for the head and a sliding piece made locally. The QUAC stick is a graduated stick for height measurement, with a scale attached for arm circumference categories according to height. It was successfully used in emergencies (Arnhold, 1969) and its relevance is being recognized (Brozek, 1974). The arm circumference, because of convenience, is a measure particularly well suited to low cost equipment, such as X-ray strips (Shakir, 1975).

It should be noted that the cost of the instrument is not the only financial

Table II. Comparative Costs of Instruments

Cost	Weight	Height/length	Circumference	Fatfold
		Measurement		
High	Heavy duty scale $200	Stadiometer $250 (250)[a]	Metal tape $6.00 (600)	Precision caliper $100 (200)
Middle	Salter scale $30	Custom-made board $40 (40)	Fiberglass tape $0.50 (50)	Plastic anthrogauge $5.00 (10)
Low	—	Graduated stick $1.00 (1)	X-ray strip $0.01 (1)	Cardboard anthrogauge $0.50 (1)

[a] Proportional compared with low cost.

factor to be considered. Others include maintenance and replacement of the instruments, the salary of the measurer, and the "correction cost." The last refers to the extra time and expense required for action taken to rectify an incorrect measure.

Both absolute and relative costs must be taken into account. To arbitrarily allow a few hundred dollars for instruments in a survey costing up to a hundred times that amount can be counterproductive.

4. Application of Anthropometry with Other Measures

Anthropometry alone provides a profile of growth or body size attained, or of change over time. It reflects nutritional status in terms of the effects of protein–calorie malnutrition (PCM)—its location, extent, severity, and duration. It indicates what, where, and how great is the problem, but used alone does not identify associations or causes.

Usually anthropometry is considered with other nutrition-related information for complete assessment purposes. One practical field methodology for nutrition surveys limits measurements to body length and weight, age, hemoglobin concentration, and pretibial edema, to which can be added other measures, if appropriate and feasible (Miller *et al.*, 1976). This core approach emphasizes the need for a valid sample selection and minimization of the number and complexity of survey items so that sufficient resources are available for training and supervision. Studies may be phased: first, to identify the scope of the problem, using a cross-sectional survey, and later to clarify associations or causes, using more in-depth studies. This has the advantage of delineating more clearly the populations which require further investigation.

Alternatively, a judicious choice of selected indicators, such as those related to infections and the diet, can be added to the basic assessment (Blankhart, 1971). Where extensive in-depth information is included in a cross-sectional survey on a large scale, the effort involved may compromise resources, the validity of survey items, and the sample size. In addition, proven causal information is possible only in a careful longitudinal study.

There are circumstances, such as in a famine, when the main causal factors are obvious and the urgent objective is to determine the numbers at risk for mortality and morbidity (Lowenstein and Phillips, 1973; U.S. Department of Health, Education and Welfare, 1975).

Sometimes it is possible to "piggyback" a limited number of body measurements in an ongoing survey in a nutrition-related field, such as an agricultural or household expenditure survey. This procedure can be efficient and economical. It seems to have been little tried in the past, and provides a ready-made opportunity to link key nutritional indicators.

Flexibility is needed to maximize the usefulness of applied anthropometric assessment, according to the special needs in any given situation.

5. Anthropometric Measures

Simplicity and relevance are the keys to efficient assessment and rapid application of results. Relevance pertains to the measurer's capabilities, the essential information required, the seting in which the measurements are being made, and the availability of appropriate instruments. Table III, showing adaptive anthropometric measurement techniques, indicates that decreased costs and reduced complexity of the measurement techniques allow investigators to increase their sample size and the utilization of trained measurers, with little or no professional background. This action, however, may have the negative effect of also reducing the reliability of measurements. Clearly, then, there can be no uniform pattern for all situations, and, once again, flexibility is required.

It is important to apply rigorous procedures to the measurement process. Essential are proper standardization, training, and supervision of measurement taking. One of anthropometry's major weaknesses relates to poor measurement techniques. These should not be tolerated, even among nonprofessionals. A general approach to this problem is presented at the end of this section. It is acknowledged that an obsession with some ideal "standard of measuring" is unrealistic. There is little reason now not to record in the metric system, providing suitably calibrated instruments are available.

5.1. Weight

Weight is the most traditional and popular measurement for assessing health and nutritional status in all but the least sophisticated setting. Weight remains almost the only rough indicator of nutrition at birth. Weight, with height or body frame, is the basis for actuarial estimates of life expectancy— at least in developed countries. Weight measurement spans science, industry, and marketing.

Table III. Adaptive Anthropometric Measurement

Site/worker	Expected training	Instrument complexity, cost	Reliance on[a] Weight	Height length	MUAC	TF[b]	Numbers reached
Institute	Most	Most	+	+	+/−	+	Least
Hospital	↑	↑	+	+	+/−	+/−	
Health center			+	+/−	+/−	−	↓
Auxiliary[c]			+/−	+/−	+	−	
Community-based worker	Least	Least	−	+/−	+[d]	−	Most

[a] Highest +, medium +/−, lowest −.
[b] Triceps fatfold.
[c] Auxiliary workers may be better suited for training than many other groups listed.
[d] For longitudinal assessment in the community only the tape measure or its equivalent, such as an X-ray strip, may be readily available.

Despite the recognition of the many factors which cause nonnutritional variability in weight estimation, such as food, feces, and fluid, weight remains the most popular method of monitoring changes due to dehydration and overhydration, as well as nutrition. The claim that this measure is more sensitive to soft tissue alterations than other measures, such as arm circumference (WHO, 1970), should be reviewed in the particular situation concerned. Thus, Yarbrough and colleagues (1974) from INCAP point out that the large day-to-day variability in weight measurement amounted to as much as a standard deviation of 0.2 kg, where there was no control for nonnutritional variables. Such a control, particularly in the young children with whom this study was done, is usually impractical in field situations. This variability is far greater (at least ten times) than that of the instrument or measurement. Weight is unique for all measures in this regard. The importance of this would be less in older children and adults.

Yarbrough *et al.* (1974) comment that the variability problem of weight can be overcome by using large sample numbers. Although weight is a good index for extremes of nutritional status, it is not sufficiently sensitive to detect all but gross changes in nutrition for population studies or for screening the same child. In their studies, most of the differences in weight were "due" to nonnutritional factors and to changes in height. It would take at least 6 months to demonstrate significant differences in weight, "using practical sample sizes" for control and fed groups. They also could not demonstrate any improvement in weight for height in the supplemented children, contrary to the expectations of Seoane and Latham (1971). Because of day-to-day weight variability, they recommend that weighing scale precision may be sacrificed somewhat for improved robustness in field work.

Falkner (1961) also questions the reliance on weight, which "has possibly been misinterpreted and given more exalted status than any other. It is a highly unstable measure." He continues with height "is the standby and most stable. When combined with weight, a good idea of size is obtained."

These comments must not detract from the proven special value of serial weight estimation with growth charts. In addition, an "unusual" weight, compared with a prior recording in longitudinal assessment, may be readily cross-checked.

5.1.1. Instruments

Jelliffe (1968b) analyzed replies from over 50 pediatric nutrition centers throughout the world regarding an "ideal scale." These are briefly noted: (a) Low cost was the major limiting factor (at least no more than the present UNICEF models); (b) "accuracy" to 0.1 kg; (c) sturdiness-durability, ease of cleaning, and repair; (d) easy transportability; (e) clear readability; (f) weight range—for children under 5 years up to 25 kg; (g) nonfrightening appearance; and (h) appropriate weighing surface.

The Salter scale (Portable Baby Weigher Model 235) satisfies virtually all of these requirements (Fig. 1). It is a hanging scale with a stable Teflonized

Fig. 1. Salter scale.

spring. Its application to growth charts is well described (King *et al.*, 1972). Either a basket or trousers may be used for the children and can readily be made locally. It is suggested that where trousers are used, several should be provided for waiting children, to speed up the procedure. The scale markings may be in 0.25- or 0.5-kg increments for the growth chart and 0.1- or 0.2-kg units for regular use. Scales are available for 25-kg and 50-kg ranges.

In the field, there is usually little problem in using a horizontal branch, beam, or structure for hanging the scale. Even in desert or similar treeless conditions, two assistants can support a length of bamboo or wood on their shoulders for the purpose.

The Detecto baby scale is a beam balance scale used for weighing up to 16 kg at intervals of 0.02 kg. Its cost is similar to that of the Salter. It is light, but inconvenient for field work where home visits are involved. This scale also compares unfavorably with the Salter because (a) the child must be properly centered in the pan and be perfectly still, (b) knife edges require checking, and (c) it has no direct readout dial.

There remains a need for inexpensive, sturdy, portable scales for older children and adults—both for cross-sectional and longitudinal purposes.

The adult Detecto scale is suitable for use in health centers and similar facilities. Its range is 140 kg at intervals of 0.1 kg. However, where frequent transport is required, problems of portability and precision arise. If it is used for community level field work, checking and calibration before each weighing session site and after each journey is required. It also contains a sliding measuring rod for height (Fig. 2).

The Homs hand carrying scale (No. 300 or 150k) is more expensive. It is sturdy, compact, but very heavy, weighing 16 kg. The metric model (150k) measures up to 150 kg by 5-kg increments on the main indicator scale and 50-g intervals on the fractional bar. The scale is near ground level and requires repeated stooping to read.

Spring bathroom-type scales are not recommended apart from exceptional circumstances for only cross-sectional surveys in adults. These were used in the Nigerian emergency where portability and availability were the most desirable requirements. Frequent calibration and checking with known weights are essential during the weighing session, at least after every 10 or 20 measures. A firm flat surface, checked with a level, is also essential. The "better" bathroom scales are not necessarily cheap; some cost more than $20.

Commercially available weights of 5 or 10 kg should be used for calibration. A container of known size (e.g., 10 or 20 liters) is a convenient alternative. This can be filled with water in the field.

In the country concerned, locally available scales are often preferred, if available and otherwise suitable. They will usually be cheaper and easier to replace than imported models.

5.2. Height and Length

Height (or length) remains the standby for measurement of skeletal growth. It probably correlates better with socioeconomic status in most age groups than soft-tissue measures or weight (Beaton and Bengoa, 1973). It is relatively insensitive to short-term nutrition deficits, but significant measurable

Fig. 2. Adult Detecto scale. Reading areas.

catch-up height can be estimated (Yarbrough *et al.,* 1974). Length measurements have been recommended for children up to 36 months of age and height measurements thereafter (Center for Disease Control, 1976).

5.2.1. Instruments

The instruments range from a graduated stick to expensive, precise, digital readout anthropometers. A graduated stick is the cheapest instrument and has particular value where approximate information is required. Locally made wooden length boards have the advantage of being relatively inexpensive and adapted to the needs of the assessment. The usual problem is to have the local craftsman understand the prototype. A flat steel or protected metric tape can be used as the scale.

Several varieties of prototype boards, usually of wood, have been developed, none yet on a commercial basis. The UCLA Nutrition Assessment Unit has used a portable board for height and length measurement, which has been made in several African countries for national nutrition surveys (Fig. 3). The Center for Disease Control, Atlanta, Georgia has also developed a portable board for both measures. This uses a hinge mechanism, so that the headboard becomes the base of the instrument when height is desired. Another board for length is illustrated by Fomon (1974).

Commercial instruments are costly but may be preferred when small differences in stature require assessing. Quality control will also become more important. Stadiometers (boards with a scale attached) with a digital readout and counterbalanced movable vertical boards are available. A model specifically for neonates and infants has an advantage in that a ratchet mechanism prevents the upright for the feed end from slipping during the measuring (Davies and Holding, 1972).

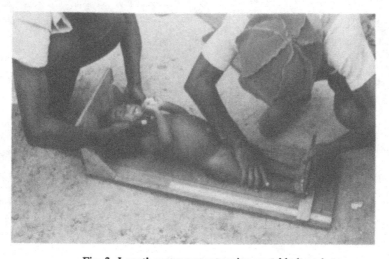

Fig. 3. Length measurements using portable board.

Anthropometers are scaled measuring rods. Although more readily portable than boards, they probably encourage more measuring errors, especially in young children. The height attachment on certain weighing scales, such as the Detecto, is not recommended for use, as the headpiece is too wobbly and narrow.

5.3. Mid-Upper-Arm Circumference

Mid-upper-arm circumference (MUAC) is an objective counterpart to the clinical sign of thinness (Jelliffe and Jelliffe, 1969). When combined with triceps fatfold it may be used to estimate muscle bulk (Gurney and Jelliffe, 1973). It has particular value when scales are unavailable for reasons of cost or convenience (King *et al.*, 1972). Moreover, some studies suggest that MUAC may be more sensitive than weight for detecting nutritional change. Because MUAC alters little in normal children from 1 to 5 years, the measurement does not require a precise age estimate over this age range (Gurney *et al.*, 1972). This is a great advantage when ages are difficult to determine.

5.3.1. Instruments

Tape measures should not be wider than 12 mm nor thicker than 0.2 mm. Metal and firm plastic tapes are more likely to retain a circular shape during measurement, while less rigid tapes tend more to follow the actual surface of the arm. Cloth or paper tapes are unsuitable as they stretch in length and can tear. Numeral design on commercial tapes varies greatly and should be clearly understood by the measurer. The numbers are best printed along the long axis of the tape and not crosswise, as the latter can confuse.

Fiberglass, plastic, or laminated tapes are suitable. Permanent creasing may occur with plastic tapes. Laminated tapes may adhere to the skin surface in humid conditions and should be smoothed during measurement.

The recently developed "insertion tape" (Fig. 4) has a window for reading the measurement and a slot for stabilizing at one end (Zerfas, 1975). Modifications of this tape may be utilized to plot results from a survey on the body of the tape itself, so that recording of results on a form is unnecessary.

Simple and inexpensive methods for assessing arm circumference are based on the "precise-age-independent" property of this measure in children aged 1–5 years. They include an X-ray strip (Shakir, 1975) and a bangle (Laugeusen, 1975).

5.4. Fatfold Thickness

The fatfold measure is the only convenient field method available to assess body fat bulk directly and objectively. Caliper results have shown a good correlation with ultrasonic and electrical conductivity measures of fatfold thickness (Booth *et al.*, 1966), as well with total body fat (Parizkova, 1961).

Fatfold thickness is superior to other measures, such as weight, in as-

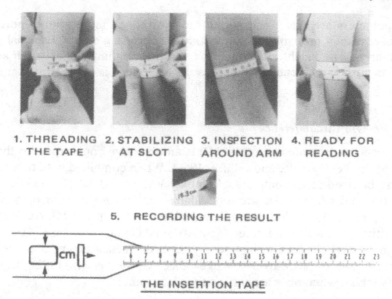

Fig. 4. Measurement procedure using the insertion tape.

sessing obesity. Where possible one limb measurement (left triceps) and one body measurement (left subscapular) are the minimum required to account for differing distribution of subcutaneous fat (American Academy of Pediatrics, 1968; Owen, 1973). Triceps fatfold has been favored for adolescents (Seltzer and Mayer, 1967) and may be combined with arm circumference for arm muscle and arm fat area estimations (Gurney and Jelliffe, 1973).

Adequate training and practice is essential for fatfold measures to be reliable. Where possible, the same person should perform all measurements, so that practice is not wasted, variability is kept to a minimum, and consistent results are obtained at follow-up.

5.4.1. Instruments

Precision calipers are designed so that their jaws exert a constant tension. However, this varies according to fatfold compressibility, which in turn varies with age, sex, hydration, and fatfold site. Calipers measure a double layer of fatfold; due to their tension the caliper reading is approximately 1.3 times the uncompressed single-layer fatfold value, depending on compressibility (Garn and Gorman, 1956).

Precision calipers are expensive; they include the Lange (Lange and Brozek, 1961), the Harpenden (Tanner and Whitehouse, 1955), and Holtain–Tanner–Whitehouse calipers (Fig. 5). The Lange is available in the United States; the others are made in England.

Certain plastic calipers have inadequate springs and incorrect tensions. They are not recommended, especially for children. The "anthrogauge" is based on the wire gauge principle. It is a flat plastic sheet ⅛-inch thick with

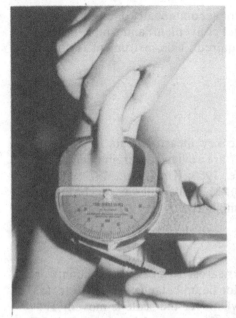
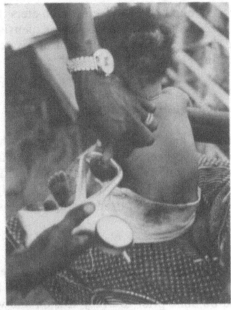
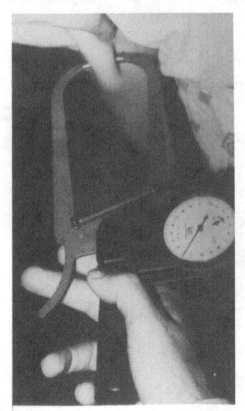

Fig. 5. Measurement of triceps fatfold using the (a) Lange, (b) Holtain, and (c) Harpenden calipers.

smoothed notches of selected diameters to accommodate different fatfold thicknesses (Fig. 6). It may be an inexpensive, convenient, and simple method when an approximate fatfold measure is required. This instrument is still in the testing phase.

6. Methods of Measurement

Weight and height or length remain the key measurements in children. Methods have been described elsewhere but are reviewed herein (Center for Disease Control, 1976; Falkner, 1961; Jelliffe, 1966; King *et al.*, 1972; Owen, 1973; Weiner and Lourie, 1969; WHO, 1970). Methods will vary according to the type of instrument used and age of the subject.

6.1. Weight

Body weight should be measured to the nearest 10 g or $\frac{1}{2}$ oz for infants or 100 g or $\frac{1}{4}$ lb for children. Checking at zero for beam balance scales (Detecto) should be done prior to every session and whenever the scale is moved. The scale should be frequently calibrated using reference weights (e.g., Toledo Scales Company, Toledo, Ohio).

6.1.1. Procedural Steps

1. Have the mother completely undress her child. Request the mother to hold the child.

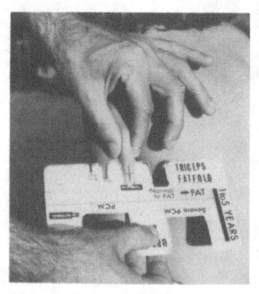

Fig. 6. Measurement of triceps fatfold using the anthrogauge.

2. Calibrate the scale to the zero point, using the scale screw.
3. Have the mother place the child on the scale.
4. If the child is weighed on an upright scale, be certain that he/she is not holding on elsewhere for support.
5. Read the scale face to the nearest graduation only when the child is reasonably still and the needle is stationary. It may be necessary to wait.
6. When you are satisfied that the value is stable, record the weight.
7. If it is impossible for the child to be fully undressed, weigh an equivalent amount of clothing and subtract this value.

6.2. Length

Whether a table or special board is used, one or two wood or metal scales should be calibrated in millimeters and ⅛ inch. A board has the advantage in that one or two grooves or tracks will ensure that the movable footpiece is kept vertical to the scale and is easier to control. Alternatively, each length scale (one is metric, the other is imperial units) may be fixed at opposite sides of the board and the footpiece wedged between.

6.2.1. Procedural Steps

A valid measurement for length depends on proper positioning of the whole child during the reading and recording phase for at least 2–3 sec.

1. Lay the measuring board horizontally on a flat, firm surface or preferably on a table top.
2. Instruct the mother to remove any footwear or headgear on the child and carry the child over to the board.
3. With the help of the mother, place the child on the board with the crown of the head against the fixed end of the board.
4. Position the child so that shoulders, back, and buttocks are flat along the center of the board. Have the mother on the opposite side of the board face you and keep her child flat and central.
5. Stand or kneel behind the fixed end of the board and position the child's head so that the eyes are pointing directly upward and the Frankfort plane is perpendicular to the main board.
6. Use your right hand to bring the movable foot board firmly against the child's heels so that the feet are perpendicular to the main board.
7. Keeping the movable board snugly against the child's flattened feet, grasp the knees and depress them firmly against the main board.

6.3. Height

Readings may be recorded to the nearest ½ cm or ⅛ inch. Recording to the nearest ½–1 inch is too crude, especially for children at borderline levels of low stature. The height attachment on weighing scales should not be used as the headpiece is too wobbly and narrow. The child should not step aside when the reading is made, otherwise downward slippage of the headpiece may occur and introduces a source of error.

6.3.1. Procedural Steps (Using a Board)

1. Place the board where there is a hard flat surface leading to an upright structure (e.g., wall or door). Stand the measuring board vertically.
2. Instruct the mother to remove any footwear or headgear on the child and lead the child to the board.
3. With the mother's help place the child on the horizontal platform so that the feet are together and central.
4. Be certain that
   a. the child stands flatfooted with the knees fully extended;
   b. the back of the heels, calves, buttocks, trunk, and shoulders all touch the vertical surface of the measuring board (or the wall if not board is used);
   c. the mother keeps the buttocks, trunk, and shoulders flat and central.
5. Bring the movable sliding piece headboard to rest firmly on the crown of the child's head. Position the child's head so that the eyes are pointing directly forward (i.e., Frankfort's plane is now parallel to the headboard).

6.4. Mid-Upper-Arm Circumference

To determine the midpoint of the upper arm, a small mark should be made on the left upper arm halfway between the tip of the olecranon and the tip of the acromial process with the arm flexed at a right angle. The arm hangs loosely by the side at the time of measuring. Correct tape tension is the critical factor for adequate measurement; the tape must be applied to the skin surface around the whole arm, but not too tightly to cause an indentation. Because of problems of arm size and cooperation in infants, it is recommended that the arm be held out by the mother or assistant while the child sits on the mother's lap.

6.4.1. Procedural Steps

1. The mother sits comfortably on a chair, and the child sits on the mother's lap or stands. The measurer sits or crouches to face the child's left side.
2. Get the mother to bare the child's left arm and shoulder.
3. Ensure suitable positioning of child, mother, and measurer. The child's left arm should be bent at a right angle for midpoint measure.

4. Identify the posterior tip of the acromion process of the left scapula. Straighten the tape down the arm and note the distance to the nearest centimeter at the elbow point. Halve the distance and mark a horizontal straight line across the arm corresponding to your result.
5. Straighten the child's arm so that it is relaxed and hanging loosely.
6. Loop the tape around the left upper arm.
7. Pull the tape until firm and gentle and uniform contact is made with the arm circumference skin surface at the marked midpoint. Do not compress the soft tissues of the area.
8. Read the value indicated (between the opposing arrows for the insertion tape) to the nearest 0.1 cm and record.
   Note: Arm circumference of babies (under 6 months):
      a. The arm tissues are often flabby and it is difficult to maintain suitable tension around the arm.
      b. The upper arm is difficult to keep still. Thus modifications with these children are as follows:
        1. Have the mother hold out the baby's hand so that the arm is away from the side and steady.
        2. Check tape tension around the arm at least twice before reading the measure.

6.5. Triceps Fatfold

The correct site and depth of placement of caliper jaws is essential for adequate results in the measurement of triceps fatfold. The midpoint of the upper arm is marked. A posterior plane is drawn vertically above the olecranon process to cross the midpoint.

The arm should be hanging straight and relaxed at the time of measurement. The left thumb and forefinger grasp and lift the triceps fatfold about $\frac{1}{2}$ inch proximal to the marked cross. If anchoring of the deep fatfold to the intermuscular septum is apparent or the main bulk of the triceps muscle is not directly felt with the thumb and forefinger deep to the fatfold, recheck the position—it is likely to be too medial or lateral. The grasp should readily identify the muscle–subcutaneous interface, particularly in infants where definition is less clear-cut and the caliper prone to be placed too deeply. Two readings should be made at the same site to ensure agreement and derive an average.

6.5.1. Procedural Steps

1. Measure the mid-upper arm point as for arm circumference. Then straighten the child's arm so that it is relaxed and hanging loosely.
2. Align a long pencil or equivalent directly up the back of the upper arm from the elbow point. Mark along this line at the region of the

mid-upper mark previously made for the arm circumference. The two lines should cross at a right angle.

3. Standardize your caliper to zero if indicated and pick it up with your right hand. Do not show the open jaws to the child. Place the left thumb and index finger about 2 cm apart* so they are 1 cm above the mid-upper arm point and in line with the just completed midline mark. Gently but firmly pick up a fold of fat at this site (see Fig. 5). Do not pinch underlying muscle nor only the skin (it hurts!). Check the child's expression.

4. Lift the fatfold enough to clear it from underlying tissue felt deeply with your fingertips.

5. Apply caliper jaws just below the pinch to the part of the fatfold at the mid-upper arm point with the jaws meeting in the plane of the marked line. (The upper margin of the jaws should be no closer than $\frac{1}{2}$ cm below the finger and thumb, the jaws placed at the same depth as the pinch but about 1 cm down the arm.)

6. From the onset of jaw application wait for 2 sec. At the same time you will note the dial indicator decreasing a little. Exactly after these 2 sec read and call out the value to the nearest 0.1 mm.

7. Remove caliper, keeping the left thumb and index finger gently in position.

8. Record the value. Check that it is correct.

9. Reapply the caliper at the same site and repeat steps 6, 7, and 8. Do not do this if the child is noticeably uncomfortable due to the procedure.

10. *Note:* Depending on your ease of identifying the fatfold, confidence in jaw positioning, and child cooperation, step 6 may be repeated two or three times before a consistent reading is called out; afterward an average is taken of consecutive measurements.

6.6. Quality Control of Measurements

Measurements poorly performed are usually not worth doing and the results can be dangerously misleading. Errors may be due to poor instruments or techniques, especially those of reading and recording (Center for Disease Control, 1975).

Common errors of measurement are listed in Table IV. Whenever practicable, all measurements should be performed by the same person, so that practice is not wasted, variability is kept to a minimum, and consistent results for the same child are obtained at follow-up.

A suggested training schedule for measuring is as follows:

1. Procedures and instruments demonstrated.
2. Instrument practice under supervision.

---

* The distance apart may be adjusted to accommodate the fatfold.

Table IV. Common Errors of Measurement

All measurements	Inadequate instrument Restless child (procedure should be postponed) Reading part of instrument not fixed when value taken Reading Recording errors
Length	Incorrect age for instrument Footwear or headwear not removed Head not in correct plane Head not firmly against fixed end of board Child not straight along board Body arched Knees bent Feet not vertical to movable board Board not firmly against heels
Height	Incorrect age for instrument Footwear or headwear not removed Feet not straight nor flat on vertical platform or wall Knees bent Body arched or buttocks forward (body not straight) Shoulders not straight on board Head not in correct plane Headboard not firmly on crown of child's head
Weight	Room cold, no privacy Scale not calibrated to zero Child wearing unreasonable amount of clothing Child moving or anxious due to prior misregard

Head circumference	Occipital protuberance/supraorbital landmarks poorly
	    defined
	Hair crushed inadequately, ears under tape
	Tape tension and position poorly maintained by time of
	    reading
	Headwear not removed

Triceps fatfold	Wrong arm (should be left arm)
	Mid-arm point or posterior plane incorrectly measured or
	    marked
	Arm not loose by side during measurement
	Examiner not comfortable nor level with child
	Finger–thumb pinch or caliper placement too deep (muscle)
	    or too superficial (skin)
	Caliper jaws not at marked site
	Reading done too early, pinch not maintained, caliper
	    handle not fully released

Arm circumference	Tape too thick, stretched, or creased Wrong arm (should be left arm) Mid-arm point incorrectly measured or marked Arm not loosely hanging by side during measurement Examiner not comfortable nor level with child Tape around arm, not at midpoint; too tight (causing skin contour indentation, too loose (inadequately opposed)

3. Trial practice review.
4. Standardization test.
5. Review and further practice under simulated field conditions.
6. Further standardization test if required.
7. Field trials under supervision.
8. Review.

Standardization exercises simply compare the results for the same measurements on the same child (Martorell, 1975). For example, if the result by the "supervisor" is subtracted from that by the "trainee," the difference gives an estimate of performance by the trainee. This assumes the supervisor is competent. A single exercise should be performed for at least eight children initially. As an approximate guide, it is suggested that there should be no difference *in any of the eight measurements* greater than the following: 100 g (3 oz) for weight; 1.2 cm ($\frac{1}{2}$ inch) for length; 2.5 cm (1 inch) for height; 1.2 cm ($\frac{1}{2}$ inch) for head or arm circumference; 1.5 mm for triceps fatfold. These figures apply to subjects up to 5 years of age. The guides may be altered according to the age of the child and the type of study required.

During the actual survey, some form of field check should be made by the supervisor.

ACKNOWLEDGMENT. The author acknowledges the assistance of D. B. Jelliffe and Jane Kurtzman in the preparation of this chapter.

7. References

Aall, C., 1970, Relief, nutrition and health problems in the Nigerian/Biafran War, *J. Trop. Pediatr.* **16:**70.

American Academy of Pediatrics, Committee on Nutrition, 1968, Measurement of skinfold thickness in childhood, *Pediatrics* **42:**538.

Arnhold, R., 1969, The QUAC stick: A field measure used by the Quaker service team, Nigeria, *J. Trop. Pediatr.* **15:**243.

Ashworth, A., 1969, Growth rates in children recovering from protein–calorie malnutrition, *Br. J. Nutr.* **23:**835.

Beaton, G. H., and Bengoa, J. M., 1973, Practical population indicators of health and nutrition, in: *Nutrition and Preventive Medicine,* WHO, Geneva.

Behar, M., 1975, The appraisal of nutritional status of population groups, in: *Nutrition in Preventive Medicine* (G. H. Beaton and J. M. Bengoa, eds.), WHO, Geneva.

Berg, A., 1973, *The Nutrition Factor: Its Role in National Development,* The Brookings Institution, Washington, D.C.

Billewicz, W. Z., 1974, Some remarks on cross-sectional and longitudinal studies in relation to malnutrition and its consequences, in: *Methodology in Studies of Early Malnutrition and Mental Development,* p. 49, WHO Workshop, Saltsjobaden, Sweden, 1973.

Blankhart, D. M., 1971, Outline for a survey of the feeding and nutritional status of children under three years of age and their mothers, *J. Trop. Pediatr. Environ. Child Health* **4:**176.

Booth, R. A. D., Goddard, B. A., and Paton, A., 1966, Measurement of fat thickness in man: A comparison of ultrasound, Harpenden calipers and electrical conductivity, *J. Nutr.* **20:**719.

Brozek, J., 1974, From a QUAC stick to a compositional assessment of man's nutritional status, in: *Nutrition and Malnutrition* (A. F. Roche and F. Falkner, eds.), Plenum Press, New York.

Center for Disease Control, 1975, *Nutrition Surveillance: Sources of Error in Weighing and Measuring Children,* United States Department of Health, Education and Welfare, Washington, D.C.

Center for Disease Control and Health Services Administration, 1976, *Evaluation of Body Size and Physical Growth of Children,* Public Health Service, U.S. Department of Health, Education and Welfare, Washington, D.C.

Cochran, W. G., 1963, *Sampling Techniques,* pp. 157, 247–248, Wiley, New York.

Cochrane, A. L., and Holland, W. W., 1971, Validation of screening procedures, *Br. Med. Bull.* 27:341.

Davies, D. P., and Holding, R. E., 1972, Neonatometer: A new infant length measurer, *Arch. Dis. Child.* 47:256.

Falkner, F., 1961, Office measurement of physical growth, *Pediatr. Clin. North Am.* 8:13.

Fomon, S. J., 1974, *Infant Nutrition,* 2nd ed., pp. 35–37, Saunders, Philadelphia.

Garn, S., and Clark, D. C., 1976, Trends in fatness and the origins of obesity, Ad Hoc Committee to Review the Ten-State Survey, *Pediatrics* 57:455.

Garn, S. M., and Gorman, E. L., 1956, Comparison of pinch-caliper and teleroentgenogrammetric measurements of subcutaneous fat, *Human Biol.* 28:407.

Gordon, J. E., Wyon, J. B., and Asch, W., 1967, The second year death rate in less developed countries, *Am. J. Med. Sci.* 254:121.

Graham, G. G., and Adriansen, B. T., 1972, Late "catch-up" growth after severe infantile malnutrition, *Johns Hopkins Med. J.* 131:204.

Gurney, M., and Jelliffe, D. B., 1973, Nomogram for arm muscle and arm fat area, *Am. J. Clin. Nutr.* 26:912.

Gurney, M., Jelliffe, D. B., and Neill, J., 1972, Anthropometry in the differential diagnosis of protein–calorie malnutrition, *J. Trop. Pediatr. Environ. Child Health* 18:1.

Habicht, P., Martorell, R., and Yarbrough, C., 1974, Height and weight standards for preschool children, *Lancet* 1:611.

ICNND, 1963, *Manual for Nutrition Surveys,* U.S. Gov. Printing Office, Washington, D.C.

IUNS (International Union of Nutritional Sciences), 1972, The creation of growth standards: A committee report, *Am. J. Clin. Nutr.* 25:218.

IUNS, 1977, Guidelines on the at-risk concept and the health and nutrition of young children, *Am. J. Clin. Nutr.* 30:242.

Jelliffe, D. B., 1966, Assessment of Nutritional Status of the Community, WHO Monograph No. 53, Geneva.

Jelliffe, D. B., 1968a, The pre-school child as a bio-cultural transitional, *J. Trop. Pediatr.* 14:217.

Jelliffe, D. B., 1968b, Weight scales for developing regions, Letter to Ed., *Lancet* 2:359.

Jelliffe, D. B., 1973, Nutrition in early childhood, *World Rev. Nutr. Diet.* 16:1.

Jelliffe, D. B., and Jelliffe, E. F. P., 1969, The arm circumference as a public health index of PCM in childhood: Current conclusions, *J. Trop. Pediatr.* 15:253.

Jelliffe, D. B., and Jelliffe, E. F. P., 1971, Age independent anthropometry, *Am. J. Clin. Nutr.* 24:1377.

Jelliffe, D. B., and Jelliffe, E. F. P., 1973, The at-risk concept as related to young child nutrition programs, *Clin. Pediatr.* 12:65.

King, M., King, F., Burgess, L., and Burgess, A., 1972, *Nutrition for Developing Countries,* Oxford University Press, London.

Kish, L., 1965, *Survey Sampling,* pp. 3–30, Wiley, New York.

Lange, K. O., and Brozek, J., 1961, A new model of skinfold calipers, *Am. J. Phys. Anthropol.* 19:98.

Laugeusen, M., 1975, Child's bangle for the diagnosis of undernutrition, *Indian J. Pediatr.* 12:126.

Lowenstein, M. S., and Phillips, J. F., 1973, Evaluation of arm circumference measurement for determining nutritional status of children and its use in an epidemic of malnutrition, *Am. J. Clin. Nutr.* 26:226.

Martorell, R., 1975, Identification and evaluation of measurement variability in the anthropometry of preschool children, *Am. J. Phys. Anthropometry* 43:347.

McLaren, D. S., and Burman, D., 1976, *Textbook of Paediatric Nutrition*, pp. 106–112, Churchill-Livingstone, London.

Miller, D., Nichaman, M., and Lane, M., 1976, Simplified Field Assessment of Nutritional Status in Early Childhood (submitted for publication), Center for Disease Control, Atlanta, Georgia.

Morley, D., 1970, National nutrition planning, *Br. Med. J.* **4:**85.

Morley, D., 1973, *Paediatric Priorities in the Developing World*, Butterworths, London.

Owen, G. M., 1973, The assessment and recording of measurements of growth of children: Report of a small conference, *Pediatrics* **51:**46.

Parizkova, J., 1961, Total body fat and skinfold thickness in children, *Metabolism* **10:**794.

Rosa, F. W., and Turshen, M., 1970, Fetal nutrition, *Bull. WHO* **43:**785.

Seltzer, C. C., and Mayer, J., 1967, Greater reliability of the triceps skinfold as an index of obesity, *Am. J. Clin. Nutr.* **20:**950.

Seoane, N., and Latham, M. C., 1971, Nutritional anthropometry in the identification of malnutrition in childhood, *J. Trop. Pediatr. Environ. Child Heath* **17:**98.

Shakir, A., 1975, The surveillance of protein–calorie malnutrition by simple and economic means, *J. Trop. Pediatr. Environ. Child Health* **21:**69.

Tanner, J. M., 1959, The measurement of body fat in man, *Proc. Nutr. Soc.* **18:**148.

Tanner, J. M., 1962, *Growth at Adolescence*, 2nd ed., Blackwell, Oxford.

Tanner, J. M., and Whitehouse, R. H., 1955, The Harpenden skinfold caliper, *Am. J. Phys. Anthropol.* **13:**743.

U.S. Department of Health, Education and Welfare, 1972, Ten-State Nutrition Survey, 1968–1970, DHEW Publication No. (HSM) 72-8132.

U.S. Department of Health, Education and Welfare, Public Health Service, Center for Disease Control in Cooperation with U.S. Agency for International Development, 1975, Sahel Nutrition Surveys—1974 and 1975.

Weiner, J. S., and Lourie, J. A., 1969, *Human Biology: A Guide to Field Methods*, International Biological Program Handbook No. 9, Blackwell, Oxford.

World Health Organization, 1970, Nutritional Status of Populations: A Manual on Anthropometric Appraisal of Trends, Geneva, *WHO/NUTR* **70:**129.

World Health Organization, 1976a, Methodology of Nutritional Surveillance, Report of a Joint FAO/UNICEF/WHO Expert Committee, WHO Technical Report Series No. 593, Geneva.

World Health Organization, 1976b, Anthropometry in Nutritional Surveillance: An Overview, *UN Protein Advisory Group Bull.* **6:**2.

Yarbrough, C., Habicht, J. P., Martorell, R., and Klein, R. E., 1974, Anthropometry as an index of nutritional status, in: *Nutrition and Malnutrition* (A. F. Roche and F. Falkner, eds.), Plenum Press, New York.

Zerfas, A. J., 1975, The insertion tape: A new circumference tape for use in nutritional assessment, *Am. J. Clin. Nutr.* **28:**782.

Anthropometric Field Methods: Criteria for Selection

Jean-Pierre Habicht, Charles Yarbrough, and Reynaldo Martorell

1. Introduction

Since "nutrition" cannot be measured directly, one must rely on measurements of indicators of nutritional status for inferential statements. One such statement addresses the issue of whether or not the nutrition of an individual has changed. This involves examining changes in the measurements of an indicator of nutritional status. In this context the indicator is dealt with as a continuous variable and as such it may also be used to identify differences in nutrition between individuals or groups of individuals. For such purposes one needs answers to the following questions: Does the indicator identify changes or differences in nutrition? Can the sensitivity of the indicator be improved? These concerns are addressed in Sections 2–7.

Indicators of nutritional status are also used to identify individuals who require nutritional intervention. In this case the indicator is used as a diagnostic tool, and for this purpose one must address the question: What proportion of those diagnosed will benefit from a particular nutrition intervention? This also involves answering: How many who could have benefited from such an intervention are correctly diagnosed? These concerns are addressed in Section 8.

The use of the indicator both to measure changes or differences in nutrition and for diagnostic purposes is seriously impaired if the indicator is affected by nonnutritional influences. How one deals with such a lack of specificity is addressed in Section 9. Finally, the fact that the measurement of the indicator must be ethically acceptable and operationally feasible is discussed in Section 10.

Jean-Pierre Habicht • Division of Nutritional Sciences, Cornell University, Ithaca, New York. *Charles Yarbrough* • Computers for Marketing Corporation, Kenwood, California. *Reynaldo Martorell* • Food Research Institute, Stanford University, Stanford, California.

More emphasis is placed in this chapter on the sensitivity of the indicator as it relates to individuals, because this seems to be an area of major concern in nutritional anthropometry. Our discussion is based on usual statistical theory, but, to preserve continuity, calculations and formulas have been relegated to notes collected in an Appendix (Section 11) starting on page 382 (see Note 1).

Definitions are provided for various crucial concepts. Where possible we have tried to retain the usual terminology. In some cases concepts are redefined because of lack of clarity in the literature (e.g., nutritional status, precision, dependability, and accuracy).

These discussions distinguish between three related concepts: nutrition, nutriture, and nutritional status. *Nutrition* is "the process by which the organism uses food, or anything normally ingested through digestion, absorption, transport, storage, metabolism and elimination for purposes of maintenance of life, growth, normal functioning of organs and the production of energy" (McLaren, 1976). *Nutriture* is "the state resulting from the balance between supply of nutrition on the one hand and the expenditure of the organism on the other" (McLaren, 1976). We feel this is tantamount to defining nutriture as the physiological state which results from the cellular availability of nutrients.

Finally, *nutritional status* is the expression of nutriture in a specific variable. Therefore one must always specify the variable or variables when referring to nutritional status (i.e., nutritional status as reflected in height, one variable, or in growth, a number of variables). Thus, better nutriture may be expressed by better growth in height. Accordingly, growth in height becomes an *indicator* of nutritional status. The difference between nutritional status and its indicator is that the indicator also reflects the influences of nonnutritional factors, and thus the indicator is not specific for nutritional status.

2. Components of Sensitivity in Individuals

The sensitivity of an indicator of nutritional status is a function of the extent to which it reflects or predicts changes in nutriture. Small height for age is indicative of a history of chronic malnutrition whereas low enzymatic activity related to growth is suggestive of current malnutrition. Both of these indicators reflect events of past and present nutriture and hence are *reflective* indicators of nutritional status. A second kind of indicator of nutritional status is *predictive* in that it is used to predict future outcomes, such as improved performance, or death due to malnutrition. Some indicators may be both reflective and predictive at the same time. Weight for age, for example, reflects chronic malnutrition and predicts mortality (Gómez *et al.*, 1956). Clearly, the purpose of the indicator must be defined prior to evaluating its sensitivity or specificity. Indicators that are both reflective and predictive may have very different sensitivities depending on how they are used.

In defining sensitivity, we would like it to be larger when a small change

Table I. Sensitivity of Arm and Calf Circumference to Differentiate between Preschool Boys with and without Recent Protein–Calorie Malnutrition[a]

Indicator of nutritional status	Random nonnutritional influence[b] = ν (cm)	Difference between two groups[c] (cm)	Nutritional effect[c] = n (cm)	Sensitivity of measured indicator[d]	t[e]	p[f]
Arm circumference	±1.28	1.99	0.98	0.59	5.7	<0.01
Calf circumference	±1.69	1.49	0.71	0.18	3.2	<0.01

[a] Data described in Cheek et al. (1977).
[b] Standard deviation within two comparison groups, corresponds to $\sqrt{\nu^2}$ of Note 2.
[c] Difference = Δ. Δ^2 corresponds to $2(2n^2 + \nu^2/N)$, where N is the sample size in each group [here = 26.96 = 2/(1/28) + (1/26)], n^2 and ν^2 as in Note 2.
[d] Sensitivity = $(n/\nu)^2$ as in footnote c.
[e] $t = (1 + 2 \times 26.96 \times \text{sensitivity})^{1/2}$ as per Note 12.
[f] Probability that the difference found was only due to chance—obtained from t tables in Snedecor and Cochran (1967b).

in nutriture produces a large change in the indicator relative to the effect of random nonnutritional influences (see Note 2). Thus, Table I indicates that in a certain population arm circumference differentiates between children with and without recent protein–calorie malnutrition with a sensitivity of 0.59 while another indicator, calf circumference, does so with a sensitivity of only 0.18. In this population, therefore, arm circumference is a much more sensitive indicator of recent protein–calorie malnutrition.

In general, we perceive three steps in relating nutriture to measurement and each component of sensitivity is related to one of these steps. The first step is to estimate the vagaries of measurement, which we call *imprecision*. Another influence is the myriad of nonnutritional factors which affect the indicator, and render it *undependable* as an indicator of nutritional status. Stature, for instance, can be affected by heredity or by the time of day it is measured. Finally, a same change in nutritional status may reflect quite different changes in nutriture at different levels of nutriture. To use an extreme case, a 5% increase in weight for height is not the same in an emaciated child with only 60% of the weight he should have for his height as in an overweight child with 150% weight for height. It is thus *inaccurate* to assume that equal changes in nutritional status reflect equal amounts of change in nutriture. Thus there are three subcomponents of sensitivity (see Note 3)—precision, dependability, and accuracy. These are discussed in greater detail in the following four sections.

3. Accuracy of Nutritional Status as a Description of Nutriture

Changes in nutriture are accurately reflected by changes in nutritional status when one can infer from the change in nutritional status what the change in nutriture is. For instance, weight for height is often claimed to be a sensitive indicator of recent protein–calorie malnutrition in preschool children. This is true of severe malnutrition but not of mild to moderate malnutrition because weight for height changes very little as protein–calorie nutriture begins to deteriorate, as shown in Fig. 1. It is not until there is a sudden rapid loss of weight a week or so before symptomatic kwashiorkor develops that weight becomes a sensitive indicator of malnutrition. Extrapolating the sensitivity of weight found in severe malnutrition to moderate malnutrition is therefore inaccurate because it assumes a linear relationship between nutriture and the nutritional status, weight. Thus, improving accuracy means understanding and quantifying the relationship between nutritional status and nutriture at different levels of nutriture. This relationship is probably one of decreasing changes in nutritional status for similar changes in nutriture as nutriture improves, as suggested by Fig. 2, which shows the relationship between caloric supplementation and growth in weight in moderately malnourished children.

Similarly, growth in head circumference seems to be as sensitive an indicator of protein–calorie nutritional status as is growth in height during the first 2 years of life, but growth in head circumference may have no relationship

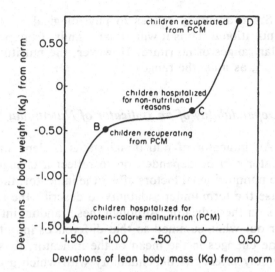

Fig. 1. Relationship of body weight to protein-calorie nutrition. (Data from Cheek *et al.*, 1977.)

to nutritional status thereafter. Assuming a constant relationship between growth in head circumference and protein-calorie nutritional status at different ages is therefore inaccurate. On the other hand, knowledge of the relationships between growth in head circumference, growth in height, age, and nutritional status may permit the identification of the timing of past malnutrition in populations of older preschool children (Malina *et al.*, 1975).

Quantifying the relationships between nutritional status indicators and nurture improves accuracy because it permits one to specify the magnitude of change in the indicator that one expects at various levels of nurture. This is particularly important if one is using indicators of nutritional status to diagnose malnutrition in individuals. In such a case one should ascertain and improve the accuracy of nutritional status at the level of nurture one has defined as warranting nutrition intervention.

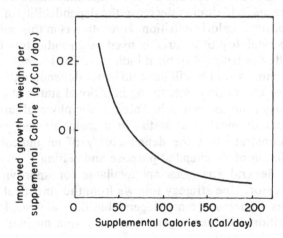

Fig. 2. Relationship of improvements in growth with increasing calorie intake. (Data from Yarbrough *et al.*, 1977.)

Since accuracy relates to physiological relationships between nutriture and nutritional status it will remain similar from population to population over similar ranges of nutriture. However, the nutriture in question must be well defined, as must the range.

4. Dependability of an Indicator of Nutritional Status

An indicator of nutritional status is dependable when a change in the indicator can be depended on to reflect a change in nutritional status. The more nonnutritional factors affect the indicator, the less dependable it is. Thus we use the term undependability to describe the random nonnutritional influences on the indicator. Although these random influences tend to cancel each other out when summed so that differences in nutritional status are reflected by the changes in the mean of the indicator, these nonnutritional influences adversely affect the dependability with which a change in nutritional status can be ascertained, in that one cannot be sure that any single individual change in the indicator is due to a change in nutritional status.

The magnitude of undependability can be ascertained and related to all influences on the measurement of the indicator, including nutritional influences so as to deliver a "dependability" score (see Note 4). This score can range from a totally undependable score of zero to a perfectly dependable score of 100%. This perfect score is attained when the nonnutritional influences on the indicator are negligible relative to the variability in nutritional status and measurement errors.

Therefore the nutritional influence of interest must be carefully specified, for this will also define by exclusion the "nonnutritional influences." For example, the dependability of body weight as an indicator of hydration (water nutrition) is adversely affected by short-term changes in stomach content. Although changes in stomach content are a nutritional process, it is not the nutritional influence of interest in measuring hydration, and would be considered in such a context a "nonnutritional influence." Analogously, short-term changes in hydration decrease the dependability of body weight as an indicator of protein–calorie nutrition. Hydration is in this context "nonnutritional." The dependability of a variable used as an indicator, therefore, depends on what kind of nutritional status it indicates. Furthermore, even for identical purposes the same variable will have different dependabilities as an indicator depending on whether one is measuring nutritional status at one point in time or changes across time as shown in Table II for physical anthropometries used as indicators of nutritional status in a malnourished population. These examples demonstrate that the dependability of an indicator cannot be evaluated in isolation of its intended purpose and setting.

Several strategies are available for improving the dependability of an indicator. One strategy follows from the statistical fact that the mean of many measurements is more dependable than any single measurement. If the nonnutritional influences are random within an individual over a span of time,

Table II. Upper Limits of Dependability of Some Anthropometric Indicators of Protein–Calorie Nutritional Status[a]

Variable used as indicator	Nonnutritional variability of indicator[b] = undependability	Total variability of indicator between children[c]		Dependability of indicator[d] (%)	
		Attained	Incremental	Attained	Incremental
Body weight (kg)	0.29	1.23	0.64	94	59
Supine length (cm)	0.25	4.00	1.54	100	95
Crown–rump length (cm)	0.50	2.30	1.25	95	68
Total arm length (cm)	0.27	1.85	1.23	98	90
Head circumference (cm)	0.30	1.40	0.51	95	31
Chest circumference (cm)	1.23	2.00	1.74	62	28
Mid-arm circumference (cm)	0.16	1.00	0.84	97	92
Thigh circumference (cm)	0.54	2.02	1.40	93	70
Calf circumference (cm)	0.00	1.40	0.68	100	100
Wrist breadth (cm)	0.06	0.23	0.24	93	88
Knee breadth (cm)	0.00	0.39	0.34	100	100
Triceps skinfold (mm)	0.36	1.65	1.54	95	89
Subscapular skinfold (mm)	0.15	1.15	1.03	98	96
Biceps skinfold (mm)	0.47	1.10	0.98	82	54
Midaxillary skinfold (mm)	0.34	1.20	0.93	92	73
Anterior thigh skinfold (mm)	0.30	1.86	1.69	97	94
Lateral thigh skinfold (mm)	0.00	2.07	0.94	100	100
Medial thigh skinfold (mm)	0.58	1.30	1.19	80	52

[a] Data from Martorell et al. (1975c) in 30-month-old malnourished boys and girls.

[b] $d = \sqrt{\bar{d}^2}$; d^2 is defined as in the footnote on page 382; d^2 was estimated by remeasuring the children 1–2 weeks apart, finding the variance around each child's means of the two measurements, and subtracting the variance due to measurement imprecision and the variance over this short period due to the children's mean growth. d^2 is therefore made up of all short-term influences on the anthropometric variables. We think that d^2 is a good approximation of the variance of all nonnutritional influences on differences in growth in this population, because these children are sufficiently malnourished (Yarbrough et al., 1977) so that we believe that all differences in growth between the children are due to differences in nutriture. If this assumption is in substantial error and there are, for instance, strong nonnutritional influences on growth in this population (i.e., genetic), the tabled estimates of d^2 will be too small.

[c] This is a standard deviation, the square root of the variance which is made up of $(n^2 + a^2 + d^2 + p^2) = (n^2 + v^2)$ as in Notes 2 and 3 on page 382. It was estimated by measuring attained size in children aged 30 months or the 6-month incremental growth of children from 24 to 36 months.

[d] Dependability $= 1 - [d'^2/(n^2 + v^2)]$ where $d'^2 = d^2$ for attained growth, and for incremental growth $d'^2 = 2d^2$; d^2 as in footnote b.

repeated measurements over that span of time can be averaged to increase the dependability by using their mean as the indicator. If the nonnutritional influences are distributed at random across many individuals, the dependability of the indicator mean will increase as the number of individuals measured increases (see Note 5). Section 7 discusses this strategy in greater detail.

A second strategy for improving dependability is to exclude or take into account disturbing nonnutritional influences. One can exclude disturbing influences by experimental design. For instance, the effect of age can be excluded by comparing the growth between groups of children all of the same age, or the effect of diurnal variations can be excluded by always measuring at the same time of day, or the effects of disease can be excluded by preventing disease.

Alternatively, where disturbing influences cannot be controlled through the design, the investigator can measure the factors and attempt to exclude their effects from the analysis. The procedure has an additional payoff in that it allows relative comparisons of the factors influencing nutritional status (see Note 6). This in turn may lead to practical recommendations for therapy or public health interventions. For instance, in an area of moderate malnutrition, malnutrition accounted for only half of the stunting in growth in preschool children (Yarbrough *et al.*, 1977), while diarrheal disease accounted for another tenth (Martorell *et al.*, 1975a), and other common diseases had no marked adverse effect on growth (Martorell *et al.*, 1975b).

If the indicator of nutritional status is to be used for diagnostic purposes, one should ascertain and improve the dependability of the indicator within the range, which will be considered diagnostic.

5. Precision of Measurement

Precision is the degree to which repeated measurements yield the same value. The precision of measurement is lowered by random imperfections in the measuring instruments or in the measuring and the recording techniques. These imperfections are inevitable and claims of perfect precision in measurement reflect ignorance of the causes of imprecision.

Thus imprecision describes the random imperfections of measurement, and can be related to all the influences on the indicator of nutritional status including nutriture, inaccuracies, and undependabilities so as to deliver a "precision" (see Note 7) score. This score, like "dependability," can range from zero to 100%. A perfect score occurs when measurement errors are negligible relative to the influences on the indicator of nutritional status. Table III presents the precision of measurement of some physical anthropometries in a malnourished population.

The value of imprecision depends only on the vagaries of measurement over a specific range of a variable, and therefore depends on measurement technique. In contrast to undependability, imprecision is independent of the

*Table III. Precision of Some Anthropometric Indicators of Protein-
Calorie Nutritional Status[a]*

Variable used as indicator	Measurement variability[b] = imprecision	Between groups of children (%)	
		Attained growth[c]	Incremental growth[d]
Body weight	±0.02 kg	100	100
Supine length	±0.34 cm	99	90
Crown–rump length	±0.33 cm	98	86
Total arm length	±0.57 cm	91	57
Head circumference	±0.14 cm	99	85
Chest circumference	±0.79 cm	84	59
Mid-arm circumference	±0.18 cm	97	91
Thigh circumference	±0.40 cm	96	84
Calf circumference	±0.23 cm	97	77
Wrist breadth	±0.09 cm	85	72
Bicondylar femur	±0.11 cm	92	79
Triceps skinfold	±0.66 mm	84	63
Subscapular skinfold	±0.38 mm	89	73
Biceps skinfold	±0.37 mm	89	72
Midaxillary skinfold	±0.30 mm	94	79
Anterior thigh skinfold	±0.88 mm	78	46
Lateral thigh skinfold	±1.34 mm	58	0
Medial thigh skinfold	±0.38 mm	91	80

[a] Data from Martorell *et al.* (1975c) in 30-month-old malnourished boys and girls.
[b] $p = \sqrt{p^2}$; p^2 as defined in Note 7 on page 383. p^2 was estimated by repeating the measurement twice within an hour on each child and finding the variance around the mean of the two measurements.
[c] $1 - (p^2/(n^2 + v^2))$; $(n^2 + v^2)$ from Table II, second column and footnote *c*.
[d] $1 - (2p^2/(n^2 + v^2))$; $(n^2 + v^2)$ from Table II, third column and footnote *c*.

population and setting except as these affect the range of the variable. However, the precision score does include undependability and therefore the score is dependent on the population and setting.

There are two different basic strategies to improve precision. One improves the technology, the second takes advantage of the statistical fact that means of measurement are more precise than each individual measurement.

As indicated before, imprecision is a function of several factors. One such factor which is relatively easy to ascertain is instrument error (see Note 8). Elsewhere (Habicht, 1973) simple hand calculation techniques are presented for estimating the imprecision within a measurer or between measurers as compared to a standard, and for using this standard to determine the appropriate truncation of significant digits for recording the data. Inspection of the relative contributions to imprecision of the various components can help one decide which component is most limiting to precision, and in greatest need of improvement.

The second strategy for improving precision consists of using a mean of multiple measurements. Since this discussion is intimately related to sample size estimations, it is more appropriately discussed in Section 7.

6. Interrelationships between Accuracy, Dependability, Precision, and Reliability

The preceding sections have outlined how one may judge the accuracy with which nutritional status reflects nutriture, the dependability with which the indicator reflects nutritional status, and the precision with which the measurement reflects nutritional status. All of these positive attributes are calculated from estimates of the respective negative attributes, inaccuracy, undependability, and imprecision. Both the positive and negative attributes can be used, but for different purposes.

Because the negative attributes are independent (see Note 3) of each other, their behavior at different values of an indicator and in different populations and settings is easier to understand than is the behavior of the positive attributes. The negative attributes should therefore be used in comparisons across studies. At present, however, certain precautions should be observed when making these comparisons. Imprecision may depend on the magnitude of the value measured. This is apparently not true for height and weight but is the case for skinfolds. Undependability may also depend on the value of the indicator and it certainly depends on the number and the extent of nonnutritional influences, which may change from study to study. As these variations of imprecision and undependability are studied across different circumstances it is likely that those variations will prove to be predictable by simple algebraic functions (see Note 9), which will probably not be the case for the positive attributes of precision and dependability because the positive attributes are not independent of each other.

However, the negative attributes, inaccuracy, undependability, and imprecision, are calculated in units of measurement. When different indicators of nutritional status are measured in different units it is clear that their negative attributes cannot be compared. Even when the units are the same, the negative attributes cannot be compared usefully across indicators. For instance, equal imprecisions of ±0.66 mm have quite different implications for skinfold measurements over the triceps than under the scapula.

For comparisons of attributes among different indicators the positive attributes of accuracy, dependability, and precision should be used, because these are independent of the units of measurement. A combination of dependability and precision, called reliability (see Note 10), is already used widely in the social sciences (Rummel, 1970) to compare variables among themselves. Separating out the components of dependability and precision has implications for technical improvements in the quality of data collection, but reliability as presented in Table IV is useful for the considerations dealt with in Section 7. Where accuracy may be presumed to be similar as in skinfolds, the variable with the higher reliability, such as subscapular skinfolds relative to all other skinfolds, should be more sensitive to changes or differences in nutriture.

Unreliability is therefore made up of undependability and imprecision (see Note 11). A low ratio of undependability to unreliability, as in Table V, identifies which indicators will most benefit from improvements in precision.

Table IV. *Reliability of Some Anthropometric Indicators of Protein–Calorie Nutritional Status*

Variable used as indicator	Attained[a] (%)	Incremental[b] (%)
Body weight	95	59
Supine length	99	85
Crown–rump length	93	54
Total arm length	88	48
Head circumference	94	16
Chest circumference	47	0
Mid-arm circumference	94	84
Thigh circumference	89	54
Calf circumference	97	77
Wrist breadth	78	59
Bicondylar femur	92	79
Triceps skinfold	79	52
Subscapular skinfold	87	69
Biceps skinfold	70	25
Midaxillary skinfold	86	52
Anterior thigh skinfold	75	39
Lateral thigh skinfold	58	0
Medial thigh skinfold	71	32

[a] $1 - (p^2 + d^2)/(n^2 + v^2)$; d and $(n^2 + v^2)$ from Table II, p from Table III.
[b] $1 - (2p^2 + 2d^2)/(n^2 + v^2)$.

Thus improvements in weighing technique will not improve the quality of weight data collected to identify weight changes due to protein–calorie nutrition. On the other hand, subscapular skinfolds could be improved markedly by better measuring techniques, which would increase subscapular skinfold reliability still more relative to bicepital skinfold measurements for instance. Such considerations would suggest that bicepital skinfold measurements are not useful if subscapular skinfolds are available and both skinfolds are equally accurate.

7. Sensitivity and Sample Size

Data on indicators of nutritional status are sometimes collected to identify differences or changes in nutriture. By appropriate statistical testing, the investigator calculates whether or not the differences are statistically significant at specified levels of probability. The larger the differences, the more likely it is, of course, that the findings will be statistically significant. On the other hand, the larger the variability of the difference, which is composed of all the variabilities due to inaccuracy, unreliability, and imprecision which have been described earlier, the less likely it is that significance levels will be reached. A third factor, sample size, is also important. The larger the sample size, the more likely it is that an actual difference will be identified as significant (see Note 12). These three factors, the size of the difference, the variance of the

Table V. The Ratio of Undependability to
Unreliability of Some Anthropometric
Indicators of Protein–Calorie Nutritional
Status[a]

Variable used as indicator	Ratio (%)
Body weight	99
Supine length	34
Crown–rump length	70
Total arm length	28
Head circumference	82
Chest circumference	71
Mid-arm circumference	44
Thigh circumference	64
Calf circumference	0
Wrist breadth	33
Bicondylar femur	0
Triceps skinfold	23
Subscapular skinfold	14
Biceps skinfold	60
Midaxillary skinfold	56
Anterior thigh skinfold	10
Lateral thigh skinfold	0
Medial thigh skinfold	70

[a] $d^2/(d^2 + p^2) = d^2/r^2$; d from Table II, p from Table III,
r is unreliability.

difference, and the sample sizes, determine, therefore, the likelihood of finding statistically significant differences as in a simple t-test of group means.

Estimating the sample size required in an investigation is just as important as improving the accuracy, reliability, and precision of measurements. Unfortunately, so little work has been done on "accuracy" that the issues of sample sizes and strategies for optimizing sample stratification must regretfully be discussed on the basis of expected differences in nutritional status rather than on the basis of differences in nutriture.

Weight will be used as an example of an indicator of nutritional status. In usual clinical practice weight is measured with an imprecision of ±300 g as contrasted to ±20 g (Table III) in a carefully controlled scientific study. If weight is used as an indicator of hydration in a child with diarrhea, single measures ±300 g taken at the beginning and at the end of the day would not be able to detect a life-threatening loss of half a liter of fluids with any certainty (see Note 13). However, if the child were to be measured 36 times at the beginning of the day (each measurement taken "independently," in ignorance of the previous measurements) and 36 times at the end of the day, the imprecision of the means at the beginning and end of the day would be ±50 g, and one would easily detect a 500 g loss (see Note 14). Improving the precision of weighing techniques to ±50 g per single measurement is just as effective as 36 measurements. In this case there is a clear trade-off between the cost of improving the quality of the instrument and technique and the cost of carrying out the multiple measures necessary to attain the same level of precision.

If weight is used as an indicator of protein–calorie nutriture in a single child, even perfectly precise weighings at the beginning and at the end of the observation period would not identify a 500 g weight loss or gain, because the day-to-day undependability is ±290 g. This level of undependability could be reduced slightly by timing the weighings to take account of diurnal rhythms. On the other hand, 36 independent weighings over 36 days at the beginning and end of the observation period would permit easy identification of the 500 g weight change.

The variability which obscures identifying a meaningful difference between measurements is made up of inaccuracy, unreliability, and imprecision. One should therefore consider them all together when trying to estimate desirable sample sizes. Figures such as the ones presented in Tables II (dependability) and III (precision) are therefore essential for sample size estimates. A comparable table on accuracy is lacking, but combining the data in Tables II and III as in Table IV, or calculating reliability directly (see Note 10), will permit estimates of *minimum* sample sizes required to detect a given nutritional effect (see Note 15). The actual sample sizes needed may, however, be considerably larger if the accuracy of the nutritional status is low.

Table V, the ratio of undependability to unreliability (the d^2/r^2 ratio), permits one to evaluate whether the major variability decreasing sensitivity is undependability, in which case the ratio is high, or imprecision, in which case the ratio is low. Taking again the example of identifying a 500 g loss in weight due to protein–calorie malnutrition, one would need 36 daily weighings at the beginning and again at the end of the period if there were no imprecision because the undependability is ±290 g from day to day. Adding an imprecision of ±300 g to the ±290 g undependability results in a d^2/r^2 ratio of 48% (see Note 16) and would result in having to weigh daily for 75 days (36/0.48) before and after the observation period to achieve an ease of identification equal to that of 36 daily weighings with a perfect precision for each weighing. An imprecision of ±20 g, on the other hand, corresponds to a 99% ratio and hardly affects sample size requirements.

Increasing the number of measurements in each session to obtain greater precision is a sampling strategy that may be less expensive than increasing the number of subjects or measuring sessions (see Note 17). The least expensive strategy can be ascertained exactly (Snedecor and Cochran, 1967a, Sec. 17.12) if the cost of each measurement can be separated from the cost of each measuring session.

Closely allied to the effect of precision and dependability on sample size is the effect of these two components of sensitivity on regression analysis of data. As indicated earlier precision and dependability can be combined to deliver reliability (see Note 10). Reliability can be used to estimate the maximum association that could be expected from a measured indicator of nutritional status and some other variable (see Note 18). For example, the relationship in adults between obesity and hypertension (Blair *et al.,* 1977) is greater when obesity is measured by subscapular skinfolds than by tricepital skinfolds (see Note 19). In this case the "accuracy" is presumed to be the same in that both skinfolds reflect local fat deposition equally well. Nevertheless, before one can infer that subscapular fat reflects the hypertensive effects of obesity better than does tricepital fat, one must take the reliability of the indicator into account (see Note 19).

Just as the correlation coefficient will be reduced if reliability is low, so will the regression coefficient describing the relationship between an indicator and some outcome be underestimated. Whereas unreliability in either the indicator of nutritional status or in the other variable affects the correlation coefficient, only unreliability of the determinant variable affects the regression coefficient. The determinant is nutritional status if the other variable is a consequence of nutrition, but the determinant is the variable which affects nutriture if the outcome is nutritional status. If the reliability of the appropriate variables is known, these underestimates may be corrected (see Note 20).

Taking account of precision and reliability is particularly important in any systematic attempt to infer or test hypotheses about the relative strengths of association between nutritional status and causes or outcomes, such as is done by stepwise multiple regression procedures, or by path analysis. This can be done quantitatively as we have described. These statistical manipulations do not, however, correct for inaccuracies of nutritional status. Therefore, inferences made from such analyses must be supported, if only qualitatively, by sound biological models.

8. Sensitivity in Populations

The sensitivity of an indicator when used to characterize a population has different components depending on whether one is using the indicator as a continuous variable to calculate means or to enumerate the proportion of malnourished persons in the population. The sensitivity of population means is a composite of the sensitivity of the indicator in individuals (see Note 21). When the indicator is less sensitive to a given change in nutriture among better than among poorer nourished individuals, it is clear that the sensitivity of an indicator in a population will be different depending on whether or not malnutrition in that population has a high or low prevalence. This probably explains, for instance, the lesser sensitivity of birth weight to reflect maternal nutrition in New York City (Susser and Stein, 1977), where the sensitivity of response was almost imperceptible, than in rural Guatemala (Lechtig *et al.*, 1975), where the sensitivity of response was much higher. Therefore methods which select for those most likely to benefit from an intervention will increase the sensitivity of the indicator. Such, for instance, would be a study of the effects of malnutrition on birth weight restricted to mothers who have previously born small-for-date babies (Habicht *et al.*, 1974). In practice, this population sensitivity also depends on the distribution of changes in nutriture within a population, so that a same population's response to nutrition intervention may be quite different depending on how that intervention is distributed within the population (see Note 21).

When an indicator is used to enumerate changes in the proportions of malnourished persons in a population, the sensitivity depends not only on the sensitivity of the indicator at the cutoff point (WHO, 1976) used for diagnosis, but also on the proportion of the population just above or below the cutoff point. For instance, compare two populations with the same proportion diagnosed at the same cutoff point as malnourished. In one population many of the malnourished are more severely malnourished than in the other population. An equivalent change in nutriture in individuals of the first population will shift less of their indicators across the cutoff point than in the second population, so that the indicator will appear less sensitive in the first more severely malnourished population than in the second population.

Sometimes indicators are used to select malnourished persons for treatment. In this context the sensitivity of the indicator used for diagnosis is the proportion of truly ill who are so identified (see Note 22) and is directly related to the sensitivity of the indicator in the individual at the cutoff point. The converse is the specificity of the diagnostic indicator, which is the proportion of all those correctly identified as not ill who were not ill (see Note 22). The specificity is directly related to the reliability of the indicator at the cutoff point.

In scientific investigations screening is often used to select those who will respond most to test the intervention. Similar considerations underlie the screening of populations to "efficiently" intervene among those most in need. Unfortunately, diagnostic sensitivity and specificity tell us little about how responsive the indicator will be in populations chosen for intervention on the basis of a diagnostic indicator. For this one must know the proportion of those diagnosed as malnourished who are really malnourished, which is termed the positive "predictive value" of the diagnostic indicator (see Note 22) (Vecchio, 1966). Persons misclassified as malnourished will, of course, not respond to improved nutrition. Low diagnostic specificity and low prevalence of malnutrition both tend to result in a greater proportion of such misclassifications, and such low predictive value cannot be improved much by increasing the sensitivity. In fact, even low diagnostic sensitivity is compatible with a high predictive value if diagnostic specificity and prevalence are high. Detailed descriptions of how predictive value changes with changes in sensitivity, specificity, and prevalence may be found in the book by Galen and Gambino (1975).

Just as poor reliability in an indicator will lead to poor diagnostic specificity and thus to poor predictive value and poor population sensitivity to an intervention, so will poor reliability result paradoxically in a spurious response to the intervention. This happens when a person is classified as malnourished on the basis of a temporary unreliability. The nutrition intervention is then applied and the person is remeasured to ascertain his response. If this time the unreliability is not present or is in the other direction, the individual's indicator will have "improved" resulting in a spurious response. At the time of selection, more of those will tend to be classified as malnourished whose unreliabilities are in one direction, the pathological direction, than in the other. This results upon remeasurement in a "spurious" mean improvement (Davis, 1976) which must be calculated and taken into account.

9. Specificity of Response

A poor specificity of response not only affects the sensitivity of the indicator in populations by adversely affecting the indicator's diagnostic predictive value (see Note 22), but the unreliabilities which poor specificity may reflect also reduce the indicator's sensitivity in individuals (see Note 23). These random nonspecificities can result in missing a response to a nutritional change, a false negative result. However, we have shown how to avoid such false negative results by appropriate measuring strategies and population screening.

Equally important, if not more important in scientific inquiry, is avoiding false positive results. These may arise because of the coincidental effects of nonnutritional factors which mimic nutritional effects. For this to happen the factor must have affected the indicator in a nonrandom or biased fashion so that there was an identifiable cumulative change in the indicator. If the indicator were absolutely specific for the nutritional influence under consideration, such false positive results, of course, would never occur. This, however, is unattainable in practice.

The classic strategy for avoiding false positive results is to compare individuals, some of whom are documented as being affected by the nutritional influence under consideration and some of whom are documented as not being so affected, in such a way that both groups of individuals are similarly affected by the nonnutritional (nonspecific) influences. This *randomization* of nonnutritional influences should also include the randomization of measurers across both groups, because no amount of measurer training or standardization can remedy biases due to inadequate randomization of measurers. Randomization does not improve the specificity of the indicator, but it converts most problems related to specificity into problems of reliability (see Note 23), which can be dealt with as discussed in the preceding sections. This strategy is supported by extensive statistical research and practice (see Note 1) on the benefits of randomizing nonspecific influences among the groups to be compared. Similar results are achieved in clinical practice by repeating therapeutic trials on patients with recurring nutritional problems often enough to be sure that the responses are due to the therapy and not to other factors.

A second strategy useful in determining whether an indicator is changing because of the nutritional factor under consideration is to measure different indicators of nutritional status, which will only change congruently if nutriture were truly changed. This will often require measurements of other than anthropometric variables. For instance, marked changes in vitamin D nutrition not only result in changes in bone morphology and roentgenology but also in biochemical findings, all of which must change *congruently* to substantiate a change in vitamin D nutriture.

A third strategy for improving specificity consists of measuring the non-specific influences and their effects so as to *take them into account* in the analysis of the results. This strategy has the advantage that it does not decrease sensitivity by randomizing the nonspecific factors. It also has the advantage of permitting a comparison between nutritional and nonnutritional effects on the indicator (see Note 7). But experience shows that this is a dangerous strategy that frequently results in biased outcomes. It is impossible to predict all possible causes of nonspecific changes in an indicator before an investigation begins. Rather this strategy should be used in conjunction with a design that randomizes nonspecific elements as discussed earlier. Used in this manner, it is a useful device for both increasing sensitivity and preventing biases (Snedecor and Cochran, 1967a; MacMahon and Pugh, 1970). For instance, the effects of parity on birth weight are similar to those of some nutritional factors. Therefore, if primiparas are kept separate from multiparas when comparing test and control groups, potential biases due to differences in the distribution of primiparas in the various nutritional groups will be obviated and the sensitivity of the comparison will be increased.

The three strategies proposed for improving specificity, namely, randomization, congruity of response among several indicators, and statistical control of nonnutritional influences, are discussed at greater length elsewhere in the context of intervention trials in populations (Habicht and Butz, 1979).

10. Feasibility

An indicator can be highly sensitive and specific yet inappropriate or impossible in a particular setting for a variety of reasons. Equipment costs are an obvious constraint. Less obvious are the costs of maintenance and continuous calibration. Even when these costs can be met, the need for spare parts and specialized maintenance or calibration may entail long periods when no measurements can be made, and this may be unacceptable if data collection is rigidly scheduled. Inconvenience to the subject is sometimes apparent to the researcher or clinician because of complaints and outright refusals to participate. These instrument and subject costs can be weighed against the costs of strategies outlined in this chapter to improve the data arising from less sensitive or specific methods. However, some measurements should be precluded for ethical reasons, even though cost efficiency considerations would favor them.

A cost which is often neglected but which may be substantial, is that of training and standardizing measurers. The cost of improvements in precision can, of course, be traded off against the cost of increasing the number of measurements, as described in Sections 5, 6, and 7. When measurer turnover is high or standardization between measurers is impracticable, increasing the number of measurements may be the only feasible alternative.

Often inadequate attention to the considerations outlined in this chapter has resulted in data that are useless, or very expensive to analyze. On other occasions the precision of the measuring equipment and of the measurers has far exceeded that which was useful relative to the reliability of the indicator or necessary relative to the uses of the data. In fact, only when the specific data analyses are outlined before data collection begins can the required sensitivity and specificity of the indicator measurements be described and the consequent measurement equipment and sampling strategies be specified. These are essential in assuring useful data to answer scientific or clinical questions of substance.

11. Appendix

Note 1. We present in these notes the numerical definitions and statistical considerations underlying the text. The mathematical level of the notes requires only a basic understanding of statistics at the level of *Facts from Figures* (Maroney, 1956). This excellent book requires no previous knowledge of more than elementary algebra. A more comprehensive textbook on practical statistics is that of Snedecor and Cochran (1967b).

Note 2. Sensitivity $= n^2/v^2$, variance due to nutrition divided by the variance due to nonnutritional influences. n^2 is the variance due to changes in nutriture and v^2 is the variance due to nonnutritional influences (see Note 3). In principle the sensitivity can vary from zero to infinity. Depending on how it is estimated, for example by the method of moments as used in Table I, it may in practice be a negative value but not significantly different from zero.

Note 3. $v^2 = a^2 + d^2 + p^2$, where the variance due to inaccuracy is a^2, the variance due to undependability is d^2, and the variance due to imprecision is p^2. In practice p^2 can be estimated with ease, and d^2 with more difficulty; a^2 has not yet been estimated. Inaccuracy is usually called "bias" in statistics texts and describes discrepancies between a mathematical function and the true state it is supposed to describe. Systematic bias is described in Section 8. The components of sensitivity are precision P (as in Note 7), dependability D (as in Note 4) and accuracy A (by analogy to P and D), such that sensitivity $= (A + D + P - 2)/(3 - A - D - P)$. This description of the components of sensitivity assumes no correlations between the different components of variance v^2. We believe that assumption to be generally true.

Note 4. Using the notation of Notes 2 and 3, dependability $= (n^2 + a^2 + p^2)/(n^2 + a^2 + d^2 + p^2) = D = 1 - [d^2/(n^2 + v^2)]$; d^2 and $n^2 + v^2$ can be calculated as in footnotes b and c, respectively, of Table II. Because $n^2 + v^2$ depends on the situation, the persons and times sampled must be representative of the larger study. When D is referred to as a percentage, this formula is multiplied by 100.

Note 5. The dependability of the mean ($\bar{D}$) of N estimates of the indicator is $\bar{D} = 1 - (1 - D)/N$, where D refers to the dependability of the indicator as derived from single estimates. If $\bar{D}$ is sought across individuals, N refers to the number of individuals measured. If $\bar{D}$ is sought within an individual, N refers to the number of times the individual is measured. However, within an individual the periods of time over which the N estimates are made must be long enough to include variability in all short-term nonnutritional changes.

Note 6. This is often done by multivariate analysis such as analysis of variance and regression analysis. Such techniques also can be used to partition d^2 into its different nonnutritional effects, which can then be compared to n^2, the nutritional effect.

Note 7. Using the notation of Notes 2 and 3, precision $= (d^2 + a^2 + n^2)/ (p^2 + d^2 + a^2 + n^2) = P = 1 - [p^2/(n^2 + v^2)]; p^2$ and $n^2 + v^2$ may be calculated as in footnote b, Table III, and footnote c, Table II. When P is referred to as a percentage, this formula is multiplied by 100. See also Note 4.

Note 8. This is done by using calibrated inanimate inert objects for testing the instruments, such as metal blocks for scales, steel rods for anthropometers, and small steel blocks for calipers. Instrument testing should be periodic and cover the whole range for which the instrument is used. Many instruments lose their calibration in different ways at different points in the range of measurement.

Note 9. For instance, if the relationship between imprecision p and the size of the measurement x is found to be $p = f(x)$, then p can be better expressed as a function of x. For instance, if $p = ax$, then the coefficient of variation p/x is a more meaningful description of imprecision than is $\pm p$.

Note 10. Reliability, R, equals precision $+$ dependability $- 1$, where dependability is calculated as in Table II and precision is calculated as in Table III. Reliability within individuals is usually estimated directly by calculating the square of the correlation coefficient between two measurements, the second taken some time after the first. This is the "test–retest" reliability.

Note 11. $r^2 = d^2 + p^2$, where r is unreliability, d undependability, and p imprecision.

Note 12. Significance is related to $\sqrt{1 + 2N\, n^2/v^2} = \sqrt{1 + 2N\, \text{Sensitivity}} = t$. N refers to the sample size in each group, t is obtained from tables for specified significance levels (Snedecor and Cochran, 1967b), other notation as in Note 2. n^2 increases as the difference among means due to differences in nutriture increases (see Table I for example).

Note 13. Using the notation of Note 12, $N = 1$, $n^2 = (250 \text{ g})^2$, $v^2 = (300 \text{ g})^2$, sensitivity $= 0.69$; $t = [1 + 2 (250)^2/(300)^2]^{1/2} = 1.55 = [1 + (2) (0.69)]^{1/2}$. This t value can occur more often than not even when there is, in fact, no loss of weight between the two measurements (see "How to Be a Good Judge" in Mahoney, 1956)—the statistical significance is low. In other words, one would treat for dehydration more children who were, in fact, not dehydrated than were dehydrated.

The preceding text of this note refers to how often in this circumstance one would treat a child improperly for dehydration given a 500 g difference, if one measures to an imprecision of ±300 g. In the treatment of dehydration it is even more important to know how often one would miss picking up a true weight loss of 500 g. The statistical power of the test refers to how often one would identify a weight loss of 500 g if it occurred. For equal sensitivity the greater the significance desired, the lower will be the power and vice versa. In the diagnosis of dehydration the power should be higher than the significance because the consequences of not treating severe dehydration are more serious than treating a child who does not need it. On the other hand, in scientific inquiry significance is usually set higher than power (Cohen, 1969). However one wishes to allocate power and significance, the higher the t the higher will be the power for a same significance and vice versa. This section is concerned with how to increase the t value. Sensitivity calculated on a difference between two comparison groups, as in Table I, permits direct access to power tables.

Note 14. $t = [1 + 36(250)^2 2/(300)^2]^{1/2} = 7.14 = (1 + 36 \times 2 \times 0.69)^{1/2}$, where $N = 36$ as per notation of Note 12, which has a power of over 99.5% at a significance of 99% ($p \leq 0.01$). In fact, the power in this case is in excess of usual needs. A sample size of 16 in each measuring period would deliver a power of 95% at a significance level of 95%.

Note 15. $N = (t^2/2\text{sensitivity}) - 1$ (from Note 12). If $a^2 = 0$, then sensitivity $= (1 - \text{reliability})/\text{reliability}$ and $\hat{N} = t^2/2(1 - \text{reliability})/\text{reliability} - 1$. However, if $a^2 > 0$, then $\hat{N} < N$.

Note 16. $(290)^2/(290^2 + 300^2) = 48\%$.

Note 17. The precision (P_m) will then rise as the number of measurements (m) per session increases: $P_m = 1 - (1 - P)/m$, where P is the precision as in Table III.

Note 18. The greatest correlation coefficient that can be expected between a measured indicator of nutritional status and another variable is $\leq$ (reliability of the indicator times reliability of other variable)$^{1/2}$, where reliability is from Note 10.

Note 19. Correlation coefficients between systolic blood pressure and skinfolds:

	Triceps	Subscapular
Maximum expected correlation coefficients, $\acute{r}$ (Johnston et al., 1973; Blair et al., 1977)	.988	.988
Correlation coefficients found, r	.115	.289
Corrected correlation coefficients $r/\acute{r}$ (Blair et al., 1977)	.116	.292

Note 20. The better estimate of the regression coefficient b of nutritional status with outcome is ($b'/\text{reliability of nutritional status}$), where b' is the regression coefficient calculated from the data using the measured indicator of nutritional status, and reliability is as in Note 10.

Note 21. The sensitivity of population means to changes or differences in nutriture is the mean of the individual sensitivities. Each individual's sensitivity must be estimated at that individual's nutriture. If the distribution of changes in nutriture within the population is included in the definition of population sensitivity, then it is the mean of each individual's product of his sensitivity times his share in the population's change in nutriture.

Note 22.

	"Truth"	
	Malnourished	Not malnourished
Diagnosis of:		
Malnourished	M	m
Not malnourished	n	N

where M represents those correctly diagnosed as malnourished, N those correctly diagnosed as not malnourished, n those falsely diagnosed as not malnourished, and m those falsely diagnosed as malnourished.

Sensitivity of diagnosis = $M/(M + n)$ = percentage of malnourished who are so diagnosed = Se.

Specificity of diagnosis = $N/(N + m)$ = percentage of not malnourished who are diagnosed as not malnourished = Sp.

Predictive value of diagnosis = $M/(M + m)$ = percentage of those who are diagnosed as malnourished who are indeed malnourished = V.

True prevalence of malnutrition = $(M + n)/(M + n + m + N)$ = percentage of total population who are "truly" malnourished = P.

Apparent prevalence of malnutrition = $(M + m)/(M + n + m + N)$ = percentage of total population who are diagnosed as malnourished = p.

True prevalence P is related to apparent prevalence p as $P = (p + Sp - 1)/(Sp + Se - 1)$.

Predictive value V is related to sensitivity Se, specificity Sp, true prevalence P, and apparent prevalence p as

$$V = SeP/[SeP + (1 - Sp)(1 - P)]$$
$$= SeP/p = Se(p + Sp - 1)/p(Sp + Se - 1).$$

Note 23. The more frequently and more severely nonnutritional influences randomly affect an indicator of nutritional status, the larger will be d^2 (as in Note 3) and the lower is the sensitivity. The same holds true of random imprecision for measurement, p^2, and of inaccuracies for nutritional status, a^2. The result of all these random nonspecific influences is to increase v^2 relative to n^2 (see Note 2), so that the quantitative expression of "random specificity" and sensitivity are identical.

ACKNOWLEDGMENT. This chapter is a report of research of the Cornell University Agricultural Experiment Station Division of Nutritional Sciences.

12. References

Blair, D., Habicht, J.-P., Sims, E., and Sylwester, D., 1977, Example of an analysis of the relationship between the distribution of body fat and the risk of certain morbid events, unpublished.

Cheek, D. B., Habicht, J.-P., Berall, J., and Holt, A. B., 1977, Protein–calorie malnutrition and the significance of cell mass relative to body weight, *Am. J. Clin. Nutr.* **30:**851–860.

Cohen, J., 1969, *Statistical Power Analysis for the Behavioral Sciences,* Academic Press, New York.

Davis, C. E., 1976, The effect of regression to the mean in epidemiologic and clinical studies, *Am. J. Epidemiol.* **104:**493–498.

Galen, R. S., and Gambino, S. R., 1975, *Beyond Normality: The Predictive Value and Efficiency of Medical Diagnoses,* Wiley, New York.

Gómez, F., Ramos-Galván, R., Frenk, S., Cravioto-Múnoz, J., Chavez, R., and Vasquez, J., 1956, Mortality in second and third degree malnutrition, *J. Trop. Pediatr.* **2:**77–83.

Habicht, J.-P., 1973, Standardization procedures for quantitative epidemiologic field methods, in: *Manual of Internationally Comparable Growth Studies in Latin America and the Caribbean,* Chap. 7, Pan American Health Organization—or in Spanish in revised form: 1974, *Boletin de la Oficina Sanitaria Panamericana* **76:**375–384.

Habicht, J.-P., and Butz, W., 1979, Measurement of health and nutrition effects of large-scale nutrition intervention projects, in: *Measurement of Impact of Nutrition and Related Health Programs in Latin America, PAHO,* in press.

Habicht, J.-P., Lechtig, A., Yarbrough, C., and Klein, R. E., 1974, Maternal nutrition, birth weight and infant mortality, in: *Size at Birth, CIBA Foundation Symposium 27,* Elsevier, Amsterdam.

Johnston, F. E., Hamill, P. V. V., and Lemeshow, S., 1973, Skinfold thickness of children 6–11 years, United States Department of Health and Welfare Publication No. (HSN)73-1602, Series 11, No. 120.

Lechtig, A., Habicht, J.-P., Delgado, H., Klein, R. E., Yarbrough, C., and Martorell, R., 1975, Effect of food supplementation during pregnancy on birth weight, *Pediatrics* **56:**508–520.

MacMahon, B., and Pugh, T. F., 1970, *Epidemiology: Principles and Practice,* Little, Brown, Boston.

Malina, R. M., Habicht, J.-P., Martorell, R., Lechtig, A., Yarbrough, C., and Klein, R. E., 1975, Head and chest circumference in rural Guatemalan Ladino children birth to seven years of age, *Am. J. Clin. Nutr.* **28:**1061–1070.

Maroney, M. J., 1956, *Facts from Figures,* Penguin, Baltimore.

Martorell, R., Yarbrough, C., Lechtig, A., Habicht, J.-P., and Klein, R. E., 1975a, Diarrheal diseases and growth retardation in preschool Guatemalan children, *Am. J. Phys. Anthropol.* **43:**341–346.

Martorell, R., Habicht, J.-P., Yarbrough, C., Lechtig, A., Klein, R., and Western, K. A., 1975b, Acute morbidity and physical growth in preschool Guatemalan children, *Am. J. Dis. Child.* **129:**1296–1301.

Martorell, R., Habicht, J.-P., Yarbrough, C., Guzman, G., and Klein, R. E., 1975c, The identification and evaluation of measurement variability in the anthropometry of preschool children, *Am. J. Phys. Anthropometry* **43:**347–352.

McLaren, D. S., 1976, *Nutrition in the Community,* Wiley, New York.

Rummel, R. J., 1970, *Applied Factor Analysis,* Northwestern University Press, Evanston.

Snedecor, G. W., and Cochran, W. G., 1967a, Various sections on multivariate analyses and stratified sampling, in: *Statistical Methods,* Iowa State University Press, Ames.

Snedecor, G. W., and Cochran, W. G., 1967b, Table A.4. The distribution of *t* (two-tailed) tests, in: *Statistical Methods,* Iowa State University Press, Ames.

Susser, M., and Stein, Z., 1977, Prenatal nutrition and subsequent development, in: *The Epidemiology of Prematurity* (D. M. Read and F. J. Stanley, eds.), Urban & Schwarzenberg, Munich.

Vecchio, T. J., 1966, Predictive value of a single diagnostic test in unselected populations, *N. Engl. J. Med.* **274:**1171–1173.

WHO, 1976, World Health Organization Report of a Joint FAO/UNICEF/WHO Expert Committee, Methodology of Nutritional Surveillance, WHO Technical Report Series 593.

Yarbrough, C. Y., Habicht, J.-P., Klein, R. E., Martorell, R., Lechtig, A., and Guzman, G., 1977, Responses of indicators of nutritional status to nutrition intervention in populations and individuals, in: *Evaluation of Child Health Services—the Interface between Research and Medical Practice*, U.S. Gov't. Printing Office, Washington, D.C., DHEW publication #(NIH) 78-1066, 1978.

18

Anthropometric Field Methods: Simplified Methods

Adnan Shakir

1. Discussion

Protein–calorie malnutrition (PCM) is a widespread problem of infants and young children in the developing world and to a lesser extent in poverty areas in the industrialized nations.

There is extensive evidences that malnutrition, especially the severe forms, may permanently impair the mental and physical development if started early in life. Even mild forms can increase the morbidity and mortality of infectious diseases in young children. The greater frequency and severity of some infectious diseases is related to poor nutrition. The early detection of children with mild to moderate as well as severe PCM (by periodic surveillance of the population at risk) is important for quick, economical, and effective treatment and prevention.

Although severe forms of PCM can of course be detected clinically, milder degrees of malnutrition or growth failure are easily missed (McLaren and Kanawati, 1970; Shakir *et al.*, 1972).

Nutritional anthropometry provides indices of growth failure and bodily disproportion, which are characteristic of PCM, and thus can be of great help in detecting the early forms. Many different anthropometric measurements are used in nutritional assessment and surveillance. These include weight, height, head circumference, mid-upper arm circumference, and skinfold thickness. All these measurements (with the exception of mid-upper arm circumference) are more or less dependent on knowing the precise age (Jelliffe and Jelliffe, 1971).

Weight for age is perhaps the most generally useful and widely used single measurement. A continuing weight curve of every child in the community,

Adnan Shakir • Department of Pediatrics and Child Health, Baghdad University Medical College, Baghdad, Iraq.

using simple weight charts, is the best method of continuing assessment at present (Chapter 16). A weight curve which shows faltering or flattening remains the most sensitive measure of impaired nutritional states.

Weight for age has the major disadvantage of needing an accurate knowledge of the child's age. In most developing countries, accurate ages are usually not known, and several methods have been suggested for use when the precise ages are not available. They are mostly based either on (a) a single measurement, which supposedly changes little with age, but is affected early in malnutrition, such as the mid-upper arm circumference and the skinfold thickness over a certain age range (1–5 years old), or (b) a ratio of two measurements other than age, on the assumption that one of the measurements reflects readily the recent nutritional state, such as weight or arm circumference, and the other, being a skeletal measurement, such as height or head circumference, is not affected readily by acute malnutrition, and can therefore be used as an approximate indication.

It is extremely important to know how these various indices work and how they compare with the weight for age, and try to find the most practical method, using the simplest equipment that can be employed even by junior paramedics for nutritional screening in community field surveys or similar situations.

To be practical for field use, the method should be (a) precise-age independent, (b) capable of standardization, (c) objective (not subject to observer variation), (d) usable by paramedicals with minimum special training, (e) rapid, (f) not antagonizing to parents, (g) repeatable at intervals on the same child, (h) applicable to both prevalence surveys and screening programs, (i) capable of misclassifying only a minimum of malnourished children, (j) useful for communities with both acute and/or chronic PCM, and (k) reliable in defining levels of severity of PCM in terms of weight for age or an agreed alternative.

The equipment used to carry out the method should be (a) locally made or locally purchasable, (b) easily transportable and durable, (c) reliable, requiring little or no adjustment during period of use, (d) cheap, (e) usable by junior paramedics, with the minimum possible reading and recording errors, and (f) giving ready classification of the child, without the need for reference tables, nomograms, or complicated calculations.

Methods based on weight and height include (a) weight for height (or length), which has been advocated as an age-independent measure of PCM of recent onset (WHO, 1971); (b) the weight over height squared ratio (Rao and Singh, 1970); (c) the weight over height to the power of 1.6 (Dugdale, 1971).

All these have the disadvantages that they need a weighing scale and trained workers. They cannot be used for screening in areas in which chronic malnutrition is prevalent, when stunted height is also likely to be found, and both weight and height may be reduced proportionately (nutritional dwarfing), resulting in normal weight for height. This is particularly so if international standards are used. Subjects have to have two measurements taken, including height (or length), which with small children, under field conditions, and in the hand of paramedicals is likely to be inaccurate. Finally, reference tables or calculations are needed before children can be classified.

The difficulties encountered in our Baghdad study (Shakir, 1975a), which found 73% of the children with PCM being stunted, illustrate some of these problems. Some children with severe PCM were misclassified as normals, and some normals were misclassified as severely malnourished, according to the weight-for-age definition and on the presence or absence of edema. It was difficult to define clear cutoff points for the different degrees of PCM; even the suggested upper-borderline limits showed these misclassifications.

The weight-for-head circumference (Jelliffe, 1966) has the same disadvantages mentioned earlier. The chest/head ratio (Jelliffe, 1966) is simple and easy to carry out, needing only a string or tape. Results with this measurement in our Baghdad study (Shakir, 1975a) were very disappointing, however; it correctly classified only 6% of the underweight, 50% of the marasmic kwashiorkor, and 25% of the kwashiorkor groups.

The arm for height has been developed under the name of the "QUAC stick" (Arnhold, 1969). The name was taken from the Quaker arm circumference. The child's mid-upper arm measurement is taken and he then stands up against a stick on which either one or three lines are used (Fig. 1). On the lines are marked 85, 75, and 70% of the standard arm circumference for height. If three lines are used, the children can be separated into four different groups. This method obviates the need for correct age estimation (Anderson, 1975), and does not require weighing scales, but it has the following problems: (a)It does not show the relative contributions of height (chronic PCM); (b) it may have different interpretations for different age groups and it is simpler to describe a selected age group; (c) special training is required for its estimation; and (d) two measurements are required, one (height) requiring an additional piece of lengthy equipment.

The results of using the QUAC stick in Baghdad (Shakir, 1973), when the children were measured by experts under study conditions, gave a good indication of its value for prevalence surveys. It also had an acceptable usefulness as a screening tool. However, reports on its use under field conditions, using paramedics from a study in Bangladesh (Sprague and Foster, 1972), showed that it could be used efficiently in prevalence surveys, but its value in screening children was not accurate, since slightly more than one-fourth of malnourished children were reclassified into another category on repeat surveys.

Kanawati and McLaren (1970) had advocated the arm/head circumference ratio for both surveys and screening programs. The method is based on the same principle as that for weight and height. The authors suggested the ratio of 0.310 as the upper limit for mild-moderate PCM, and 0.250 as the lower limit defining severe PCM.

The Baghdad study (Shakir, 1974), using an arm/head ratio of 0.290, showed 39.3% to have PCM, which is comparable to the 34.5% found by weight for age. The 0.310 ratio included all cases with PCM and the 0.290 ratio missed 6.3% of the children with mild-moderate PCM, but the former includes

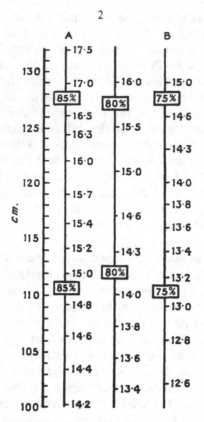

Fig. 1. QUAC stick, diagram of upper half of height range.

48% of the normal children in the malnourished group. The 0.250 level seemed a suitable upper limit for defining severe PCM; it missed none of the severe and included none of the normal cases.

The arm/head ratio could be used for both surveys and screening PCM, but it has all the disadvantages of the QUAC stick, mentioned earlier. It also has the problem of necessary calculations before the child can be classified, and that its use is limited to children under 5 years of age.

In order to facilitate nutritional surveys and screening of children, and to provide a ready classification under field conditions when using the arm/head ratio, Shakir (1974) suggested using a ruler (Fig. 2) on which the stated arm/head ratios are incorporated. Zerfas (1975) devised a ratio insertion tape using the same principle.

The use of a single measurement, such as the triceps or subscapular fatfold thickness, is based on the fact that these measurements readily reflect the body mass changes of malnutrition, and that they change very little with age during a stated interval. The standard values of the triceps and subscapular skinfold measurements vary less than 2 mm between the ages of 1 and 5 years (Jelliffe, 1966). The triceps skinfold measurements seemed to correlate well with the percent weight for age, with a coefficient of correlation of .83 ($p <$ 0.001). The percent weight for age could be predicted from the percent skinfold standard with a confidence of ±15% (Shakir, 1975a). The fatfold measure is the only convenient method, both direct and indirect, to estimate body fat content. It indicates calorie stores and, together with mid-upper arm circumference (MUAC), can be used as a measure of underlying muscle mass.

As a tool for survey work and for screening PCM under field conditions, fatfold thickness has many major limitations, such as (a) significant ethnic differences in the distribution of subcutaneous fat (Robson *et al.*, 1971); (b) variations of fat composition may occur with different diets and environmental temperature, possibly associated with atypical physical properties, such as compressibility (McLaren and Read, 1962); (c) measuring the skinfold thickness, usually the triceps, by one of the usual calipers has to take into account the high cost of the instrument, the need for frequent checking, the need for thorough training, the unreliability between observers and the fact that the difference between the well-nourished and malnourished levels of measurements is very small and requires a measurement validity rarely obtained in the field; (d) the age groups to be selected would depend on the particular country's needs; for example, where extensive, unregulated school feeding programs exist, the value of the measure to monitor possible changes in obesity rates could be limited to this situation. Otherwise, adults may best reflect trends in obesity, although it is difficult to relate the measure from one age group to the total population.

The mid-upper or maximal upper arm circumference (MUAC) measurement has been increasingly used for children aged 1–5 years since 1959 (Jelliffe

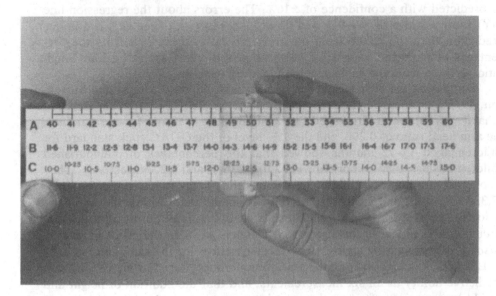

Fig. 2. The ruler suggested for use to give immediate classification of nutritional states of children on the basis of arm/head ratio.

and Jelliffe, 1961). Its value in predicting mortality has been validated in a postwar situation in Bangladesh (Sommer and Lowenstein, 1975). In young children, MUAC alone generally correlates well with weight-for-age estimate (Jelliffe and Jelliffe, 1969). In school age children, most studies have compared MUAC for height with weight for age or height, and variable correlations found (Sastry and Vijayaraghavan, 1973). MUAC is not tissue specific, since the arm is composed mainly of subcutaneous fat as well as muscle. Where fatfolds are relatively thin, as is often the case in 7-year-old males in less technically developed countries, MUAC may parallel muscle mass better than weight for age (Gurney and Jelliffe, 1973).

The mid-upper arm circumference standard increases 5.4 cm during the first year of life. From 1 to 6 years the increase is only 1.5 cm (Burgess and Burgess, 1969). The precise-age independence of the MUAC for 1- to 5-year-olds has been shown in industrialized countries, e.g., Polish children. Shakir (1975b) has shown that the same is true for Baghdad children aged 1–6 years. A mid-upper arm circumference measurement of 16.5 cm is considered a constant standard for children aged 1–6 years and can be used when the precise ages of children are unknown. Results appear similar to those obtained using age-specific arm standards. When applied to children under 1 year of age, MUAC tends to "overdiagnose" a larger number of normal children as malnourished. This misclassification is in favor of the child, and could be usefule in this critical age interval, especially in crisis situations.

The arm circumference measurements (MUAC) gave the highest coefficient correlation, i.e., .92 ($p \leq 0.001$), to the percentage of weight for age than all other measurements and ratios mentioned. It also gave the best prediction of percent weight for age from the percentage of arm standard, which could be predicted with a confidence of ±16%. The errors about the regression line (SY) show a range of 4.9–9.5, and do not increase with age. This is a better prediction than the ±20% from the arm/head ratio, the ±24% from the percentages of expected weight for height, and the ±28% from the Rao–Singh ratio, derived from the same group of children (Shakir, 1975b).

Comparing the 80% weight-for-age Harvard standard (Stuart and Stevenson, 1971), as an upper limit for PCM, with the 85% constant arm standard (Shakir, 1975b) for children aged 13–72 months, the 85% arm for age standard, the arm/head ratio of 0.290, and the 85% level of the QUAC stick standard for children aged 13–72 months showed that the percentages of malnourished children misclassified as normal were 2.0, 2.2, 6.3, and 1.2, respectively. They all overdiagnose normal children as malnourished in the following percentages: 17.3, 17.3, 10.6, and 11.0, respectively. It seems that neither the arm/head ratio nor the arm for height yields results significantly different from the arm circumference alone, in this age group. If this is so, then the ratios have little *raison d'etre*, requiring almost twice as much effort for the same information. El-Lozy (1974) has found that the arm circumference alone has almost as much predictive ability as all six measurements, and that the addition of height and weight to arm circumference does not increase appreciably the multiple correlation. On the other hand, the addition of the arm circumference to the height and weight leads to a considerable increase in predictive ability. It must be noted that the MUAC of older children will probably not mirror that of younger age target groups. Its level may be less sensitive to environmental effects (illness, food intake) and may reflect hereditary and maturation patterns more. Soft-tissue measures (such as MUAC) in adults are probably less sensitive than in younger children, although the raw values are numerically larger, and hence easier to measure accurately. They may have use with regard to (a) the MUAC of pregnant women which may reflect fetal nutrition status; (b) an estimate of manual work capacity; and (c) a simple measure to compare the nutritional status between vulnerable (mothers and young children) and less vulnerable groups (adult males); thus at times both groups may be below acceptable levels during periods of prolonged food shortage (A. J. Zerfas and D. B. Jelliffe, personal communication).

Simple methods for categorizing MUAC have recently been suggested, and include (a) a strip treated X-ray film, colored to define categories of PCM derived from the constant standard (Shakir, 1975b; Morley, 1976) (Figs. 3 and 4); (b) a quipu, a string with different colored knots at 10% increments below a 100% standard value (Jelliffe and Jelliffe, 1975); (c) a further simplification has been the use of bangles in communities where children regularly wear these on their arms. Investigation has shown that the size of these bangles is remarkably constant when they are produced by different manufacturers. One that is widely available has a diameter of 4 cm with an internal circumference of 12.6 cm. In children who have recently become malnourished, the bangle will not pass over the elbow. However, in those who have been malnourished for some time the bangle will pass over the elbow and up to the mid-upper arm. Provisional studies in India suggest that this may well prove to be a particularly acceptable method of assessing malnutrition using schoolchildren and other community groups (Morley, 1976). (d) An anthrogauge, which is a widened plastic or cardboard ruler with large notches to indicate specified diameters of the mid-upper arm; (e) a more refined arm measurement has been developed in Papua New Guinea and Rhodesia which can be used up to the age of 12 years (Morley, 1976).

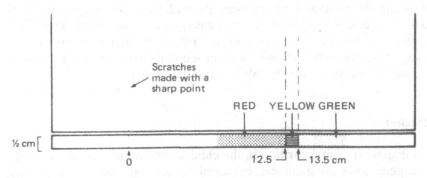

Fig. 3. The Shakir strip: .5–1 cm strip of washed X-ray film colored with a felt-tip pen.

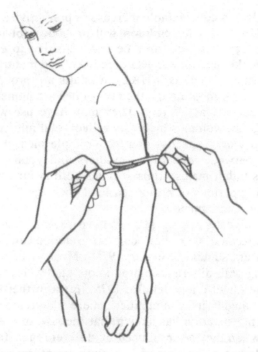

Fig. 4. The Shakir strip, in use measuring arm circumference.

All these methods are crude and, although of limited value for individual assessment, they may be especially suited for group evaluation, provided no systematic bias occurs with measuring. The Shakir's X-ray film strip has been used in both Africa and India by schoolboys with only a primary education. Following limited instruction, these boys arrived at a remarkably good estimate of the nutritional state of young children in their community (Morely, 1976). A similar method is being explored for use with antenatal mothers in India using a strip of X-ray film with markings at 21 and 23.5 cm (Shah, 1974).

Any child found with edema of nutritional origin is to be considered as severely malnourished, irrespective of the arm circumference measurement. Those with an arm measurement of more than 12.5 cm are to be classed as kwashiorkor. This will include all the rare cases of so-called acute kwashiorkor, which is caused by some acute severe infection such as measles, and results in a rapid fall in serum albumin and subsequent edema, without affecting physical measurements materially (McLaren *et al.*, 1968).

2. Conclusion

Each diagnosis of malnutrition in the child should be treated individually and assessment made using clinical, biochemical, and various anthropometric measurements. In longitudinal assessment of growth, serial increments in body dimensions are extremely useful.

In those countries in which ages of children are known, and the weight and height charts are used as a national document, the possibility arises of a month-by-month surveillance of all the children in the community. This is achieved by recording the date when measurement is taken and the age of the child. In this way it is possible for those at district and national headquarters to identify the nutritional states of young children. Each child is assessed and these children will reflect the nutritional state of the population. Since children are the most sensitive indicators of poor nutrition in the population, this may prove to be a system that can be generally adopted, once weight charts are more widely used. With this information those in a decision-making position will be warned of the approach of a period of food shortage some time before it becomes serious. They will also be able to identify those areas which are particularly affected.

In most developing countries weight charts are not in use. There is the added difficulty that with few exceptions the parents will not know the age of their children. Further, attempting to estimate the age of children is time-consuming and can be inaccurate. Even in these difficult conditions a great deal could be done by using one of the simplified methods for measuring MUAC and identifying nutritional edema. This is particularly practical between the ages of 1 and 5 years, since this is the period in which the syndromes of protein–calorie malnutrition, marasmus, and kwashiorkor are most frequent. After the age of 5 years, the arm circumference alone may not be quite so effective. The use of arm circumference in combination with height (i.e., the QUAC stick) seems to give a better estimate of the child's nutritional state.

In crisis situations, such as a limited food supply, the methods discussed herein can identify the geographical areas, groups, and individuals in greatest need of services. In noncrisis situations, these methods can give a community diagnosis of PCM by type and prevalence at one point in time, and when repeated (and it is possible to do so cheaply), they provide information on the changing incidence and pattern of PCM in the community. These techniques can also be used, with due caution, to assess the individual child, both at one time and over a period of time to gauge progress and response to treatment.

3. References

Anderson, M. A., 1975, Use of height–arm circumference measurement of nutritional selectivity in Sri Lanka School feeding, *Am. J. Clin. Nutr.* **28:**775.

Arnhold, R., 1969, The "QUAC stick": A field measure used by the Quaker service team in Nigeria, *J. Trop. Pediatr.* **15:**243.

Burgess, H. J. L., and Burgess, A. P., 1969, A modified standard for mid-upper arm circumference in young children, *J. Trop. Pediatr.* **15:**189.

Dugdale, A. E., 1971, An age independent anthropometric index of nutritional status, *Am. J. Clin. Nutr.* **24:**174.

El-Lozy, M., 1974, Letter to the editor, *Lancet* **2**:175.

Gurney, J. M., and Jelliffe, D. B., 1973, Arm anthropometry in nutritional assessment: Monogram for rapid calculations of muscle circumference and cross-sectional muscle and fat areas, *Am. J. Clin. Nutr.* **26**:912.

Jelliffe, D. B., 1966, The Assessment of the Nutritional Status of the Community, Geneva, WHO (Monograph Series No. 53).

Jelliffe, D. B., and Jelliffe, E. F. P., 1961, The nutritional status of Haitian preschool children, *Acta Trop.* **18**:1.

Jelliffe, D. B, and Jelliffe, E. F. P., 1969, The arm circumference as a public health index of protein–calorie malnutrition in early childhood. Current conclusions, *J. Trop. Pediatr.* **15**:253.

Jelliffe, D. B., and Jelliffe, E. F. P., 1971, Age-independent anthropometry, *Am. J. Clin. Nutr.* **24**:1377.

Jelliffe, D. B., and Jelliffe, E. F. P., 1975, The quipu in measuring malnutrition, *Am. J. Clin. Nutr.* **28**:203.

Kanawati, A. A., and McLaren, D. S., 1970, Assessment of marginal malnutrition, *Nature* **228**:573.

McLaren, D. S., and Kanawati, A. A., 1970, The epidemiology of protein–calorie malnutrition in Jordan: Part I. Application of a simple scoring system, *Trans. Roy. Soc. Trop. Med. Hyg.* **64**:754.

McLaren, D. S., and Read, W. W. C., 1962, Fatty acid composition of adipose tissue: A study of three races in East Africa, *Clin. Sci.* **23**:247.

McLaren, D. S., Faris, R., and Zekian, B., 1968, The liver during recovery from protein–calorie malnutrition, *J. Trop. Med. Hyg.* **71**:271.

Morley, D., 1976, Nutritional surveillance of young children in developing countries, *Int. J. Epidemiol.* **5**:51.

Rao, K. W., and Singh, D., 1970, An evaluation of the relationship between nutritional status and anthropometric measurements, *Am. J. Clin. Nutr.* **23**:83.

Robson, J. R. K., Bazin, M., and Soderstrom, R., 1971, Ethnic differences in skinfold thickness, *Am. J. Clin. Nutr.* **24**:864.

Sastry, J., and Vijayaraghavan, K., 1973, Use of anthropometry in grading malnutrition in children, *Indian J. Med. Res.* **61**:1225.

Shah, P. M., 1974, *Early detection and Prevention of Protein Calorie Malnutrition*, Popular Prakasan, Bombay.

Shakir, A., 1973, QUAC stick in the assessment of protein–calorie malnutrition in Baghdad, *Lancet* **1**:762.

Shakir, A., 1974, The arm/head ratio in the assessment of protein–calorie malnutrition in Baghdad, *Trop. Pediatr.* **20**:122.

Shakir, A., 1975a, The surveillance of protein–calorie malnutrition by simple and economic means (A report to UNICEF), *J. Trop. Pediatr Environ. Child Health* **21**:69.

Shakir, A., 1975b, Arm circumference in the surveillance of protein–calorie malnutrition in Baghdad, *Am. J. Clin. Nutr.* **28**:661.

Shakir, A., Demarchi, M., and El-Milli, N., 1972, Pattern of protein–calorie malnutrition in young children attending the outpatient clinic in Baghdad, *Lancet* **2**:143.

Sommer, A., and Lowenstein, M. S., 1975, Nutritional status and mortality: A prospective validation of the QUAC stick, *Am. J. Clin. Nutr.* **28**:287.

Sprague, J. S., and Foster, S., 1972, UNROD Information Paper 21 (October 1972).

Stuart, H. C., and Stevenson, S., 1971, Growth charts, in: *Textbook of Pediatrics* (W. Nelson, ed.), p. 39, Saunders, Philadelphia.

WHO, 1971, Joint Expert Committee on Nutrition (FAO/WHO) 8th Report, Food Fortification, Protein–Calorie Malnutrition, WHO Technical Report Series No. 577, Geneva.

Zerfas, A. J., 1975, The insertion tape: A new circumference tape for nutritional assessment, *Am. J. Clin. Nutr.* **28**:782.

19

Presentation of Data

Miguel A. Guzmán, Charles Yarbrough, and Reynaldo Martorell

Growth is the result of interplay between genetic and environmental factors. Though the matter is still debatable, evidence to date indicates that most of the differences in growth (i.e., height, weight) between children from developed and developing nations, result from differences related to the environment and not to racial origins (Guzmán, 1968; Habicht *et al.*, 1974). Consequently, growth reflects the adequacy of dietary, principally proteins and calories, and health conditions; or more generally, it reflects nutritional status.

This chapter is limited to a discussion of the kinds of growth assessment surveys and their uses, with attention being devoted to techniques and manners of presenting data. The presentation is organized around three kinds of growth assessments: cross-sectional surveys, longitudinal studies, and surveillance. As Table I shows, the basic distinction between them has to do with who is being measured and with what frequency. Though the actual collection procedures are essentially the same in all cases, the sampling procedures as shown in Table I are unique in each case because fundamentally different questions are being asked.

1. Cross-Sectional Surveys

Cross-sectional surveys are one-time approaches in which populations are measured usually over as short a period as possible. Such surveys are cheap, easy to conduct, and, of course, take less time to execute and analyze than longitudinal studies.

A matter of concern for all growth evaluations, but particularly with cross-sectional surveys because of time constraints, is the selection of the study

---

Miguel A. Guzmán • Institute of Nutrition for Central America and Panama, Guatemala City, Guatemala. *Charles Yarbrough* • Computers for Marketing Corporation, Kenwood, California. *Reynaldo Martorell* • Food Research Institute, Stanford University, Stanford, California.

Table 1. Types of Growth Assessment

Cross-sectional survey	All or a representative sample of a population are measured once
Longitudinal studies	The same individuals are measured at two or more points in time
Surveillance	Selected members of a population, not necessarily the same individuals, are measured at two or more points in time

sample. The aim is often not to measure the universe, be it a region or a town, but a given portion of it. Needless to say, if one is to generalize, the sample measured must be representative, as communities selected for reasons of ease and convenience frequently are not. If not all individuals within a town are to be measured, a random sample should be selected from a proper frame. This ideal situation contrasts with a frequent pattern in which those who are measured are those who "show up" to be measured.

Cross-sectional surveys have a variety of potential uses, the most common one being to evaluate the nutritional status of populations. One example of this is the INCAP/OIR nutritional status assessment of Central America (INCAP, 1971). In this survey the investigators were interested principally in two general aspects: (a) relative comparisons between groups to identify communities, regions, or countries with better or worse nutritional status, and (b) comparison of each of the groups of concern with reference standards to quantify deviations from normal. Both of these concerns require comparison of two or more populations.

There are many approaches to comparing two populations: direct comparison of means through *t*-tests or other appropriate tests when the distribution is skewed; direct plotting of cases against a percentile distribution; and calculation of the percentage of cases below a certain value. It is the dictates of the survey which determine what techniques to use. For instance, a government may be interested in estimating the extent of severe growth retardation in various regions of the country, in which case the percentage of children falling below an unacceptable value would be the desired statistic.

A second usage of cross-sectional surveys is to screen individuals to select those at greatest immediate risk of death. Recent history has seen the application of such surveys in the war-devastated areas of Biafra and Bangladesh, where international relief organizations wanted to direct their resources to individuals, particularly children, most depleted and in need of food (Sommer and Lowenstein, 1975). In both regions, the QUAC stick indicator, which consists of comparing the arm circumference of a test child to that of a normal child of the same height, was utilized. This indicator, like other indicators derived from the comparison of test and normal children in terms of labile measurements such as weight, muscle circumference, and skinfold relative to stable measurements of length or width of the skeleton, evaluates current nutritional status and is age independent. The basic assumption in the rationale of these indicators is that in normal children of a given skeletal length or width, there is a tolerable range in labile tissue and that marked deviations from this represent abnormality, where severe body depletion and obesity represent the extremes. Data of this kind are, therefore, usually reported in terms of how many subjects fall above or below a chosen criterion.

Cross-sectional surveys may also be utilized to construct standards of growth if children are measured at fixed ages (Luna-Jaspe *et al.*, 1970). Such data are ideal for estimating means and variances or percentiles of growth at specific ages. Moreover, subtracting means at two different ages will yield adequate estimates of the mean growth increments. To estimate variability of increments, however, longitudinal data must be utilized. Cross-sectional data on children not measured at specific ages are not recommended for constructing standards of growth, for these data do not allow for adequate estimates of variability at fixed ages. For the same reason, percentile charts should not be constructed if children are not measured at the same age.

A fourth general category of studies for which cross-sectional surveys are useful is the investigation of factors related to growth. Because of the very nature of surveys, such a research design can at best only inform about associations and not about cause–effect relationships. Examples of such studies include the relationship between familial and socioeconomic factors related to growth (Rawson and Valverde, 1976). In some cases, the investigators are not interested in causes but the indicators of risk. Examples of these latter studies include the identification of the maternal anthropometric characteristics that best predict low birth weight or infant mortality (Lechtig *et al.*, 1976) with the aim of selecting women with such traits as recipients of more than normal attention.

2. Longitudinal Evaluations

By convention "longitudinal" has come to mean "during (part of) the life of." Thus, longitudinal evaluations are those in which data are collected at various times on the same individuals. As with simple surveys, the intent of this may be to evaluate the health status of an individual, to establish norms of growth, or to study the growth process itself. By far the most widespread collection of longitudinal data is for purposes of monitoring the growth of an individual child in view of two major types of questions.

The first question, and the most significant medically, is whether the child is growing adequately. In particular, the interest is upon successive changes, with the idea of identifying a period of relatively higher stress through a faltering in the rate of growth (Marsden and Marsden, 1964). This in turn implies some standard for comparison and a way to look for differences. The most common clinical technique is the plotting of serial measures of either height or weight against age on a chart which already shows the standard.

Another common procedure is to calculate differences (i.e., the change itself) and to make a direct numerical comparison with standards of change. In both instances, comparisons may be absolute, "the child's growth is faltering," or relative, "he's always been short for his age; he's even shorter than you'd expect given his parents height," but in both cases the outcome is essentially a yes/no answer to the question of whether the child is currently doing "ok." If the answer is no, then some immediate cause is sought.

The second question, though less important medically, is one very frequently asked, namely, "what will the child's adult stature be." This can be done roughly by obtaining a child's current percentile, assume that his adult percentile will be the same, and then convert the latter percentile to absolute height. Alternatively, predictive regression equations that utilize age, anthropometric measurements, and skeletal age may be used if more precision is required (Walker, 1974).

Deriving standards for change is a somewhat tricky business since it cannot be done from cross-sectional data: the rate of growth is not necessarily smooth and, unlike the mean change, the distribution of changes cannot be inferred from the separate distributions of attained values. The most obvious example is the adolescent spurt, which every child undergoes, but which is not reflected in the steady rise of average height during the adolescent years (Tanner, 1962). Thus longitudinal standards imply longitudinal studies from which to derive the estimates of change.

Longitudinal assessments are preferred for investigations of the growth process itself. Measurements of the rate of growth, for instance, can be related to whatever other factors are of interest (Gopalan et al., 1973). As in the case of assessing individual histories, the comparison may be absolute, "the better fed children grew more," or relative, "the impact of supplemental feeding on growth is greater between the ages of 1 and 3 years than in later ages." Finally, for some studies the appropriate variable may be more truly longitudinal, as for example the age of maximal growth spurt of parameters derived from fitting a curve (Johnston et al., 1976).

3. Surveillance

The term "surveillance" originates from the French surveiller, meaning to keep a very close watch with the intent of control. Thus, for governments and technical agencies which have in recent years become increasingly concerned with the nutritional status as an indicator of social well-being, surveillance is not an isolated activity but an integral part of the formulation and execution (which includes evaluation) of policy. Because of the diversity of interests and nonuniformity of the conditions in different areas, this calls for different approaches and methodologies. In this context surveillance relates to measurements on sequential samplings of populations. The information collected in the initial point in time in the observation process may be used as a reference baseline for evaluation in the time sequence.

A joint FAO/UNICEF/WHO Expert Committee (WHO, 1976) has defined nutritional surveillance as a continuous process with the following objectives:

1. To describe changes in the nutritional status of populations with particular references to subgroups identified as being at risk.
2. To provide information that will contribute to the analyses of causes and associated factors for the selection of preventive measures.
3. To promote decisions by governments concerning priorities and allocation of resources to meet the needs of normal development and emergencies.
4. To enable predictions of the probable evolution of nutritional problems.
5. To monitor nutritional programs and evaluate their effectiveness.

Clearly, information from different fields (production, consumption, income, ecology, and health status, among others) is needed to fulfill these objectives. In this chapter, however, we limit considerations to growth data as it may contribute to a surveillance effort. Furthermore, considerations of growth data are limited to the use and presentation of selected anthropometric measurements, in the context of nutritional surveillance within an epidemiological frame, and for the evaluation of the impact of interventions. For these purposes, the data in a given time sequence may or may not originate from the same individuals, although to be of real value, it must originate from the same population—the target of surveillance.

The use of anthropometric measurements in nutritional surveillance within an epidemiological frame, in general, relates to the evaluation of shifts in the distribution of anthropometric characteristics in a population. In turn, the population groups for surveillance may be selected intentionally because of ease of access and high risk to malnutrition; the behavior through time of results derived from observations made in the selected population then serves as indicators of change in the nutritional situation of an area, country, or region. In this context, shifts in the distribution of a given anthropometric characteristic or index can be documented and identified using a time series of means and standard deviations presented in a set of conventional tables and/or graphs. In Japan, for example, graphs of the age–sex-specific average heights and weights of schoolchildren over a sequence of years (Nakayama and Arima, 1965) have been used successfully to depict secular trends (increasing) in height and weight and have also served to document changes in such trends (decreasing), particularly in the case of weight, during World War II.

When the distribution of the measurements of the characteristics used in the surveillance exercise is approximately normal, and interest is on the population as a whole, corresponding means and standard deviations can be used successfully as appropriate indicators for evaluating change over time. However, the distribution of the measurements used in surveillance very often is not normal (e.g., skinfolds) or interest is focused on a particular group (such as low-birth-weight babies) within the population. In this case, the conceptually appropriate presentation is a sequential set of frequency distributions or, equivalently, a time series of tables of selected percentile values of the variable of interest. This approach can be cumbersome and difficult. The important point, however, is that the evaluation of shifts in time of a complete frequency distribution can be made by being sensitive to the shifts in points of greatest interest in the distribution. The procedure may be simplified by selecting arbitrary points in the original distribution of measurements and following through time the behavior of the study population in terms of these selected reference points. One example of this approach is the sequential calculation of the percentage of newborn children with low birth weight (<2.5 kg), or else the percentage of children who fall below a reference line traced on a growth chart. The former is useful in predicting changes in mortality in children, and the latter helps in estimating changes in the magnitude of the malnutrition problem (Lechtig *et al.*, 1976; INCAP, 1971). The difficult feature in such schemes relates to the definition of appropriate reference points for calculation of the percent figures and the establishment of the levels of the percent figures required to trigger action.

The procedures described can be used also for the purpose of evaluating changes in nutritional status as a result of the natural process of development. An example from Guatemala (Guzmán, 1975) illustrates the presentation of data in both tabular and graphic forms. In this case, observations on the height and weight of children from a specific public school and an orphanage in metropolitan Guatemala were compared in the time extremes of a 20-year time interval. In other words, the same locations, but with different individuals, were compared at the beginning and end of the 20-year time lapse considered. A numerical table (Table II) and graph (Fig. 1) prepared with data from the study in reference, illustrate the usefulness of this scheme of presenting anthropometric data in the detection of changes over long periods of time and contrasting such changes in different population groups.

The sequential recording of indices, such as we have described, will define

Table II. Mean Differences in Height for Male Children from an Orphanage and a Public School in Guatemala City, 1952–1972[a]

Age (years)	Orphanage		Public school	
	Δ Height (cm)	Percent change	Δ Height (cm)	Percent change
8	−1.07 ± 1.99	−0.9	6.17 ± 1.86	5.3
9	−1.90 ± 1.58	−1.5	10.91 ± 1.35	9.2
10	−0.40 ± 1.86	−0.3	10.96 ± 1.37	8.9
11	−0.95 ± 1.97	−0.7	4.80 ± 1.47	3.6
12	0.91 ± 2.78	0.7	7.76 ± 1.74	5.8
13	−1.40 ± 2.08	1.0	6.95 ± 1.58	5.0
14	−3.50 ± 2.07	−2.4	3.38 ± 2.77	2.3
All ages	−1.19 ± 0.80	−0.9	7.28 ± 0.60	5.7

[a] Δ Height = mean height 1972 − mean height 1952. Percent change relative to age-sex-specific height in 1952.

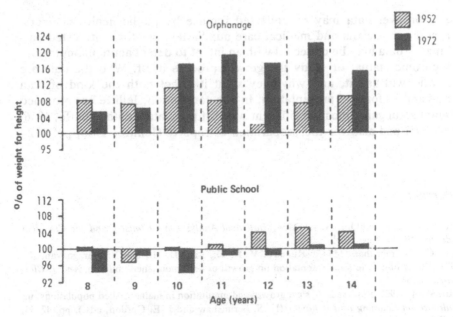

Fig. 1. Evolution of weight for height in male children from an orphanage and a public school in Guatemala City, 1952 and 1972.

trends, and reasonable predictions of expected behavior of the indices are possible from such trends, provided they do not project too far into the future.

Of particular usefulness in the evaluation of the impact of interventions is an adaptation of techniques used for industrial quality control (Bennett and Franklin, 1954). Using this scheme control charts with expected steady state behavior of anthropometric characteristics in a given population can be constructed. These charts can then be used for plotting and displaying graphically the behavior through time of selected anthropometric characteristics. Departures from the steady state can be detected in the time intervals specified for recording measurements. This scheme is particularly suited for application in centers that provide health services—mother and child care centers and maternity wards in hospitals, for example. Additionally, the control charts may provide a basis for an early warning system for deteriorating nutritional conditions. The value of control charts in surveillance systems merits research and exploration of practicability.

4. General Comment

We have now completed a review of the analysis and presentation of the data arising from the three major types of growth monitoring: survey, longitudinal study, and surveillance. Such growth data are utilized for nutritional assessment and are collected to assess either individuals or populations as a whole. Moreover data may be collected for use by public health planners, researchers, physicians and medical care auxiliaries, or other interested parties. Finally, data may be collected with an intent to detect acute malnutrition, assess chronic status, or study the growth process itself. Who the data are about, who will use it, and why they want it affect both the kind of data collected and its form of presentation. In this connection, Jelliffe has provided clear and useful guidelines for different ways of presenting data (Jelliffe, 1966, pp. 198–204) and also for the preparation of reports (Jelliffe, 1966, pp. 172–175).

5. References

Bennett, C. A., and Franklin, N. L., 1954, *Statistical Analysis in Chemistry and the Chemical Industry*, Wiley, New York.

Gopalan, C., Swaminathan, M. C., Kumary, V. K. K., Rao, D. H., and Vijayaraghavan, K., 1973, Effect of calorie supplementation on growth of undernourished children, *Am. J. Clin. Nutr.* **26**:563.

Guzmán, M. A., 1968, Impaired physical growth and maturation in malnourished populations, in: *Malnutrition Learning and Behavior* (N. S. Scrimshaw and J. E. Gordon, eds.), pp. 42–54, MIT Press, Cambridge, Mass.

Guzmán, M. A., 1975, Secular trends in height and weight as indicators of the evolution of nutritional status, in: *Proceedings of the 9th International Congress of Nutrition* (A. Chavez, H. Bourges, and S. Basta, eds.), Vol. 4, pp. 76–81, Karger, Basel.

Habicht, J. P., Martorell, R., Yarbrough, C., Malina, R. M., and Klein, R. E., 1974, Height and weight standards for preschool children: Are there really ethnic differences in growth potential? *Lancet* **1**:611.

Institute of Nutrition of Central America and Panama (INCAP) and Interdepartmental Committee on Nutrition for National Development (ICNND), 1971, Nutritional Evaluation of the Population of Central America and Panama, U.S. Department of Health, Education and Welfare, Publication No. (HAS) 72-8120, Washington, D.C.

Jelliffe, D. B., 1966, The Assessment of the Nutritional Status of the Community, WHO Monograph Series No. 53, World Health Organization, Geneva.

Johnston, F. E., Wainer, H., Thissen, D., and MacVean, R., 1976, Hereditary and environmental determinants of growth in height in a longitudinal sample of children and youth of Guatemalan and European ancestry, *Am. J. Phys. Anthropol.* **44**:469.

Lechtig, A., Delgado, H., Yarbrough, C., Habicht, J. P., Martorell, R., and Klein, R. E., 1976, A simple assessment of the risk of low birth weight to select women for nutritional intervention, *Am. J. Obstet. Gynecol.* **125**:25.

Luna-Jaspe, H., Ariza-Macías, J., Rueda-Williamson, R., Mora-Parra, J. O., and Pardo-Téllez, F., 1970, Estudio seccional de crecimiento, desarrollo y nutrición en 12,138 niños de Bogotá, Colombia. II. El crecimiento de niños de dos clases socio-económicas durante los primeros seis años de vida, *Arch. Latinoam. Nutr.* **20**(2):151.

Marsden, P. D., and Marsden, S. A., 1964, A pattern of weight gain in Gambian babies during the first 18 months of life, *J. Trop. Pediatr.* **10**:89.

Martorell, R., Habicht, J. P., Yarbrough, C., Lechtig, A., Klein, R. E., and Western, K. A., 1975, Acute morbidity and physical growth in rural Guatemalan children, *Am. J. Dis. Child.* **129**:1296.

Nakayama, K., and Arima, M. (eds.), 1965, Child Health in Japan, An Exhibition at the XI International Congress of Pediatrics, November 7–13, Tokyo.

Rawson, I. G., and Valverde, V., 1976, The etiology of malnutrition among school children in rural Costa Rica, *J. Trop. Pediatr.* **22**:12.

Sommer, A., and Lowenstein, M. S., 1975, Nutritional status and mortality: A prospective validation of the QUAC stick, *Am. J. Clin. Nutr.* **28:**287.

Tanner, J. M., 1962, *Growth at Adolescence*, 2nd ed., Blackwell, Oxford.

Walker, R. N., 1974, Standards for somatotyping children. I. The prediction of young adult height from children's growth data, *Ann. Hum. Biol.* **1:**149.

World Health Organization, 1976, Methodology of Nutritional Surveillance, Report of a Joint FAO/UNICEF/WHO Expert Committee, World Health Organization, Technical Report Series No. 593, Geneva.

Nutrition Surveillance in Developed Countries: The United States Experience

Milton Z. Nichaman and J. Michael Lane

1. Introduction

Nutrition surveillance can be defined as the continuous monitoring of both the nutritional status of population groups and the economic- and food-related variables which may be predictive of the possibility of change in nutritional status. Nutrition surveillance differs from nutrition surveys in that surveillance data are collected, analyzed, and utilized in an ongoing manner while nutrition status surveys are most frequently conducted at single or multiple discrete points in time. Surveillance can also be interpreted as the analysis of time trends obtained from the results of periodic nutrition status surveys. We would suggest that this type of activity be termed monitoring. This chapter deals with surveillance in the context of a continuous ongoing system of data collection, analysis, and utilization. Although the mechanics of implementing and carrying out a nutrition surveillance system in developed countries may differ from those in developing countries, the principles remain the same.

Before discussing the details of development and implementation and the results of the nutrition surveillance program in the United States, it is useful to ask the question: why nutrition surveillance and why is it necessary in developed countries? Many industrialized countries, where there should be no classical nutritional deficiency syndromes or protein–calorie malnutrition on the basis of either national food balance sheets, food availability, or overall socioeconomic status, have identified certain subgroups of their populations with evidence of nutritional deficiencies. Moreover, as might be expected in these affluent societies, there is evidence of nutritional excesses as well. Studies have also shown a relationship between nutritional status and various

Milton Z. Nichaman and J. Michael Lane • Nutrition Division, Center for Disease Control, Atlanta, Georgia.

socioeconomic and demographic population characteristics (Ten-State Nutrition Survey, 1972; Health and Nutrition Examination Survey, 1976; Owen *et al.*, 1974). For example, in the United States, there is a direct relationship between certain indices of nutritional status in children, such as attained growth and vitamin and mineral status, with the level of poverty as measured by income, family size, rural or urban living conditions, and the educational level of the mother. These conclusions, particularly in the United States, have resulted in the funding of large numbers of feeding programs as well as programs designed to improve the health and nutritional status of groups identified as being at nutritional risk.

The large amount of public funds involved in carrying out these programs demands the specific identification of those segments of the population most in need of intervention activities, as well as identification of those individuals who are eligible for these programs but who are not taking advantage of them. There is also need for data for the evaluation of the efficiency and effectiveness of those programs specifically designed to improve the nutritional status of program participants. Nutrition surveillance should ideally provide information that would be useful in all of these areas.

The ultimate goal of nutrition surveillance is to enhance the health of a population by identifying and preventing the development of malnutrition. In developed countries our concerns are with both undernutrition and overnutrition. More immediate objectives of nutrition surveillance are to specifically identify population groups showing evidence of malnutrition and to bring these groups to the attention of intervention programs designed to correct nutritional abnormalities.

Nutrition surveillance in affluent countries has been reviewed by Berry (1972). He suggested that the main aim and justification of national nutrition surveillance was prediction. He further indicated that many of the indices used in nutrition surveillance were not specifically nutritional. Some examples of these indices are economic status, changes in population demography, specific cultural attitudes, and certain climatic conditions. He also stressed the need for the collection of direct evidence of nutritional status in any program of nutrition surveillance. Nichaman (1974) reviewed the beginnings of a nutrition surveillance system in the United States and pointed out that, in this case, nutrition surveillance was developed out of organizational needs in deciding whether to develop nutrition action programs. Organizations involved in these types of activities needed to know where and among whom nutritional problems existed, how severe these problems were, and what resources could be applied to effectively deal with these problems. Irwig (1976) recommended the surveillance of growth, a variable directly related to nutritional status, because of its relationship to general health. WHO (1976) published methodology for carrying out nutrition surveillance that was applicable both in developed and developing countries.

The goals of nutrition surveillance are to identify pockets of poor nutritional status, monitor changes in variables which will predict deterioration of nutritional status, and ultimately to improve the health status of individuals.

In order to accomplish these goals, nutritional surveillance activities must be incorporated in health care delivery programs so that the information can be used to guide the planning of specific programs as well as for the evaluation of local intervention activities.

In this chapter we discuss the elements of nutrition surveillance with specific reference to their implementation in developed countries. Examples are taken primarily from the experience in implementing a nutrition surveillance system in the United States. It must be recognized that all developed countries are not similar and that each nutrition surveillance mechanism will have to be adapted to the medical care system in that country.

2. Development of Nutrition Surveillance Programs

Prior to the initiation of a nutrition surveillance program, agreement must be reached in regard to definitions of normal, optimal, and adverse nutritional status. Following this, the first actual step in the development of a program for nutritional surveillance is to characterize the present nutritional status of the population. The second step is to identify those indices of nutritional status which best reflect the problems in the population to be placed under surveillance. The third step is to identify the method of collection and sources of data related to these indices. The fourth and final step is the tabulation, analysis, and feedback of the data to the originating sources.

2.1. Identification of Population Nutrition Problems

Ideally the identification of the presence or absence of nutritional problems should be made from a random and representative sample of the population. Even when representativeness is an expressed goal of a baseline nutrition survey, it is often difficult to achieve because of problems with participant response. In the Ten-State Nutrition Survey, which was designed to sample representatively low income populations in preselected locations, the response rate was about 50%. This low response rate made it very difficult to extrapolate results to the total population from which the sample was drawn. The Preschool Nutrition Survey, conducted by Owen and his colleagues (1974), and the Nutrition Canada Survey (1973) also had problems with relatively poor response rate.

In spite of these deficiencies, certain general conclusions could be drawn from these surveys. The nutritional problems identified were not found throughout the entire population examined, but were most prevalent in certain specific age, sex, ethnic, economic, and geographic groups. Moreover, those problems which were found were of a rather general nature leading to deficits in linear growth, a high prevalence of overweight and potential obesity, and moderate amounts of anemia, primarily due to iron deficiency. In addition, these surveys raised the level of consciousness among administrators and deliverers of health services about problems of interpreting classical nutrition assessment data.

The Health and Nutrition Examination Survey (HANES), based on a national probability sample and with a higher response rate than the Ten-State Nutrition Survey, showed similar results. The extent of the problems found was quite different among subgroups of the population and the problems found were of a similar nature to those identified in the Ten-State Nutrition Survey and the Preschool Nutrition Survey.

The results of these surveys clearly point out that in a developed country the large representative survey can identify the general level of nutritional problems in various segments of the population. The small number of individuals actually examined in relation to the total population (in the United States the total number of individuals examined in the HANES survey between the ages of 1 and 74 was 20,700 out of a total population of 215,000,000) and the limited number of geographic areas from which these people were drawn limit the usefulness of these data in planning and implementing local nutrition action programs. In order to plan, initiate, and evaluate a nutrition action program at the local level, where all intervention programs ultimately operate, local data are needed to correspond with local needs. Individuals responsible for improving and initiating local programs at the city, county, or state level often have difficulty in relating national results to their own geographic area. The ideal solution to this problem would be to repeat the methodology of the national survey at a local level. Although representative data are the ideal, data available from various populations being provided health services are also useful. In most instances it is not possible to develop a locally representative sample because of competing priorities and the lack of fiscal and human resources to carry out an extensive survey. In such a situation an acceptable solution is to collect nutritional status information from individuals receiving nutrition and health care services. The individuals receiving such services are not a defined sample, and people are constantly entering and leaving the service programs. Such a program of collection, analysis, and utilization of data that is incorporated as a routine activity of delivery systems constitutes one aspect of nutrition surveillance.

2.2. Identification of Surveillance Indices

The decision as to which nutritional indices to incorporate into the nutrition surveillance system should be based on results from national, local, and regional surveys. Rather than attempting to introduce all of the examinations performed in large national surveys, with their inherent costs and logistical problems, those data items that are easily and inexpensively measurable at the local level and which relate to the major nutritional problems found in national surveys should be identified.

Although in many cases the national or even local surveys may not have been conducted on the specific population that will be placed under surveillance the characteristics of this population may be close enough to that of the populations surveyed so that inferences can be drawn from the results of these surveys. A major consideration in choosing the variables to be tracked is the availability of measurements at points of service delivery and the additional cost necessary to collect, analyze, and evaluate these data. In the United States, based on results from the Ten-State Nutrition Survey and the Preschool Nutrition Survey, which identified anemia, growth problems, and overweight as the most prevalent nutritional disorders in children, the ready availability of height, weight, and hemoglobin or hematocrit data at points of health care delivery indicated that these variables should be major items in a nutrition surveillance system. The additional factor of minimal increased cost necessary to make these data available for anlaysis and evaluation also was an important deciding factor in the incorporation of these variables in the system. The important variables of age, sex, and ethnic group were also generally available at health care delivery points and could be incorporated into the nutrition surveillance package with minimal additional cost and effort.

Other items that may be included in the surveillance data package are selected dietary intake data if they have been shown to be clearly predictive of potential changes in nutritional status, demographic items such as income and family size, as well as other biochemical or population variables that have been identified as risk factors for the development of nutrition-related disease. An example of such risk factor data collection is found in the program in Arizona where serum cholesterol determinations are a routine part of the nutrition screening activities in children and families receiving preventive health services as publicly supported clinics. Although hyperlipidemia was not specifically identified as a major nutritional problem in the United States surveys or the Canadian National Nutrition survey it is evident from other data that some investigators feel that the problem of hyperlipidemia and its possible risk factor relationship to the development of cardiovascular disease in adult life is a pediatric problem, and an attempt should be made to identify high risk individuals or groups of individuals at as early an age as possible. This information together with local evidence that a high percentage of pre-school and primary schoolchildren had high levels of serum cholesterol prompted Arizona to include serum cholesterol in their nutrition surveillance data package.

In order to incorporate the other important aspect of nutrition surveil-lance, the monitoring of predictive variables related to changes in nutrition status, further studies must be carried out once the initial determination of nutritional status has been made. These studies should be designed to identify the relationships between nutrition status and various population variables, for example, socioeconomic status, dietary intake, and cultural and demographic characteristics. It is in this area that information regarding food intake, food cost in relation to total income, changes in nutrient composition, and changes in lifestyle patterns becomes critical. In spite of the large numbers of dietary studies which have been made in the United States, there is little knowledge regarding the consumption of specific food items in relation to nutritional status or to changes in nutritional status. However, it does seem possible that using nutrition surveillance data as a basis, local studies can be carried out which might identify specific food items or patterns of dietary intake that are related to nutritional status as well as to changes in nutritional status. To do this in the context of a nutrition surveillance program in a health care delivery system it would be necessary to develop a mechanism that allows for record linkage so as to compare nutritional status items with dietary intake variables. From these studies inferences can then be drawn as to which variables to test further in regard to their predictive value and which variables to incorporate into the nutrition surveillance system on a theoretical and interim basis. Not only may these secondary data items be useful for predicting changes in nutrition status but also the relationship found between these factors and nutritional status can be useful as a basis for planning specific intervention programs. As a corollary to these studies the identification of possible causes of the identified nutritional problems will be made and intervention programs should be mounted in an attempt to manage and prevent these identified problems.

Other items which should be considered for study and possible inclusion in a nutrition surveillance system in developed countries are the prevalence of associated diseases and conditions which have been identified as being partly or fully caused by poor nutritional status or are themselves a cause of deterioration of nutritional status. Examples of such conditions are diarrhea and gastrointestinal diseases, cardiovascular diseases, low birth weight, and prematurity.

There are a number of blocks to adding additional items to routine surveillance data collection. These constraints are the lack of agreed upon definitions in regard to items indicative of poor nutritional status or predictive of changes in nutritional status, the lack of quality control of the diagnoses as well as of the index measurements, and the multiplicity of nonnutritional causes for many conditions.

The nutrition surveillance experience in the United States has been mainly in the area of collecting information concerning the nutritional status of population groups. Only now, based on data and relationships inferred from nutritional surveys, is consideration being given in nutrition surveillance data collection programs to including items related to food intake habits, food availability and food costs, and the prevalence of associated nutritionally related diseases.

2.3. Identification of Data Sources

The next step in developing a surveillance system is to determine where and how one can obtain data on the selected high risk segments of the population. For example, if a decision has been made that low income preschool children are at high risk for having nutritional problems, locations at which such children congregate and where information is available which relates to nutritional status must be identified. In developing the nutritional surveillance program in the United States, it was evident that nutrition problems were most prevalent in preschool children of minority groups and that these children were likely to attend a variety of medical care programs such as well-baby clinics, Headstart programs, and day care centers. As noted earlier, children from such clinics and programs are not a representative sample of the population at risk. For this reason, we suggest that these groups be called "samples of convenience." It is clear, however, that if nutrition problems are found in these "convenience samples," they constitute an appropriate surveillance universe in spite of the fact that they are not representative of the total population that is the target of the nutrition action program. The use of convenience samples as target groups for surveillance programs does not deny the fact that there are probably segments of the population who do not enter into these samples but who are still in need of nutrition and health services. With the usual situation of limited resources, it is reasonable to restrict surveillance activities to these samples of convenience. In some cases, such as in the Supplemental Food Program for Women, Infants and Children, the identification of the population served and therefore the population under surveillance may be based on administratively useful criteria for selection of individuals for the program. In these situations, the data available from such a program do not directly tell us the specificity of the program in identifying and serving all those in need.

It is often desirable to combine population-based survey data with data collected on a continuous basis from a number of convenience samples. In this case, a population group or geographic area would first be surveyed and those at-risk individuals in the group would be identified as the universe for nutrition action programs. Following this, there can be an ongoing collection of data from these intervention programs (convenience samples) to assess the progress of the action program as well as any changes that may occur or be predicted in the total target population. At various points in time repeat examinations of random samples of the total population must be undertaken to assure that the outreach program continues to function in identifying at-risk individuals and bringing them into the prevention and management system.

2.4. Handling of Nutritional Surveillance Data

Once a decision has been made regarding the population groups to be placed under nutrition surveillance, the places from which the data will be gathered, and the data items to be included, a system of reporting, analyzing, and utilizing data must be developed. In the United States the nutrition surveillance system being conducted on a pilot basis in selected states is attempting to place under surveillance at-risk preschool and minority group children who are seen on a regular or irregular basis in a variety of health care facilities such as well-baby clinics, Headstart programs, and day care centers. Data from local health and nutrition delivery services are collected in a standardized form and transmitted to the central facilities of a state health department. At this point data are edited to eliminate obviously erroneous items, converted to a computer-compatible format, and basic tabulations are performed either at the state computing facilities or at the Center for Disease Control. The results of these tabulaitons are immediately returned to the originating facility for their use in program planning and clinic management.

Although this program is theoretically complete with the development of one section logically dependent on the completion of the previous steps, in real life situations information is not always available in many of these areas and can be obtained only after the expenditure of considerable cost and time. Because of this, nutrition surveillance activities are often carried out before all of the pieces are in place and before the sensitivity, specificity, and predictive value of the nutritionally related variables have been established.

3. Examples from the Center for Disease Control Nutrition Surveillance System

The Center for Disease Control Nutrition Surveillance System was initiated in 1973 in cooperation with the public health departments of five states: Washington, Arizona, Louisiana, Kentucky, and Tennessee. Based on the results from national nutrition surveys it was decided to initiate surveillance of those items related to the nutritional status of high risk groups of children. These nutritionally related variables included height, weight, and hemoglobin or hematocrit. Selected demographic information and information regarding the type of health facility in which the data were collected was also obtained. An example of a typical data collection form is seen in Fig. 1. This form was developed in consultation with those individuals responsible for the delivery of nutrition and health services at the local level. These individuals are the ones who represent the major point of contact with those whose data will be placed in the surveillance system. Because in all such health care situations, whether in developed or developing countries, the number of professional personnel is never sufficient for the task imposed upon them, the collection and transmission of data to a central location for analyses must be kept as simple as possible. No new measurements were imposed on the clinic personnel and the only requirement was that the information be transmitted to a central location at the state level and from there transferred to the Center for Disease Control in some type of computer-compatible form.

Of paramount importance in a surveillance system is that there is a regular program of feedback of the results of the surveillance system directly to the source collecting and providing the data. This analysis and feedback should be done in such a way that the information being returned is useful and pertinent to the operation of the health care facility. Tables I, II, and III present examples of the type of information that is returned to the agencies providing the data. The information contained in these tables summarizes the number and percentage of children having nutritional index measurements below or above preselected and agreed upon cutoff points. Although it is possible to adjust cutoff points to meet specific local needs, it is a great advantage to have generally agreed upon cutoff points so that comparisons can

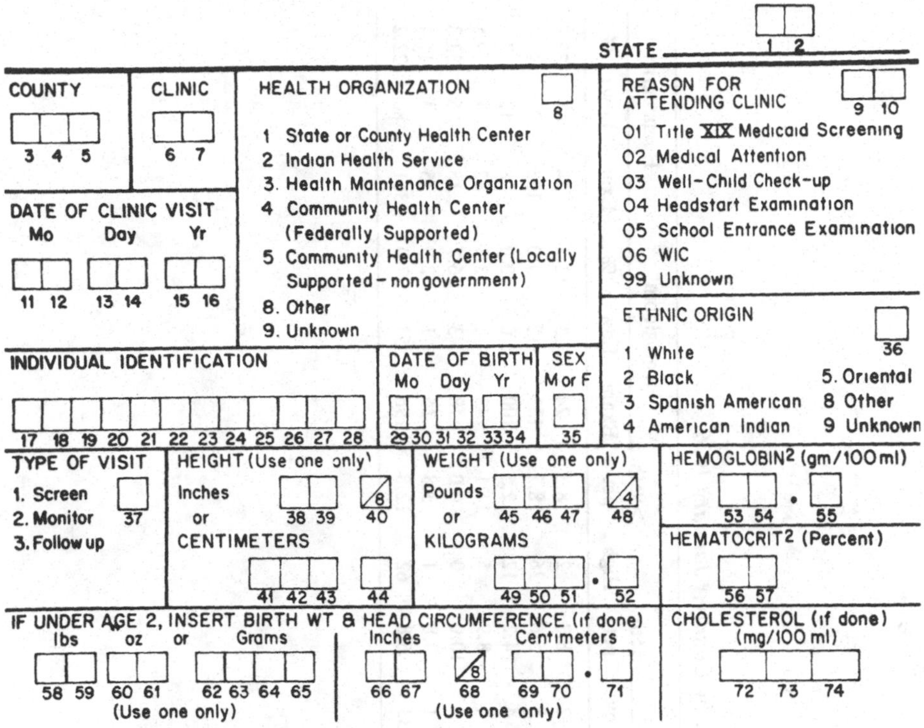

STATE _____ 1 2

COUNTY | CLINIC
3 4 5 | 6 7

HEALTH ORGANIZATION 8
1 State or County Health Center
2 Indian Health Service
3. Health Maintenance Organization
4 Community Health Center (Federally Supported)
5 Community Health Center (Locally Supported – nongovernment)
8. Other
9. Unknown

DATE OF CLINIC VISIT
Mo 11 12 Day 13 14 Yr 15 16

REASON FOR ATTENDING CLINIC 9 10
01 Title XIX Medicaid Screening
02 Medical Attention
03 Well-Child Check-up
04 Headstart Examination
05 School Entrance Examination
06 WIC
99 Unknown

ETHNIC ORIGIN 36
1 White
2 Black
3 Spanish American
4 American Indian
5. Oriental
8 Other
9 Unknown

INDIVIDUAL IDENTIFICATION
17 18 19 20 21 22 23 24 25 26 27 28

DATE OF BIRTH
Mo Day Yr 29 30 31 32 33 34

SEX M or F 35

TYPE OF VISIT
1. Screen
2. Monitor 37
3. Follow up

HEIGHT (Use one only)
Inches 38 39 /8 40
or
CENTIMETERS 41 42 43 44

WEIGHT (Use one only)
Pounds 45 46 47 /4 48
or
KILOGRAMS 49 50 51 . 52

HEMOGLOBIN[2] (gm/100 ml)
53 54 . 55

HEMATOCRIT[2] (Percent)
56 57

IF UNDER AGE 2, INSERT BIRTH WT & HEAD CIRCUMFERENCE (if done)
lbs 58 59 oz 60 61 or Grams 62 63 64 65
(Use one only)
Inches 66 67 /8 68 Centimeters 69 70 . 71
(Use one only)

CHOLESTEROL (if done) (mg/100 ml)
72 73 74

Fig. 1. Anthropometry and hematology worksheet.

be made from time to time and place to place. The cutoff points utilized in the CDC Nutrition Surveillance System have been agreed upon by all the participants in the system. The criteria for identifying individuals with low or high values are presented in Table IV.

Table I presents the summarization of data that are returned to a state listing the prevalence of abnormalities by county or region. Table II presents similar information for a single county or region by individual clinic or reporting site. The information contained in these tables can be used by people, who are responsible for the delivery of nutrition services at the central or regional level, to maximize the utilization of resources as well as to provide support for increased or new program activities. Table III lists by individual identification those children in each clinic or reporting site who were found to have one or more measurement values below or above the preselected cutoff points. Although it is clear that children identified with potential problems should be evaluated and referred for management immediately after the abnormal value is found and confirmed, the information returned to the local clinic can be effectively used by the clinic staff to assure that all those individuals needing follow-up have been appropriately managed.

Continuous monitoring of the results from individual clinics, counties, or regions can be an effective tool for improving the quality of the measurements

Table I. Nutrition Tabulation by County,[a] January 1976

County	Total exam	ht./age			wt/age			wt/ht					Hemoglobin			Hematocrit		
		Exam	Low	<5th pct[b]	Exam	Low	<5th pct	Exam	Low	±5th pct	High	>95th pct	Exam	Low	%	Exam	Low	%
01	28	27	4	14.8	26	2	7.7	24	2	8.3	2	8.3	24	3	12.5	24	5	20.8
02	150	150	14	9.3	148	6	4.1	143	8	5.6	26	18.2	133	12	9.0	133	23	17.3
03	115	113	20	17.7	105	12	11.4	96	4	4.2	12	12.5	100	0	0.0	98	12	12.2
04	48	47	3	6.4	45	2	4.4	45	4	8.9	5	11.1	43	1	2.3	42	4	9.5
05	21	21	1	4.8	21	4	19.0	20	2	10.0	0	0.0	18	3	16.7	18	6	33.3
06	39	39	3	7.7	34	0	0.0	26	1	3.8	1	3.8	35	1	2.9	35	4	11.4
07	699	692	96	13.9	676	74	10.9	483	33	6.8	62	12.8	196	53	27.0	361	80	22.2

[a] Counties are within one state.
[b] Percentile.

Table II. Nutrition Tabulation by Clinic,[a] January 1976

Clinic	Total exam	ht/age Exam	Low	<5th pct	wt/age Exam	Low	<5th pct	wt/ht Exam	Low	<5th pct	High	>95th pct	Hemoglobin Exam	Low	%	Hematocrit Exam	Low	%
01	28	28	3	10.7	27	1	3.7	27	3	11.1	3	11.1	26	0	0.0	26	4	15.4
02	11	11	0	0.0	11	0	0.0	11	1	9.1	1	9.1	9	1	11.1	8	0	0.0
03	7	7	0	0.0	6	1	16.7	6	0	0.0	1	16.7	7	0	0.0	7	0	0.0
04	2	1	0	0.0	1	0	0.0	1	0	0.0	0	0.0	1	1	0.0	1	0	0.0
Total	48	47	3	6.4	45	2	4.4	45	4	8.9	5	11.1	43	1	2.3	42	4	9.5

[a] Clinics are in county 04 (see Table I).

Table III. Nutrition Listing by Individual,[a] January 1976

Individual ID	Date of birth	Age yr	Age mo	Sex	Reason	Type visit	Date of visit	Height cm	Height in.	Weight kg	Weight lb	wt/ht (cntl)	Hgb	Hct
00401000803	10/01/72	4	0	F	06	2	10/15/76	99.1	39.0	15.9	35.1	67.6	14.6	39
00401002703	10/11/71	4	11	F	06	2	09/15/76	109.2	42.9	19.3	42.5	76.8	12.9	37
00401003603	01/18/74	2	9	M	06	2	10/15/76	90.2	35.5	13.2	29.1	49.3	12.3	35
00401007304	03/28/75	1	6	F	06	2	10/15/76	76.8	30.2	9.8	21.6	42.7	11.3	34
00401008803	03/17/73	3	7	F	06	2	10/15/76	96.5	37.9	14.9	32.8	59.9	13.3	38
00401015403	02/21/74	2	7	F	06	2	10/15/76	90.2	35.5	13.6	30.0	69.6	12.1	34[b]
00401016703	03/06/75	1	7	F	06	2	10/15/76	81.3	32.0	9.2	20.3	4.0[c]	12.5	36
00401023403	08/16/74	2	1	F	06	2	09/15/76	86.4	34.0	11.1	24.5	20.5	13.3	39
00401026204	03/28/76	0	8	M	03	S	12/13/76	68.6	27.0	9.3	20.5	91.4	11.7	34
00401031503	12/04/74	1	9	M	06	2	09/15/76	81.3	32.0	11.0	24.3	42.9	13.8	37

[a] Individuals from clinic 01. See Tables I and II.
[b] Value below specified cutoff level.
[c] Indicates weight for height value is below the 5th centile (cntl) weight for height. Values above the 95th centile weight for height are also flagged as abnormal.

Table IV. Criteria for Identifying Individuals with Low or High Values

Nutritional index	Criterion		
Low height for age	Height for age less than the 5th percentile of a person of the same sex and age in the reference population		
Low weight for age	Weight for age less than the 5th percentile of a person of the same sex and age in the reference population		
Low weight for height	Weight for height less than the 5th percentile of a person of the same sex and height in the reference population		
Low weight for height	Weight for height greater than the 95th percentile of a person of the same sex and height in the reference population		
	Age	Hemoglobin (g)	Hematocrit (%)
Low hemoglobin, low hematocrit	6–23 months	10	31
	2–5 years	11	34
	6–14 years	12	37
	15 or more years (females)	12	37
	15 or more years (males)	13	40

made. The increased awareness of local clinic staffs that others are concerned and interested in the measurements that they are making has led to improved quality and less variability in these measurements. The utilization of surveillance data to identify collection sites where there appear to be problems leading to the reporting of inaccurate measurement values has been most useful. Preparatory to developing a program to train local clinic personnel in the techniques of performing and recording anthropometric measurements a study was carried out to identify those items most related to measurement errors. Table V presents the frequency of errors in technique by a clinic staff in the weighing and measuring of children 0–5 years of age. This information is currently being utilized in the development of a training manual to teach and improve techniques in weighing and measuring children.

Another way in which these data have been utilized in the identification of technical problems and the improvement of the quality of measurements can be seen in the retrospective analysis of hematocrit data from one specific county. It was noted in this county that over a period of time the prevalence of low hematocrit values in children being initially screened fell from 47.1 to 9.5%. It was clear that the initial prevalence of 47.1% was many times higher than the prevalence seen in other areas providing data to the nutrition surveillance system. Analysis of the hematocrit data from this clinic revealed that between March 10 and June 24 there was a prevalence of low hematocrit values of 47.1%. Between June 25 and August 24 the prevalence fell to 25.8% and after August 25 the prevalence fell further to 9.5%. A closer look at these three periods of time revealed that during the first time period nurse No. 1 was doing all of the hematocrit determinations while during time period two,

*Table V. Frequency of Errors in Technique by Clinic Staff Weighing
and Measuring Children 0–5 Years Old*

Item	Frequency (and percent)	
	<2 yr	2–5 yr
Number of examinations error-free	17 (11.2)	40 (26.3)
Number of examinations in which errors observed	135 (88.8)	112 (73.7)
Total number of examinations	152 (100.0)	152 (100.0)
Specific weighing errors		
Excessive movement	47 (14.5)	10 (4.9)
Clothing left on	21 (6.5)	39 (19.0)
Specific measuring errors		
No assistance	61 (18.8)	—
Excessive movement	45 (13.8)	11 (5.3)
Not stretched or positioned correctly	33 (10.1)	61 (29.6)
Improper Frankfort plane	33 (10.1)	46 (22.3)
Headboard not placed against child's head	27 (8.3)	—
Child measured standing (<2) or recumbent (>2)	25 (7.7)	5 (2.4)
Measured with cap or shoes	12 (3.7)	12 (5.8)
Other	21 (6.5)	22 (10.7)
Total number of errors	325 (100.0)	206 (100.0)
Average errors per child	2.1	1.4

after nurse No. 1 retired, a number of different nurses did the hematocrit determinations. In time period three a new nurse was doing all of the determinations. It was clear from this analysis, as well as from retrospective interviews of the individuals concerned, that the nurse during time period one was performing the analyses incorrectly due to a misinterpretation of the directions provided by her supervisors.

The continued identification of a high prevalence of low hemoglobin and hematocrit values together with the working hypothesis that most low values were due to iron deficiency has prompted a number of local areas to review their procedures in regard to treatment and referral of these children. In some areas public health nurses, under a physician's supervision, are now allowed to initiate primary therapy. Although it might be considered advisable to refer all of these children for a complete hematologic workup to accurately establish the cause of their low hemoglobin or hematocrit, the large numbers involved preclude such a triage and in many areas it is necessary that paramedical personnel initiate certain forms of primary therapy.

Information submitted by the individual states to the CDC Nutrition Surveillance Program is periodically summarized and made available through the publication of a quarterly Nutrition Surveillance Bulletin. In this bulletin the nutrition indices are summarized by state, sex, age, and ethnic group. Although it is inappropriate to make statistical comparisons between various states or other groupings because of the lack of valid sampling, it is nevertheless useful to identify major geographic and demographic areas of need. Examples of these tabulations are presented in Tables VI, VII, and VIII. It is evident from Table VI that there is no major problem in regard to low weight for height in any of the areas submitting data. The overall prevalence of 4.9% of children having weight for height less than the 5th percentile is the same as the percentage expected if these children were similar to those in the reference population (NCHS Growth Charts, 1976). On the other hand, in all the states submitting data, there are twice as many children having high weight for height than would be expected on the basis of the reference population. Similarly there is a 5% excess of children who are short for their age indicating a continuing problem of growth retardation, most likely due to chronic mild undernutrition.

Tables VII and VIII summarize the data submitted to the CDC data bank by age, sex, and ethnic group. It is clear from these tables that there are major differences among ethnic groups and among various age groups. These data are being used to identify those groups in need of a high level of services and when used at the state and local level these data have had a major influence on the allocation of resources.

The problem in regard to probable iron deficiency anemia is well documented in all of the areas reporting. There is great variability in prevalence from area to area. It is also clear that the use of hematocrit and hemoglobin as initial screening tools identifies different percentages of children in the same area. This finding prompted us to investigate further the relationship between hemoglobin and hematocrit. The results of this investigation show that if hemoglobin is considered to be a true measure of the presence of probable anemia, the overall sensitivity and specificity of hematocrit in identifying these potentially abnormal children was not very good. Because the possibility was initially considered that the results based on the two different measurements were due to technical error, the investigation of this problem was conducted with data from the HANES I survey. The quality control of these data was such that the amount of technical error was minimal. Table IX shows the sensitivity, specificity, and the percentage of false negative and false positive identifications made on the bases of hematocrit levels less than 31% in identifying children who had hemoglobin levels less than 10 g/100 ml. Although the specificity is very high with a very small percentage of false positives, the sensitivity is only 33% with over 66% false negatives. In this population the overall prevalence of low hemoglobin values was 11.7% whereas the prevalence of low hematocrit values was only 4.2%. The predictive value of a positive screening hematocrit test was 91.7%. Results of these investigations in older age groups revealed a reverse situation in which the prevalence of low values was higher with hematocrit than with hemoglobin.

In addition to the summary tabulations made available to the states and presented in the Nutrition Surveillance Bulletins, periodic summaries of the complete distributions of the nutrition indices are made available. In many states it is necessary that data be summarized and evaluated on the basis of program funding source and administrative responsibility. The CDC Nutrition

Table VI. Nutrition Indices by State, July–September 1975: Persons Less Than 18 Years of Age[a]

State	Hemoglobin		Hematocrit		Height for age		Weight for age		Weight for height		
	Number examined	% low	Number examined	% low	Number examined	% low	Number examined	% low	Number examined	% low	% high
Arizona	1,590	13.5	1,186	10.8	2,603	15.2	2,609	9.8	2,562	6.4	9.6
Kentucky	515	13.4	426	15.0	1,303	11.1	1,317	7.4	1,270	5.6	8.3
Louisiana	2,840	20.3	1,085	20.8	4,134	11.3	4,200	7.5	4,088	5.9	8.3
Tennessee	425	6.1	6,351	21.3	7,496	10.0	7,560	6.0	6,376	4.8	9.5
Washington	245	10.2	2,367	9.9	3,400	10.2	3,391	5.7	3,359	3.1	10.3
Total	5,615	16.2	11,415	17.6	18,936	11.1	19,077	6.9	18,655	5.0	9.3

[a] Totals for Hgb and Hct include unknown sex. Incomplete reporting for Kentucky. Louisiana November data not included (not received).

Table VII. Nutrition Indices by Sex and Ethnic Group, July–September 1975: Persons Less Than 18 Years of Age

Sex and ethnic group	Hemoglobin		Hematocrit		Height for age		Weight for age		Weight for height		
	Number examined	% low	Number examined	% low	Number examined	% low	Number examined	% low	Number examined	% low	% high
Male											
Black	1,456	18.6	982	21.1	2,579	11.6	2,601	8.0	2,539	5.7	8.2
White	807	10.5	3,833	19.0	5,399	11.1	5,433	6.8	5,334	4.9	9.7
Sp. American	306	12.7	315	12.1	662	14.8	663	9.5	654	5.8	11.6
Am. Indian	158	16.5	195	11.3	412	12.4	410	7.8	406	5.9	12.3
Oriental	1	0.0	34	2.9	41	14.6	41	2.4	41	2.4	4.9
Other	29	31.0	31	19.4	72	16.7	72	16.7	71	9.9	7.0
Unknown	31	12.9	145	11.7	174	10.9	172	2.9	170	5.9	7.6
Total	2,788	15.6	5,535	18.4	9,339	11.6	9,392	7.3	9,215	5.3	9.4
Female											
Black	1,376	20.3	1,010	21.7	2,543	10.0	2,577	5.8	2,491	5.6	9.6
White	857	11.9	4,031	16.6	5,631	10.4	5,682	6.6	5,542	4.6	8.8
Sp. American	350	14.9	353	10.5	688	13.7	689	7.1	678	3.8	10.8
Am. Indian	174	14.9	214	10.7	443	11.5	442	5.4	439	3.4	12.1
Oriental	2	50.0	24	8.3	37	10.8	37	13.5	36	2.8	5.6
Other	26	42.3	51	17.6	79	13.9	81	6.2	79	6.3	10.1
Unknown	31	12.9	133	9.8	176	13.1	177	9.6	175	1.7	4.6
Total	2,816	16.9	5,816	16.7	9,597	10.6	9,685	6.5	9,440	4.7	9.2

Table VIII. Nutrition Indices by Sex and Age, July–September 1975: Persons Less Than 18 Years of Age

Sex and age group	Hemoglobin		Hematocrit		Height for age		Weight for age		Weight for height		
	Number examined	% low	Number examined	% low	Number examined	% low	Number examined	% low	Number examined	% low	% high
Male											
<1	289	11.1	603	9.6	2,861	13.6	2,901	9.0	2,784	6.3	11.0
1	388	12.6	812	9.9	1,047	14.3	1,063	7.2	1,041	6.5	14.4
2–5	853	15.5	2,246	19.5	2,679	10.8	2,686	5.1	2,666	4.0	9.5
6–9	495	19.2	872	31.2	1,142	6.9	1,137	5.5	1,136	5.7	5.7
10–12	316	18.7	436	21.3	683	10.4	679	9.3	677	5.0	6.1
13–17	447	15.0	566	13.8	927	10.9	926	9.7	911	4.5	6.3
Total	2,788	15.6	5,535	18.4	9,339	11.6	9,392	7.3	9,215	5.3	9.4
Female											
<1	271	12.2	590	9.5	2,862	12.4	2,899	6.6	2,770	4.5	12.1
1	384	18.0	864	9.8	1,077	16.0	1,091	8.3	1,069	5.7	13.4
2–5	897	16.3	2,299	16.7	2,753	10.2	2,777	5.5	2,746	3.6	7.2
6–9	461	19.1	898	28.4	1,163	5.2	1,163	4.9	1,160	6.5	5.2
10–12	319	16.0	551	19.4	763	7.9	767	6.8	754	7.4	5.6
13–17	484	18.4	614	14.0	979	9.7	988	8.3	941	3.3	9.9
Total	2,816	16.9	5,816	16.7	9,597	10.6	9,685	6.5	9,440	4.7	9.2

Table IX. Sensitivity and Specificity of Hematocrit Levels <31% in Identifying Children (12–24 Months, HANES I) with Hemoglobin Levels <10 g/100 ml

	Hemoglobin		
Hematocrit	<10 g/100 ml	10 g/100 ml+	Total
<31%	11	1	12
31%+	22	248	270
Total	33	249	282

Sensitivity: 11/33 = 33.3% Prevalence of low hemoglobin: 33/282 = 11.7%
Specificity: 248/249 = 99.6% Predictive value of positive test: 11/12 = 91.7%
False negative: 22/33 = 66.7%
False positive: 1/249 = 0.4%

Surveillance Program provides the opportunity for states to obtain such summaries on a yearly basis.

Because of concern about the problem of overweight which was identified in all the reporting areas a more detailed analysis was made of the data utilizing the cross-tabulation of height for age and weight for height recommended by Waterlow and Rutishauser (1974). In this type of analysis height for age is considered to be a reflection of long-term nutritional status while weight for height is reflective of short- or medium-term changes in nutritional intake. Table X presents data from a group of low income American children. It is evident that in this population there are a large number of children who are

Table X. Distribution of Children 12–59 Months of Age at Initial Visits According to Classifications of Nutritional Status Based on Length or Stature for Age and Weight for Length or Stature: Nutrition Surveillance 1974–1976

Weight for length or stature, Z score	Length or stature for age Z score			
	1.00–3.99	−1.99 to 0.99	−3.99 to −2.00	Class total
2.00–3.99				
Number of cases	715	2,796	591	4,102
Percent of total cases	1.36	5.31	1.12	7.79
−1.99–1.99				
Number of cases	7,013	36,564	3,806	47,383
Percent of total cases	13.32	69.45	7.23	90.00
−3.99 to −2.00				
Number of cases	251	783	126	1,160
Percent of total cases	0.48	1.49	0.24	2.20
All classes				
Number of cases	7,979	49,143	4,523	52,645
Percent of total cases	15.16	76.25	8.59	100.00

Total children considered for tabulation: 56,776 Children excluded from tabulation: 4,131 overweight. A closer look at this presentation reveals that although 7.8% of the children in the total population are overweight, in the population of short children, those children having length or stature Z scores between -2.00 and -3.99, 13.0% of the children are overweight. These data, which have been reproduced in many areas and in many population groups, have been the basis for the recommendation that the amount and types of food in the food packages provided to children by various federal and local feeding programs be adjusted on the basis of their initial anthropometric values as well as changes in these values occurring over time.

With appropriate identification, nutrition surveillance data available at the local level can be used to evaluate changes occurring in individuals who are placed in intervention programs designed to improve their nutritional status. A recent attempt at such a use of surveillance data was made by the Center for Disease Control using selected information from WIC participants in four states. The data evaluated were from children who had at least three linked records at 6-month intervals during their participation in the program. Table XI lists the criteria used for the definition of change in hemoglobin values over the three visits and Table XII summarizes these changes in two age groups. It can be readily seen that the percentage of children with increases over the period of follow-up is three times that of children with decreases. In both age groups approximately 50% of the children had no change over the 1-year period. On the basis of this type of analysis, the conclusion can be drawn that participation in the WIC program appears to benefit the children involved by raising or preventing a decrease in their hemoglobin levels. Although it is clear that this conclusion cannot be firmly supported on a statistical basis, it is nonetheless the best evidence that can be obtained utilizing data from a program that is attempting to deliver improved nutrition services to an at-risk population.

Table XI. Definitions of Change in Hemoglobin over Three Visits for Children, Age 12-24 Months: Nutrition Surveillance 1975-1976[a]

	First follow-up	Second follow-up
Increase	Increase	Increase
	Increase	No change
	No change	Increase
No change	Increase	Decrease
	No change	No change
	Decrease	Increase
Decrease	Decrease	Decrease
	Decrease	No change
	No change	Decrease

[a] For any one visit, the following magnitudes represent no change: hemoglobin, ± 0.2 g/100 ml; hematocrit, $\pm 1\%$.

Table XII. Distribution of Changes in Hemoglobin Values
over Three Visits for Children, Age 12–24 Months:
Nutrition Surveillance 1974–1976[a]

	Age			
	6–23 mo		24–47 mo	
	Number	%	Number	%
Increase	169	38	93	36
No change	229	50	137	52
Decrease	52	12	31	12
Total	450	100	260	100

[a] See Table V for definition of change in hemoglobin.

4. Summary

Nutrition surveillance is the continuous monitoring of the nutritional status as well as associated economic, demographic, and food-related variables in population groups. The goals of nutrition surveillance are to enhance the health of a population by identifying and preventing the development of malnutrition. Implementation of a nutrition surveillance program requires the development of definitions of normal, optimal, and adverse nutritional status as well as a means for collecting, analyzing, utilizing, and disseminating summaries of the collected data. First and foremost, the summaries of the collected data should be made rapidly available to the originating source in a form that is useful to them for program planning, evaluation, and management. The data available from nutrition surveillance programs can also be used to identify problem areas in regard to the quality control of measurement and for the development of the programs to improve the quality of the data.

ACKNOWLEDGMENT. We would like to thank and acknowledge the help and support of the Office of the Center Director, CDC, and the staff of the Nutrition Division in making the development of the Nutrition Surveillance System possible. Particular thanks are due to Mr. James Goldsby, Statistician, and his staff for their unstinting efforts, long hours, and good humor during the continuing emergencies associated with this effort. Finally and most importantly thanks must go to all the many individuals at the state and local level who have provided the data, without which there would be no surveillance system.

5. References

Berry, W. T. C., 1972, Nutrition surveillance in affluent nations, *Nutr. Rev.* **80:**127–131.
Health and Nutrition Examination Survey, 1976, Division of Health Examination Statistics, NCHS, DHEW Publication No. (HRA) 76-1310.

Irwig, L. M., 1976, Surveillance in developed countries with particular reference to child growth, *Int. J. Epidemiol.* 5:57–61.

NCHS Growth Charts, 1976, Monthly Vital Statistics Report, HRA Publication 76-1120, 25: No. 3 Supplement, June 22, 1976.

Nichaman, M. Z., 1974, Developing a nutrition surveillance system, *J. Am. Diet. Assoc.* 65:15–17.

Nutrition Canada National Survey, 1973, Department of National Health and Welfare.

Owen, G. M., Kram, K. M., Garry, P. J., Lowe, J. E., and Lubin, A. H., 1974, A study of nutritional status of preschool children in the United States, 1968–1970, *Pediatrics Suppl.* 53(4):Part 11, 597–646.

Ten-State Nutrition Survey, 1972, U.S. Department of Health, Education and Welfare, DHEW Publication No. (HSM) 72-8130, -8131, -8132, -8133, and -8134.

Waterlow, J. C., and Rutishauser, H. C., 1974, *Malnutrition in Man in Early Malnutrition and Mental Development* (J. Cravioto, L. Hambraeus, and B. Vahlquist, eds.), Almquist & Wiksell, Stockholm.

World Health Organization, 1976, Methodology of Nutrition Surveillance, Technical Report Series No. 593.

Nutrition Surveillance in Developing Countries, with Special Reference to Ethiopia

Mehari Gebre-Medhin

The nutritional status of population groups has been a focus of considerable interest among public health and nutrition workers for the last two decades and a number of procedures have been developed for its appraisal (Joint FAO/ WHO Expert Committee, 1963; ICNND, 1963). The publication of a WHO monograph by Jelliffe (1966) on this topic, with particular emphasis on sub-tropical and tropical areas, stimulated intensive survey activities in many Third World countries. The reports from these surveys showed that the nutrition of individuals and communities was of wide public health and socioeconomic importance. As a result of these observations many national and international organizations and technical agencies initiated applied nutrition programs, especially among infants and children.

These community nutrition surveys were usually of a cross-sectional character. Their purpose was to determine the nature and extent, and in some instances the cause, of deficiency states, to screen population groups for individual or group health care, and to establish baseline data. The assessment was traditionally carried out in the context of public health programs.

Recently, the appraisal of the state of nutrition and health of population groups has again become a matter of great concern to governments and international technical agencies. During the 1970s severe famine due to drought has ravaged many countries and caused the death of thousands of people residing in distant, isolated hamlets (de Montvalon, 1974; Gebre-Medhin and Vahlquist, 1976). In response to this, concerted efforts are currently being made to establish new approaches which, on the one hand, would monitor the determinants of food availability and distribution and, on the other, assess the effect of the food situation on the health and nutritional status of the population.

Mehari Gebre-Medhin ● Institute of Nutrition, University of Uppsala, Uppsala, Sweden.

Relevant information should be collected together as a consolidated report to give an overview of the region's food supply and nutritional situation on a regular basis. The urgency of establishing such a system was underlined by the World Food Conference in Rome in 1974, and this subject was treated exhaustively by a Joint FAO/UNICEF/WHO Expert Committee in Geneva in 1976 (FAO/UNICEF/WHO, 1976). It is strongly recommended that the interested reader study the report of the latter expert committee—Methodology of Nutritional Surveillance—which deals with the nature of a nutritional surveillance system, methods for setting up such a system, and some of the principles in its operation. It is now generally agreed that the preexistence of such a predictive surveillance system (Blix *et al.*, 1971; Rohde, 1976) would have greatly reduced human suffering and loss of the magnitude recorded in the recent disastrous food shortages, for example in the Sahel region of Africa (Berthet, 1975).

1. Principles of a Nutritional Surveillance System for Developing Countries

The subject of nutrition encompasses technological areas that are not isolated but are closely related, so that great value lies in a broad interdisciplinary approach. A change of the nutritional status of a community may follow upon a chain of events ranging across the technological fields of meteorology, agriculture, public health/nutrition, economics, and sociology, including traditional patterns and life-styles. An efficient nutritional surveillance system requires close collaboration and consultation in the planning of the operation and in the analysis of the resulting data.

In many developing countries, data collection related to food and nutrition has been carried out for a number of years by various government agencies within their mandated areas of competence. In establishing a coordinated surveillance system it is necessary to build on existing (and future) government agencies, to meet the needs of these agencies, and to strengthen their hands in the execution of their responsibilities.

Available resources for the establishment of various types of infrastructure (extension agencies in agriculture, public health, education, community development, etc.) are totally inadequate in most developing countries. The limited funds cannot be diverted to costly investigations, especially if these are carried out in isolation from existing programs. It is therefore important to keep the costs of a surveillance system at a minimal level, without compromising the reliability of resulting data. This is best achieved by carefully utilizing already existing information and by involving relevant rural institutions and field staff. The planning and operation of the system should be characterized by simplicity. The common wisdom and experience of rural people who are acquainted with the production of food should be tapped.

2. Contents of the System

The formulation of a surveillance program will have to take into account the prevailing food supply system in a given area. In most developing countries the majority of the population derive their subsistence from settled farming. Varying proportions are pastoralists who roam over wide areas in search of pasture land and water holes. They live on their cattle or barter these in exchange for grain. With increasing urbanization steadily increasing numbers are joining the wage-earning group who are dependent on others to produce their food.

The surveillance system should ensure a regular flow of quantitative, reliable information regarding the following aspects:

Climate	Timing, duration and adequacy (or excess) of rainfall
Crops	The pattern of the crop calendar, the amount of crop produced, the presence of crop pests
Livestock	Pasture conditions, herd size, production, and condition
Water	Availability and adequacy for animals and humans
Market	Prices and flows of major food commodities and livestock; migration patterns
Nutrition	Food stocks held by families, dietary intake, and the state of nutrition
Health and vital statistics	Crude death rates, infant and child mortality rates, birthrates, disease prevalence

Out of these variables, those most indicative of the food supply situation and of the consequences of that supply on nutrition and health will be chosen. The indicators, collected by the relevant agencies, will be centrally amalgamated to give a regular consolidated report from which areas vulnerable to or suffering from food shortage can be identified.

3. Operation of the System

3.1. Reporting Stations

Whenever possible the active involvement of existing stations belonging to the different national agencies should be made a matter of primary importance. However, since in most instances the infrastructural sectors of a given technical agency are developed mainly to meet the needs of a particular program, additional stations may have to be erected for the purpose of coordinated surveillance.

The reporting stations should be situated so as to provide the basis of a national framework with sufficient density to provide adequate coverage of representative areas. Particular emphasis will have to be placed on areas traditionally vulnerable to episodes of food shortage.

3.2. Personnel

Reporting stations should be staffed, wherever possible, by existing field personnel of the particular agency responsible for an aspect of the consolidated data collection system. Where some personnel expansion is called for, the possibilities of utilizing other existing agencies should be explored before new cadres are recruited specifically for the purpose of surveillance.

The school system represents a relatively well-developed rural institution in many developing countries. Teachers and students could therefore play a major role in surveillance systems. Apart from rendering valuable assistance by virtue of their education, they could provide opportunities for creating awareness among the community of problems related to food and nutrition.

3.3. Frequency of Surveillance

The frequency of surveillance will depend on a variety of factors, including the vulnerability of the area and the agricultural calendar, i.e., the period of the year when harvest occurs and food stocks are generally replenished or are at their lowest. Figure 1 depicts the periodicity of surveillance in a northeastern Ethiopian province, a predominantly arable highland area with two harvests in the year and also having a pastoral population in the lowlands.

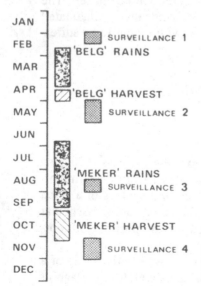

Fig. 1. Schematic presentation of periodicity of surveillance for a predominantly agricultural area in northern Ethiopia.

3.4. Data Processing

Efficient data processing for transforming periodic flows of information from widely dispersed reporting stations into manageable reports is an important requirement of a surveillance system. All data which will go to make up consolidated reports should be processed centrally. In this way cross-disciplinary analysis and interpretation will be facilitated. Up to a certain level of data flow, hand processing is an inexpensive method which can be made efficient with good organization. Although computer processing allows rapid calculation of data, there is a certain critical level of data flow below which the use of a computer is not justified. Furthermore, the setting up of a processing system is time-consuming and expensive.

3.5. Administration

Many developing countries have a ministry or a unit responsible for the overall planning of development programs, including bilateral and international cooperation. In order to facilitate national and international action in times of food crisis, the central organization of the consolidated surveillance system should be placed close to the planning machinery of the government.

4. Preliminary Experience in Nutritional Surveillance in a Developing Country—The Recent Ethiopian Famine

The history of Ethiopia is replete with accounts of disastrous food shortages (Pankhurst, 1972). Recent dietary and nutritional status surveys (Selinus *et al.*, 1971; Miller *et al.*, 1976) conducted in different communities have substantiated the existence of major food shortage situations, often outright hunger, especially during the preharvest season when food stocks traditionally reach their lowest levels. The country has a well-documented proneness to recurrent famine due to a combination of climatic, demographic, social, and political factors (Pankhurst, 1966; Cliffe, 1974).

A number of reports have dealt with the recent prolonged and severe drought which has hit larger parts of Ethiopia as well as the so-called Sahel region of Africa (Shepherd, 1975; Mason *et al.*, 1974; Miller and Holt, 1975). It is belived that in Ethiopia alone the resultant famine has caused the death of several hundred thousand people during the last few years (Shepherd, 1975). A national Relief and Rehabilitation Commission with a wide mandate has been established in the capital to coordinate national and international efforts in this field. In the past various government agencies in Ethiopia have been active in the collection of limited data related to food and nutrition as a basis for planning within their technological area of competence. However, there has been no coordinated and continuous flow of information documenting the dynamics of a changing food supply situation and nutritional status, or satisfying the planning needs of a development strategy.

   The recent drought emphasized the need for such coordination when the Relief and Rehabilitation Commission called for information on which to base its estimates of required assistance. An Interministerial Technical Working Group composed of a number of relevant national agencies was convened in order to examine the technical aspects of "a system for the regular flow of action-oriented information related to food and nutrition from drought-affected as well as other potentially vulnerable areas" (ENI, 1974). The objective of the system was to "establish a national monitor of food supply and nutritional status and early warning of deterioration as the result of climatic or other changes, mediated by a national data bank on food and nutrition matters" (ENI, 1974).

   The first systematic data collection in Ethiopia was carried out in the eastern lowlands of the country in the Ogaden area—a vast desert-like land which spreads on both sides of the Ethio-Somali border (Holt *et al.*, 1975a,b; Gebre-Medhin *et al.*, 1977). The majority of the Ogaden people are nomads and communication and other infrastructural services in the area are rudimentary. Figure 2 depicts a simplified model of the food supply system. The population roams over this vast area in search of pasture land for their cattle.

   In this area serious disturbance of the rainfall pattern occurred during 1973–1975, resulting in a severe shortage of water and inadequate pasture. In the early part of 1974 news of increased human and animal mortality reached Addis Ababa. Toward the end of 1974 the situation had deteriorated seriously, calling for immediate massive government action. Aid programs were established whereby nomads were able to stay in provisional shelters and obtain medical care (Fig. 3). The pressing involvement of the government dictated the need for the Technical Working Group to carry out surveys in order to predict any deterioration in the area and to assess the impact of ongoing assistance. For this purpose, rainfall data, estimates of livestock holdings, and prices for major grains and those fetched by animals were recorded. Data on the nutritional status of children and age-specific mortality rates were also obtained.

   Findings related to rainfall obtained from two meteorological stations in the area are shown in Fig. 4. In the period January to June 1973 and throughout 1974 there was both delay in the onset of the rains and a marked reduction in

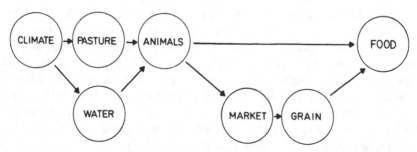

Fig. 2. Simplified model of food supply system of pastoralists in southeastern Ethiopia.

Fig. 3. Ogaden women pounding grain in a relief shelter.

the total amount of rainfall. In 1975 a delay of 1–2 months occurred, although the same maximum levels as in 1957–1970 were reached.

The situation regarding livestock holdings is shown in Fig. 5. A drastic fall is seen for practically all animals from June 1973 to June 1975. For the entire period May–June 1974 to June 1975, the average holding size per family

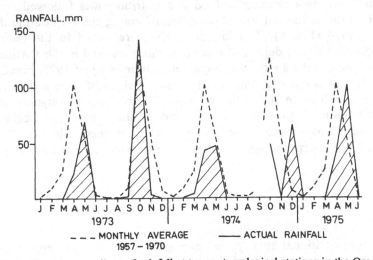

Fig. 4. Monthly recordings of rainfall at two meteorological stations in the Ogaden.

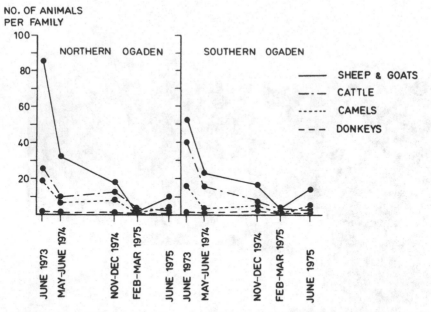

Fig. 5. Average livestock holdings per family by type of animal in the Ogaden.

was below 7.5 standard stock units (SSU), which is regarded as the minimum requirement for survival by a family. One SSU is equivalent to 500 kg live weight. Grain prices were far in excess of the predrought levels. This was mainly the result of the partial failure of the crops in the highland parts bordering this area. The data for animal prices generally reflect distress sales of cattle due to the steadily deteriorating pasture situation.

The general situation relating to food and nutrition was followed by a marked deterioration in the nutritional condition of young children during the second half of 1974 (Fig. 6). The lowest levels were noted in the period February to March 1975. A degree of recovery, more marked in the northern part of the area than in the south, was apparent in June–August 1975. Finally, the mortality figures for young children, the most vulnerable age group, remained excessively high in the entire drought-affected area throughout the period of observation. Both in the northern and southern parts, particularly the northern, there was a steady deterioration during the second half of 1974. A similar mortality pattern, though less marked, was observed in the age group 5–15 years.

5. Comments

Carrying out nutritional surveys in developing countries presents many problems. Such surveys are particularly difficult to conduct among pastoral nomads with changing migration patterns. Add to this a situation of disastrous food shortage leading to the disruption of established migration patterns, and the difficulties assume almost insurmountable proportions. In spite of this, the experience in Ethiopia has clearly shown that it is feasible to achieve a relatively extensive collection of practically useful data even under situations of great strain such as famine.

Accurate quantitative data related to rainfall are difficult to obtain in most developing countries. Meteorological stations are extremely sparse and are usually located close to towns and in geographic areas often not representative of sections that are drought-prone. Ethiopia is no exception in this respect. Nevertheless, in all its limitations the present preliminary survey has shown that the inadequacy and delay in rainfall was a major contributory factor to the food shortage problems and that periodic assessment would have given timely warning of impending disaster.

Accurate appraisal of livestock holdings in traditional societies is complicated by a number of sources of bias. Pastoralists and even settled farmers may be reluctant to report the size of their cattle stock accurately. Some tribes practice split-herd grazing so that only part of the herd is counted during a surveillance program. However, despite these sources of error, the present surveillance has shown that changes in animal holdings with time reflect the effect of pasture deterioration due to drought, although the contribution of other factors such as off-take for marketing cannot be entirely discounted. Further, by indicating average livestock holdings per family that were far below the minimum required for survival, the surveillance has pointed not

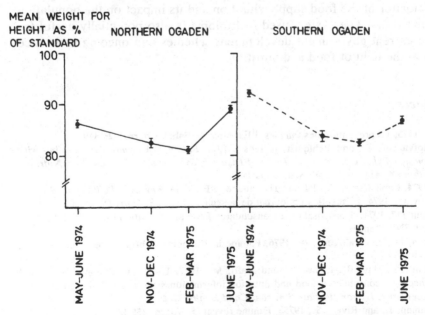

Fig. 6. Nutritional status of young children in drought-affected Ogaden. Mean values and standard error are indicated.

only to the urgency of immediate relief action but also to the need for long-term rehabilitation among a population which for a long time to come cannot expect to regain a herd size sufficiently large to support it.

The nutritional status of the population and age-specific mortality rates constitute the final links in a chain of indicators used in a surveillance system. The use of anthropometric measures in this context has been discussed in vairous publications (WHO, 1975; PAG, 1976). In the Ethiopian system weight for height was chosen as the sole indicator of nutritional status. This ratio offers a number of advantages: It is easy to obtain, is independent of age, and expresses changes in body proportions which reflect the character of nutritional deprivation. A significant deterioration in nutritional status observed in the second half of 1974 is probably accounted for by the extreme food shortage resulting from the drought superimposed on the normal long dry period between the main rains. The recovery observed in June–August 1975 could be an expression of the return of the April–May rains in combination with government relief operations.

The markedly increased excess mortality rates observed in young children are in agreement with scenes of devastating famine in a number of regions of the world in the 1970s. These figures are much higher than can be attributed to deaths which characterize "normal" times in many developing countries, including Ethiopia.

Our initial experiences have taught us that food shortage situations should no longer be regarded as inevitable events to be dealt with on an *ad hoc* basis. We have learned that it is possible to launch nutritional surveillance systems even in areas with an underdeveloped infrastructure, which would give a valuable monitor of the food supply situation and its impact on the population on a regular basis. The system used in Ethiopia is also potentially useful for evaluating current government development schemes and ongoing assistance programs in the field of food and nutrition.

6. References

Berthet, E., 1975, La prevention des famines: l'Example du Sahel, *Courrier* 25:349.

Blix, G., Hofvander, Y., and Vahlquist, B. (eds.), 1971, *Famine—A Symposium Dealing with Nutrition and Relief Operations in Times of Disaster, Symposium of the Swedish Nutrition Foundation IX*, Almqvist & Wiksell, Uppsala.

Cliffe, C., 1974, Capitalism or feudalism? The famine in Ethiopia, *Rev. Afr. Polit. Econ.* 1:34.

de Montvalon, R., 1974, Eléments d'un journal de la sécheresse, *Cah. Nutr. Diet.* 9:104.

ENI document 117, 1974, A proposal for a Consolidated Food and Nutrition Information System, RRC and ENI, Stencile.

Gebre-Medhin, M., and Vahlquist, B., 1976, Famine in Ethiopia—A brief review, *Am. J. Clin. Nutr.* 29:1016.

Gebre-Medhin, M., Hay, R., Licke, Y., and Maffi, M., 1977, Famine in Ethiopia. III. Initial experience of a consolidated food and nutrition information system. Analysis of data from the Ogaden area, *J. Trop. Pediatr. Environ. Child Health* 22:29.

Holt, J., Seaman, J., and Rivers, J., 1975a, Famine revisited, *Nature* 255:180.

Holt, J., Seaman, J., and Rivers, J., 1975b, The Ethiopian famine of 1973–74. 2. Harerge province, *Proc. Nutr. Soc.* 34:115A.

ICNND, 1963, *Manual for Nutrition Surveys,* 2nd ed., National Institutes of Health, Bethesda, Maryland.

Jelliffe, D. B., 1966, Assessment of the nutritional status of the community, WHO Monograph, Geneva.

Joint FAO/WHO Expert Committee, 1963, Expert Committee on Medical Assessment of Nutritional Status, FAO/WHO Technical Report Series No. 258, Geneva.

Joint AFO/UNICEF/WHO Expert Committee, 1976, Methodology of Nutritional Surveillance, WHO Technical Report Series No. 593, Geneva.

Mason, J. B., Hay, R., Holt, J., Seaman, J., and Bowden, M. R., 1974, Nutritional lessons from the Ethiopian drought, *Nature* **248**:646.

Miller, D. S., and Holt, J., 1975, The Ethiopian famine, *Proc. Nutr. Soc.* **34**:167.

Miller, D. S., Baker, J., Bowden, M., Evans, E., Holt, J., Mckeag, R. J., Meinertzhagen, I., Munford, P. M., Oddy, D. J., Rivers, J. P. W. R., Sevenhuysen, G., Stock, M. J., Watts, M., Kebede, A., Wolde-Gebriel, Y., and Wolde-Gebriel, Z., 1976, The Ethiopia applied nutrition project, *Proc. Roy. Soc. London B* **194**:23.

PAG, 1976, Anthropometry in nutritional surveillance: An overview, *PAG Bull.* **4**(2):12.

Pankhurst, R., 1966, Some factors depressing the standard of living of peasants in traditional Ethiopia, *J. Ethiop. Stud.* **4**:45.

Pankhurst, R., 1972, The history of famine and pestilence in Ethiopia prior to the founding of Gondar, *J. Ethiop. Stud.* **10**:37.

Rohde, J. E., 1976, The world food crisis, *Am. J. Clin. Nutr.* **29**:2.

Selinus, R., Awalom, G., and Gobezie, A., 1971, Dietary studies in Ethiopia. II. Dietary pattern in two rural communities in N. Ethiopia. A study with special attention to the situation in young children, *Acta Soc. Med. Upsalla* **76**:17.

Shepherd, J., 1975, *The Politics of Starvation,* Carnegie Endowment for International Peace, New York—Washington.

WHO document NUT/EC/75.6, 1975, Anthropometry in Nutritional Surveillance: A Review Based on Results of the WHO Collaborative Study on Nutritional Anthropometry, Nutrition Unit, WHO, Geneva.

Index

Abortions, spontaneous, 110
Absorptiometry, direct photon, 284–285
ACTH production, excessive, 72
Activity, energy requirements for, 9
Adolescence (see also Adolescent)
  growth in, 239–241
  hemoglobin concentration in, 288
  obesity in, 234, 249–251
  undernutrition in, 250
  weight gain in, 239
Adolescent (see also Adolescence), 239–251
  anemia in, 246–247
  energy needs of, 241–242
  megaloblastic anemia in, 247–248
  nutrient requirements for, 239
  pregnancy and, 248–249
  protein needs of, 242–244
  vitamin and mineral needs of, 244–246
Adopted children, resemblances to adoptive
  parents in, 39, 43
Adrenogenital syndrome, 71
Adult(s)
  calorie requirements for, 266
  nutrition effect in, 263–264
  optimizing food intake in, 285–287
Adulthood, 253–268
  interpretation of, 255
  living conditions in, 261–263
  nature of, 253–257
  nutrition and ecosensitivity in, 257–260
  nutritional requirements in, 265–267
  nutritional status in, 258–260, 267–268
  optimizing hematologic determinations in,
    287–289
  regressive changes in, 261–263
  serum and urinary vitamins in, 289–291
  sugar intake in, 264, 266
AGA, see Appropriate gestational age
Agricultural knowledge, growth and, 59
Alimento foods, 54

All Indian Council of Medical Research, 303
Amenorrhea, postpartum, see Postpartum
  amenorrhea
Amino acid disorders, in mental retardation, 68
Amino acid metabolism disorders, growth and,
  67–68
Amino acids
  essential, 12–13
  in intravenous feeding, 144
Anemia
  in adolescence, 246–247
  hypochromic, 187
  iron-deficiency, 23, 247
  megaloblastic, 247–248
Anorexia, infections and, 175
Anovulatory cycles, 109
Anthropometric field methods, 339–362,
  365–385, 289–297
  constraints in, 344–347
  goals and objectives in, 343
  height and length in, 351–353
  interrelationships among, 374–375
  measurement in, 356–362
  precision in, 370–373
  quality control of measurements in, 360–362
  response specificity in, 380–381
  scales used in, 350–351
  scope of, 343–344
  selection criteria in, 365–385
  and sensitivity in populations, 370–380
Anthropometric measures, 348–356
  instruments used in, 349–351
Anthropometry
  applications of, 347
  costs in, 346–347
  nutritional status and, 339–341
  survey duration in, 345
  surveys, screenings, and surveillance in,
    341–342
Antithyroid drugs, breast feeding and, 131

Appropriate gestational age, 155
Arm-for-height measurements (*see also* QUAC
    stick), 391
Arm circumference
    optimal, 277
    protein–calorie malnutrition and, 319–320
    reference standards for, 311–312
Arm muscle area, vs. fat area and arm fat per-
    centage, 281
Ascaris lumbricoides, 50
Ascorbic acid, 17, 22
    in adolescence, 244–245
Atole villages, in food studies, 84–112
Azotemia, 147

Bacterial infections, growth rate and (*see also*
    Infections), 48–49
Birth, multiple, *see* Multiple birth
Birth interval
    growth and, 60
    menstruating interval and, 108–110
Birth interval components, maternal nutrition
    and, 103–113
Birth weight
    dose–response relationships and, 95–99
    vs. food supplementation, 87–92
    low, *see* Low-birth-weight babies
    maternal nutrition and, 80
    nutritional interventions and, 95–99
    protein vs. calories in, 92–94
Blood, transport disorders of, 67
Blood pressure, during pregnancy, 113–115
Body composition, of young child, 156–158
Body length
    vs. gross weight, 274–275
    measurement of, 357
Body size
    congenital heart disease and, 45
    increase in, 35
Breast feeding
    advantages of, 129–130
    vs. artificial formulas, 8
    decline of, 55–56
    growth chart and, 333–334
    obesity and, 223
    technique of, 130–131
Breast milk (*see also* Milk)
    antiinfective properties of, 50–51
    ascorbic acid intake from, 22
    vs. bottle feeding, 8, 52
    composition of, 8, 132
    content of, 130
    iron in, 131–132
    iron intake from, 23
    vs. milk substitutes, 8, 53

Breast milk (*cont'd*)
    for newborn, 129
    for premature infants, 149–150
    supply of, 129
    vitamin B$_{12}$ in, 21
    vitamin intake from, 16, 19, 21
Breast milk banks, 150

Calcium, need for, 24
Calf circumference, optimal, 277
Caloric intake, by male infants, 136–138
Calories, birth weight and, 92–94
Carbohydrate metabolism disorders,
    nutrition and, 68–69
Carbohydrates, vs. low-fat diet, 15
β-Carotene (*see also* Vitamin A), 18
Celiac disease, growth and, 66
Center for Disease Control Nutrition
    Surveillance System, 416–429
    feedback in, 416–417
Central nervous system disease, growth and, 72
Cereals
    NPU of diets based on, 13
    RDA for protein and, 13
Childhood (*see also* Infant; Young child)
    energy requirements for, 10
    food practices in, 56–58
    obesity prevention in, 253–254
    protein needs for, 12
    young, *see* Young child
Cholecalciferol, 18
Cholesterol, in nutrition surveillance, 413
CIE, *see* International Children's Centre
Classification of Infantile Malnutrition, 194
Clinic assessment (*see also* Growth chart),
    329–337
Congenital heart disease
    body size and, 45
    growth and, 66
Cow's milk, composition of (*see also* Breast
    milk; Milk), 132
Cross-sectional surveys, in growth studies,
    399–401
Cushing's syndrome, 72, 225
Cystic fibrosis, growth and, 67

DBP, *see* Diastolic blood pressure
7-Dehydrocholesterol, 18
Dependability, of nutritional status indicator,
    370–372
Detecto weight scales, 350
Developing countries nutrition surveillance,
    431–440
    principles and contents in, 432–433
    system operation in, 433–435

Diabetes mellitus, nutrition and, 68–69
Diarrhea
  infection and, 50–51
  malnutrition and, 50–51
  weanling, 51–52
Diastolic blood pressures, in pregnancy,
    113–114
Direct photon absorptiometry, 284–285
Disease, growth and, 48
Dizygotic twins, 32, 38–39
Dose/time responses, in nutritional interven-
    tions, 95–99
Drought, famine and, 435–438
Dwarfism, psychosocial, 71

Early childhood, food practices in (*see also*
    Childhood; Young child), 56–58
Ecological balance, growth and, 59
"Ecosensitive" measurements, vs. "heredosen-
    sitive," 2
Edema, weight-for-age relationship and,
    315–316
End-organ failure, 72
Energy, protein-sparing effects of, 13
Energy allowances, vs. energy needs, 9
Energy requirements
  of adolescent, 241–242
  for children, 10
  vs. energy allowances, 9
  for infants, 10
Enterocolitis, necrotizing, 149
Environmental deprivation, in failure-to-
    thrive syndrome, 175–176
Ergocalciferol, 18
Escherichia coli, 52
Essential amino acids, 12–13
Essential fatty acids, 14–15
Ethiopia
  famine in, 435–438
  nutrition surveillance in, 431–440
Exterogestate fetus, 3

"Failure-to-thrive" syndrome, 171–182
  causes of, 172–173
  environmental factors in, 175–176
  infections and, 175
  investigation in, 177–179
  nutritional factors in, 174–175
  presenting complaints in, 176–177
  prevention of, 180–181
  treatment and management of, 179–180
Family food flow, 208
Family structure, growth and, 59–60
Famine
  drought and, 431, 435–438

Famine (*cont'd*)
  in Ethiopia, 435–438
FAO/WHO Expert Committee on Protein
    Requirements (*see also* Joint FAO/WHO
    Expert Committee on Nutrition),
Fat, 14–15
Fatfold (skinfold)
  optimal use of, 278–280
  triceps, 359–360
Fatfold thickness, in anthropometric field
    methods, 353–356
Fatfold values, of American population by
    age and sex, 279
Fat-free weight, 36
  formula for, 278
Fatness, genetics of, 39–40
Fat-soluble vitamins, 16–19
Fatty acids, essential, 14
Fat weight, 36
  fatfolds and, 278–280
Fels Research Institute, 306
Fetal death, socioeconomic status and, 110
Fetal growth, maternal nutrition and, 81–99
FFW, *see* Fat-free weight
Fluoride, for newborn, 133–134
Folacin, in adolescence, 245
Folic acid, 17, 21
Food(s) (*see also* Nutrition)
  body-image, 54
  "hot" vs. "cold," 54
  meaning of, 53–54
  medicinal, 53
  prestige, 53–54
Food and Agriculture Organization, 185, 199
Food chain, protein–energy malnutrition and,
    205–207
Food energy, fat and, 14–15
Food flow, family vs. national, 208–209
Food intake
  in adult, 285–287
  in early childhood, 56–58
  growth and, 47, 52–53
  in infancy, 55–56
  in pregnancy, 55
Food supplementation, vs. birth weight,
    87–92
Formula feeding, for newborn, 131
Fresco villages, in food studies, 54, 84–112
FTT, *see* "Failure-to-thrive" syndrome

Gavage, in premature infants, 140–141, 150
Genetic effect, defined, 31
Genetic research, new strategies in, 42–44
Genetics
  nutritional interactions and, 31–45

Genetics (*cont'd*)
  separation of nutrition from, 44–45
Gestation period, birth interval and, 110
Glucocorticoids
  androgen secretion and, 71
  in Cushing's syndrome, 72
Gross National Product, growth and, 34
Growth
  in adolescence, 239–241
  agricultural knowledge and, 59
  amino acid metabolism and, 67–68
  birth interval and, 60
  body composition and, 156–158
  calcium and, 24
  CNS diseases and, 72
  compensatory, 2
  congenital heart disease and, 66
  cross-sectional surveys of, 399–401
  in diabetes mellitus, 68–69
  ecological balance and, 59
  end-organ failure and, 72–73
  energy requirements for, 9
  ethnic differences and, 160–161
  failure to thrive and, 171–182
  family structure and, 59–60
  fetal, *see* Fetal growth
  food intake and utilization in, 47
  food practices and, 52–53
  food production and, 58–59
  forecasting of, 161–162
  general terms used in, 171
  infection and, 47–48
  infectious diseases and, 48
  longitudinal evaluations of, 401–402
  meaning of, 1
  multiple-birth model in, 38–39
  nutrient needs and, 7–26, 79
  obesity and, 226
  overpopulation and, 58
  poverty and, 58
  protein and, 10
  protein–energy malnutrition and, 186
  sample size in nutrition-related studies of,
    40–42
  sex factors in, 60
  social factors in, 59–60
  socioeconomic factors in, 58–60
  vitamin D deficiency and, 70
Growth assessment, surveillance in (*see also*
    Nutrition surveillance), 402–405
Growth chart
  for at-risk child, 334
  breast feeding and, 333–334
  new weighing methods and, 334–335
  objective of, 329
  in recovery, 333

Growth chart (*cont'd*)
  sample, 330
  in surveillance, 336, 402–405
  in undernutrition, 331–333
Growth failure (*see also* "Failure-to-thrive" syn-
    drome)
  causes of, 172–173
  nutritional anthropometry and, 389–397
Growth monitoring, nutritional assessment and
    (*see also* Nutrition surveillance), 273
Growth norms, population differences and, 34
Growth parameters, statistics of, 32–34
Growth pattern, low birth weight and, 155–156
Growth period, food practices during, 55–58
Growth rate, defined, 25
Growth velocity, defined, 1

Hand–wrist radiographs, 282–283
HANES I survey, 423
Head circumference
  growth in, 368–369
  optimal, 277
  reference standards for, 312–313
Health, as defined in WHO constitution, 185
Height, in anthropometric field measures,
    351–353, 357–358
Helminthic infections, growth and, 50
Hematocrit levels
  in adolescence and adulthood, 288–289
  in nutrition surveillance, 422, 427
Hematologic determinants, optimizing of,
    287–289
Hemoglobin, iron and, 22
Hemoglobin concentrations
  "deficient" or "low, " 290
  nutrition and, 287
  in nutrition surveillance, 427–429
Hemostasis, 22
Histidine, 12
Homs hand carrying scale, 351
Human milk, *see* Breast milk
Hyperlipidemia, 145
Hyperthyroidism, 71
Hypochromic anemia, in protein–energy
    malnutrition, 187
Hypophosphatasia, 72–73
Hypothyroidism, 70
  juvenile, 225

INCAP (Institute of Nutrition for Central
    America and Panama) longitudinal study,
    84–99, 110, 122
INCAP/OIR nutritional status assessment, 400
Infancy
  food practices in, 55–56
  obesity in, 219–225, 233

Infant (*see also* Children; Infancy; Newborn; Young child)
  energy needs of, 9
  nutrient requirements of, 8
  premature, *see* Premature infant
Infection
  breast milk and, 50–51
  in "failure-to-thrive" syndrome, 175
  growth and, 47–48
  helminthic, 50
  malnutrition and, 48
  protein–energy malnutrition and, 190–192
  protozoal, 50
  viral and rickettsial, 49–50
Inheritance factors, in growth parameters, 32–33
Institutional environment, for young child, 167
Internal biochemical environment, in protein–energy malnutrition, 187–188
International Children's Centre (Paris), 154
International standards, best available, 306–307
International Union of Nutritional Sciences, 299–304
Intestinal parasitism, decreased absorption and, 50
Intestinal tract, bacterial infections of, 48
Intralipid, in parenteral nutrition, 145
Intravenous feeding, in premature infants, 140–141
Iron
  need for, 22–24
  newborn and, 132–133
Iron deficiency anemia, 23
  in adolescence, 247
Isoleucine, 12

Joint FAO/WHO Expert Committee on Nutrition (*see also* FAO/WHO Expert Committee on Protein Requirements), 185, 195–197, 203, 205, 209, 431–432
Joule–protein malnutrition, 313

Kernicterus, 145
Kwashiorkor, 189–190
  growth chart in, 332
  measles and, 49
  principal signs of, 192–195
  tuberculosis and, 49

Labile protein pool, 14
Lactation, postpartum amenorrhea and, 105–106
Last menstrual period, newborn size and, 36–37
Laurence–Moon–Biedl syndrome, 255

LBW babies, *see* Low-birth-weight babies
Length/Weight optimizing, 274–277
Leucine, 12
Lineolic acid, fat and, 15
LMP, *see* Last menstrual period
Longitudinal evaluations, in growth assessment studies, 401–402
Longitudinal study, INCAP, *see* INCAP
Low birth weight, growth pattern and, 155–156
Low-birth-weight babies
  incidence of, 119–120
  number of, 118
  risk of delivery for, 119–121
Low hemoglobin concentrations, values for (*see also* Hemoglobin concentrations), 290
Lysine, 16
Lysosomal storage diseases, 73

Main puteri, 53
Malnutrition
  psychologic, 175–176
  simplified anthropometry in determination of, 389–397
Marasmus
  growth chart in, 331
  nutritional, 188–189
  principal signs of, 192–193
Maternal anthropometry
  in food studies, 86
  during pregnancy, 99–101
Maternal characteristics, nutrition and, 99–118
Maternal deprivation, failure to thrive and, 175–176
Maternal effect, defined, 31
Maternal height, birth weight and, 82
Maternal hormonal system, in pregnancy, 79–80
Maternal infection, low birth weight and, 48
Maternal nutrition (*see also* Maternofetal nutrition; Pregnancy)
  birth interval components and, 103–113
  fetal growth and, 81–99
  newborn size and, 36
Maternofetal nutrition (*see also* Nutrition), 79–122
  food supplementation vs. birth weight in, 87–92
  gestation period and, 110
  menstruation interval and, 108–110
  postpartum amenorrhea and, 105–108
  protein–calorie malnutrition and, 113–118
  variables in, 86–87
Measles, kwashiorkor and, 49

Measurements, quality control in (see also
    Anthropometric measures; Anthropometry),
    360–362
Menarche, nutrition and, 241
Menstruating interval
    factors affecting, 108–109
    fecundability and, 109–110
Mental retardation, amino acid disorders and,
    68
Metabolic anomalies
    defined, 65
    growth and, 65–73
Methionine, 12
Mid-arm muscle circumference, 310, 319
Mid-upper arm circumference, 310–311,
    319–320, 393–395
    in anthropometric field methods, 353
    measurement of, 358–359
Milk (see also Breast milk)
    caloric intake from, 136–137
    as food for first year, 135–136
    human, see Breast milk
    nutrients missing from, 136
    protein requirements and, 11
    retinol (vitamin A) content of, 16–17
Mineral needs
    in adolescence, 244–246
    of newborn, 134–135
Monozygotic twins, vs. dizygotic, 32, 38–39
Multiple-birth model, in growth studies, 38–39

National Center for Health Statistics, 303, 306
National Health Examinations, 41
National Probability Sample, 35
Nature/nurture interaction, 2
Necrotizing enterocolitis, in premature infants,
    149
Net protein utilization, 12
Newborn (see also Infant; Premature infant)
    breast feeding for, 130–131
    caloric intake by, 136–138
    fluoride for, 133–134
    formula feeding for, 131
    full-term, 129–138
    growth requirements for, 147–148
    iron for, 132–133
    marasmic, 146
    metabolic disorders in, 67–68
    minerals for, 134–135
    monitoring of, 146–148
    nongenetic determinants of size of, 36–38
    nutrition for, 129–151
    parenteral nutrition in, 142–146
    premature, see Premature infant
    "small-for-date, " 151

Newborn (cont'd)
    solid foods for, 135–138
    vitamin supplements for, 131–135
"New Growth Charts, " 35–36, 41
Niacin, 20
Nitrogen, loss of, 10, 13
Nitrogen absorption, growth and, 49
Nitrogen balance, 11
    negative, 49
Nitrogen excretion, in Q fever, 50
"Normal" child, disease pattern in (see also
    Children; Young child), 164–166
NPU, see Net protein utilization
Nutrient needs
    estimation of for infants, 8
    growth and, 7–26
Nutrients, defined, 65
Nutrition (see also Nutrient needs; Nutrition
    surveillance)
    in adults, 263–264
    birth interval components and, 103–113
    changing trends in, 3
    cystic fibrosis and, 67
    defined, 65
    in failure-to-thrive syndrome, 174–175
    growth and, 3, 65–73
    maternal, see Maternal nutrition
    maternal characteristics and, 98–118
    maternofetal, see Maternofetal nutrition
    meaning of, 1
    nondietary factors and, 47–60
    parenteral, 142–146
    postpartum amenorrhea and, 107–108
    recommended dietary allowances in, 7
    in renal diseases, 69–70
    separation of from genetic effects, 44
Nutritional assessment, optimal, see Optimal
    nutritional assessment
Nutritional development, vulnerable tissues in,
    66
Nutritional effect, defined, 31
Nutritional indicators
    feasibility for use of, 381–382
    response specificity in, 380–381
    sensitivity of in populations, 378–380
    undependability/unreliability ratios in, 376
Nutritional interventions, birth weight and,
    95–99
Nutritional maramus, (see also Marasmus),
    188–189
Nutritional status (see also Protein–calorie
    nutritional status)
    accuracy of, 368–370
    in adult, 267–268
    anthropometry and, 339–341

Nutritional status (*cont'd*)
defined, 65
dependability of indicators in, 370–372
indicators of, 365–368
sensitivity of indicators and sample size in, 375–378
Nutritional survey, "optimal" meaning in, 293–296
Nutrition indices (*see also* Nutritional indicators),
by sex and ethnic group, 425–426
by states, 424
Nutrition problems, identification of, 411–412
Nutrition surveillance, 402–405
in affluent countries, 410
anthropometry and hematology worksheet in, 417
children's statistics in, 427–429
data sources in, 414–415
defined, 409
in developed countries, 409–429
in developing countries, 431–440
feedback in, 416–417
goal of, 410–411
hemoglobin and hematocrit values in, 422
individual scores in, 421
items included in, 412–414
in United States, 409–429
Nutrition Surveillance Bulletin, 422
Nutrition surveillance data, handling of, 415–416
Nutrition surveillance indices, identification of, 412–413
Nutrition surveillance programs, development of, 411–416
Nutrition Surveillance System, of Center for Disease Control, 416–429, 432
Nutriture, nutritional status as indicator of, 368–370

Obesity
adipose tissue cellularity and, 229
in adolescence, 249–251
in anthropometry, 342
cardiorespiratory effects of, 228
in childhood and adolescence, 218–221, 249–251
diet in, 230–232
effects of, 226–229
emotional factors in, 226
energy intake in, 223
energy output in, 224–225
etiology of, 221–226
genetics of, 39–40, 222

Obesity (*cont'd*)
glucose tolerance and serum insulin in, 227–228
growth and, 226
in infancy, 217–225
metabolic effects of, 225–227
natural history of, 219–221
in older children, 224
physical activity and, 232
in prenatal period, 233
prevalence of, 217–219
prevention of, 232–234
psychiatry in, 232
treatment of, 229–232
in young child, 217–234
Ogaden area (Ethiopia), drought and famine in, 436–438
One-day dietary intake studies, 286–287
"Optimal," defined, 293–294
Optimal circumferences, in nutritional surveys, 277–278
Optimal nutritional assessment, 273–296
follow-up studies in, 293–294
optimizing hematologic determinations in, 287–289
recording, reduction, and analysis in, 294–295
screening at-risk group in, 291–293
serum and urinary vitamins in, 289–291
Ostomalacia, 70
Overnutrition, vs. protein–calorie malnutrition, 313–314
Oxytocin, 130

Parent illiteracy, growth and, 60
PCM, *see* Protein–calorie malnutrition
PEM, *see* Protein–energy malnutrition
Percent cortical area, measurement of, 282
Phenylalanine, 12
Physical activity, in young child, 162–164
Population comparisons, socioeconomic status and, 42
Population differences, national growth norms and, 34–36
Population nutrition problems, identification of, 411–412
Populations, sensitivity of nutritional indicators in, 378–380
Postpartum amenorrhea, 105–106
birth interval and, 105–111
duration of, 122
lactation and, 105–106
literature on, 110–113
nutrition and, 107–108
pregnancy outcome and, 106

Postpartum amenorrhea (*cont'd*)
  protein–energy malnutrition and, 199
  socioeconomic status and, 106–107
Postural edema, during pregnancy, 115–116
Poverty, growth and (*see also* Socioeconomic
    status), 58
Prader–Willi syndrome, 225
Pregnancy
  blood pressure during, 113–115
  diet in, 37–38
  dose–response relationships and, 96–99
  food practices during, 55
  food supplementation vs. birth weight in,
    87–92
  maternal anthropometry during, 99–101
  maternal humoral system and, 79
  maternal ingestion of food during, 86
  maternal nutritional states during, 82
  maternal nutritional states prior to, 81
  one-day dietary intake studies in,
    286–287
  postural edema during, 115–116
  proteinuria during, 116
  supplemented calories during, 113–114
Pregnancy wastage, early, 109
Premature infant (*see also* Infant; Young
    child)
  breast milk for, 149–150
  intravenous feeding of, 140–141
  monitoring of, 144–148
  necrotizing enterocolitis in, 149
  nutrition for, 138–151
  sick, 146–148
  special problems of, 138–140
Preschool children, age groups in (*see also*
    Young child), 153
Pre-School Nutrition Survey, 41
Propylethiouracil, 131
Protein
  birth weight and, 92–94
  need for, 10–14
  NPU of, 12
  RDA for, 11
  safe levels of intake for, 14
Protein–calorie malnutrition, 49, 113, 117,
    185, 299
  classification of, 317, 319–322
  indicators of, 370–373
  nutritional anthropometry and, 389–396
  vs. overnutrition, 313–314, 317
  vs. protein–energy malnutrition, 1
  reference standards and, 314–316
  reliability of indicators in, 375
  sensitivity factors in, 366–368
"Protein disaster, " growth chart and, 333

Protein–energy malnutrition, 185–212, 313
  activity and, 186
  classification of, 192–195
  cultural influences in, 200–201
  defined, 185–186
  epidemiology of, 195–198
  food chain and, 205–207
  food/nutrition policies and programs in, 209
  Gomez classification of, 193
  growth and, 186
  infections and, 190–192
  kwashiorkor and, 189–190
  mild and moderate, 186–188
  nutritional background of, 195–198
  in poverty and underdeveloped countries,
    198–200
  prevalence of, 202–204
  prevention and treatment of, 205
  secondary and tertiary prevention of,
    210–212
  severe forms of, 188
  social and economic background of, 198–200
  staple foods and, 196–197
  world infant mortality rates for, 203–204
Protein needs, in adolescence (*see also* Protein;
    Protein–calorie malnutrition), 242–244
Protein pool, "labile, " 14
Proteinuria, during pregnancy, 116
Prtozoal infections, growth and, 50
Pseudohypoparathyroidism, growth and, 72
Psychologic malnutrition, 175–176
Psychomotor development, in young child, 158

QUAC stick, in reference standards, 320,
    391–392, 400
Quinoa, 2
Q fever, 50

Radiogrammetric information, growth and,
    283
Radiography, in skeletal age determination,
    280–284
RDA, *see* Recommended dietary allowances
Recommended dietary allowances
  defined, 7
  for protein, 11
Reference standards, 299–323
  for arm circumference, 311–312
  "best available, " 322–323
  classification system in, 314–319
  factors affecting, 305–306
  genetic vs. environmental influence on,
    304–306
  for head circumference, 312–313
  ideal, 301–302

Reference standards (*cont'd*)
    international, 302–303, 306–307
    locally constructed, 303–304
    meaning of, 300–301
    sequential nutritional diagnosis and, 321–322
    for skinfold, 307–311
    uses of, 299–300
    weight for age in, 314–315
Renal disease, growth and nutrition in, 69–70
Renal transplantation, 69–70
Respiratory process, iron in, 22
Response specificity, in anthropometric field
        measurements, 380–381
Retinol, *see* Vitamin A
Riboflavin, 20
Rickets, growth and, 70
Rickettsial infections, growth and, 49–50
Rocky Mountain spotted fever, 50
"Runt" effect, 43

Sahel region (Africa), drought in, 435
Salter scale, 349–350
Sample size, parent–child similarities in, 42
Sanitation, growth and, 60
SBC, *see* Systolic blood pressure
SCDP, *see* Supplemented calories during
        pregnancy
Screening, in anthropometry, 341
Scurvy, ascorbic acid and, 22
Sensitivity, in nutritional status indicators,
        366–368
Sequential nutritional diagnosis, 321–322
SES, *see* Socioeconomic status
Sex factors, growth and, 60
Sexual differences, in young children, 158–160
Sickle cell disease, 67
Skeletal age, radiographic determination of,
        280–284
Skinfold, reference standards for, 307–311,
        320–321
Small-for-age subjects, nonnutritional size
        reductions in, 292
Smoking, newborn size and, 36
Socioeconomic status
    fetal death and, 110
    as growth factor, 41, 58–60
    nutrition and, 291
    in population comparisons, 42
    postpartum amenorrhea and, 106–107
    stillbirths and, 110
Somatomedin levels
    growth hormone and, 71
    in renal transplantation, 69–70
Spontaneous abortions, socioeconomic status
        and, 110

Stature
    inheritance of, 32
    trends in, 2
Stuart–Stevenson growth standards, 41
Subcutaneous fat, reference standards for,
        307–311
Substrate deficiency disorders, 66
Superfoods, cultural, 53
Supplemental Food Program for Women,
        Infants, and Children, 415
Supplemented calories during pregnancy,
        113–118
Surveillance
    in growth assessment, 402–405
    nutrition, *see* Nutrition surveillance
Survey, in anthropometry, 341
Systolic blood pressure, during pregnancy,
        114–115

Ten-State Nutrition Survey, 41, 307, 321, 410
Thalassemia, 67
Thiamine, 19–20
Thoracic circumference, optimal, 277
Threonine, 12
Thyroid dysfunction, growth and, 70–71
Thyroid stimulating hormone, 70
Thyroxin, 70
Total body weight, formula for, 278
Triceps fatfold, measurement of, 359–360
Triiodothyronine, 70
Tryptophan, 12
T-scores, age-specific normalized, 32
Tuberculosis, kwashiorkor and, 49
Twin studies, growth parameters and, 32–34,
        38
Tyrosine, 22

Undependability/unreliability ratios, for
        nutritional status indicators, 376–377
Undernutrition, growth chart in, 331–333
United States
    "New Growth Charts" for, 35
    nutrition surveillance in, 409–429
Uterine effect, newborn size and, 31

Valine, 12
Viral infections, growth and, 49–50
Vitamin(s), 15–22
    in adulthood, 266–267
    defined, 15
    excretion of, 16
    fat-soluble, 16–19
    need for, 15, 244–246
    for newborn, 131–133

Vitamin(s) (*cont'd*)
  serum and urinary, 289–291
  water-soluble, 19–22
Vitamin A, 16–18
Vitamin B$_6$, 21–22
  deficiency of, 21
Vitamin B$_{12}$, 17, 21
  in adolescence, 245
Vitamin D, 18–19
Vitamin D deficiency, growth and, 70
Vitamin depletion studies, 16
Vitamin needs, in adolescence, 244–246

War, growth and, 58–59
Water-soluble vitamins, 19–22
Weight
  age factor in, 389–390
  as anthropometric measure, 348–351
  measurement of, 356–357
  as nutritive status indicator, 376–377
  variability in, 376–377
Weight-for-head circumference measurements,
  391
Weight-for-height relationship, 318–319

Weight reduction, in obesity, 230
Wellcome classification, 194
Wheat gluten, intolerance to, 66
World Health Organization, 185, 431

Young child
  anorexia in, 175
  body composition of, 156–158
  ethnic differences in, 160–161
  "failure-to-thrive" syndrome and, 171–182
  forecasting growth and development for,
    161–162
  home vs. institutional environment for,
    167
  mental development in, 158
  normal, 153–167
  "normal" disease pattern in, 164–167
  obesity in, 217–234
  physical activity pattern in, 162–164
  premorbid constitution of, 166
  protein–energy malnutrition in, 185–212
  psychomotor and mental development in,
    158
  sexual differences in, 158–160